To Brian,

With Best Wishes,

Larry

Diamonds of Yakutia: Photographic Evidence for their Origin

Zdislav V. Spetsius
Institute of Diamond Industry, ALROSA Co., LTD
Mirny, Yakutia, Russia

Lawrence A. Taylor
Planetary Geosciences Institute
Department of Earth & Planetary Sciences
University of Tennessee
Knoxville, Tennessee, USA

Tranquility Base Press
Lenoir City, Tennessee
USA

PUBLISHED BY THE TRANQUILITY BASE PRESS

Library of Congress Cataloguing in Publication Data

Spetsius, Zdislav and Taylor, Lawrence

ISBN 978-0-9795835-0-6

Includes bibliographic references

1. Diamond Inclusions; 2. Diamond Morphology; 3. Origin of Diamonds;

4. Diamonds of Yakutia; 5. Diamonds of Siberia;

Library of Congress Control Number : 2008906005

ISBN: 978-0-9795835-0-6

Printed in South Korea

Table of Contents

PREFACE

Throughout history, the exploration for and exploitation of diamonds have been driven by the quest for financial rewards. It has been the blending of science and technology into exploration that has led to the continuing success in locating new diamond prospects. There is a symbiotic relationship between these contrasting disciplines involving the lab and the field, particularly as ever-more sophisticated methods are applied to exploration. However, there has always been a small contingency that is mainly interested in the science that diamonds have to reveal of the Earth's mantle, the deeper ~80 vol% portion of the Earth that largely remains unsampled. It is the academic aspect of diamond studies to which this book is mainly addressed.

Diamonds have been called "time capsules" that contain a treasure trove of scientific data about the history and evolution of our planet (e.g., Taylor and Anand, 2004), largely manifest as inclusions. It is these mineral inclusions in the diamonds that are so despised by jewelers but so prized by scientists. Detailed investigations of these inclusions received initial impetus from the development of sophisticated micro-analytical instrumentation, such as the electron microprobe in the late 1960s, and have progressed at an ever-increasing rate, especially as new geochemical instrumentation and chemical systems (e.g., Re-Os) for dating inclusions have been developed. Couple this with our increasing appreciation of the complex nature of the "stratigraphy" of the diamond's growth zones and their significance in terms of chemistry and inclusion encapsulation, and one can readily imagine the plethora of new scientific data that have been forth-coming.

With much of our data gathering, real appreciation for the "petrography" has been lost of the setting of diamonds within their host mantle rocks and the kimberlite, and of the invaluable mineral inclusions in their diamond host. It is the purpose of this book to record the invaluable photomicrographs and observations that the authors have made of the actual relationships of the diamonds to these rocks, in order that this may stimulate further excursions into the origin of diamonds.

This book has been structured into five Chapters, each addressing major topics in diamond/mantle research. The text in each Chapter is meant only as a brief introduction into the subject matter of each and is followed by a selection of photographs that further illustrate and, indeed, illuminate various aspects of the petrography of diamonds, their morphologies, mineral inclusions, and relationships to their host xenoliths and kimberlite. The reader is forewarned that the text is not intended to be all-inclusive in theory detail or even contain all the latest ideas; extensive 'literary license' has been taken in the writing of this text. We apologize herewith for any lack of proper referencing that may have occurred in our presentations. It is important to note, however, that numerous references are made throughout the text to the extensive Russian literature on Yakutian diamonds, commonly not well known or appreciated by our western comrades. It is the photographs more than the text that are of definite value to this undertaking.

At the end of this book, we have included a reprint of a paper written by Taylor and Anand in 2004, with the provocative title: "Diamonds from Yakutia: Time Capsules from the Siberian Mantle." Although a bit out-of-date in referencing, it still contains interesting and significant condensations of important topics such as the subduction origin or diamondiferous eclogites and peridotites. Permission to include this chapter, from a special edition of Chemie der Erde, was kindly given by Elsevier Publishing Company. Following this Chapter, there is a bibliography of papers, mainly about Yakutian kimberlites and diamonds. It is not intended to be all-inclusive, but should provide a starting point for many aspects of Yakutian diamond research.

Lastly, as we progress in our studies of diamonds and their host xenoliths, it has become apparent that we are narrowing the gap in our understanding of the origin(s) and formation of diamonds. However, in this continuing quest for knowledge, it seems that each investigation solves one problem but uncovers two new ones. But such is progress. Hopefully the plates in this book, and the associated 'braindroppings', will inspire some new and exciting "mental calories" for the next generation of scientists. God forbid we should ever understand everything; then, what would we do for enjoyment of the science of diamonds.

Acknowledgements: The authors are grateful to the Konstantin Kurbatov, Director of the Diamond Sorting Center of ALROSA Co Ltd. in Mirny, and Igor Lukianchikov for providing photos of sorted diamonds. Natasha Merkulova and Svetlana Banzeruk are thanked for the assistance with photography of some diamondiferous xenoliths from the Nurbinskaya pipe. Indeed, the many persons working with Yakutian diamonds who supplied the wonderful images are to be acknowledged. We particularly thank Akademician Nick Sobolev for his education of both of us about many aspects of the Yakutian kimberlite regime. In addition, we thank Darren Schnare for his assistance with proof reading and other laborious tasks. Lastly, but most significantly, Dawn Taylor is acknowledged for her enthusiastic assistance with the text, the photography, the publishing, and most importantly, her major physical- and mental-calorie input into all aspects of this book's production.

The Authors

In the open pit at Udachnaya; note the ~10 kg huge eclogite xenolith, but non-diamondiferous.
Zdislav (Slava) Spetsius - “Mr. Cool”.

In the open pit at Udachnaya; **Larry Taylor (Center)** between colleagues, Nick Sobolev (left) and Nick Pokhilenko, from the Siberian Branch of the Russian Academy of Sciences in Novosibirsk.

CHAPER 1

Introduction to Kimberlites

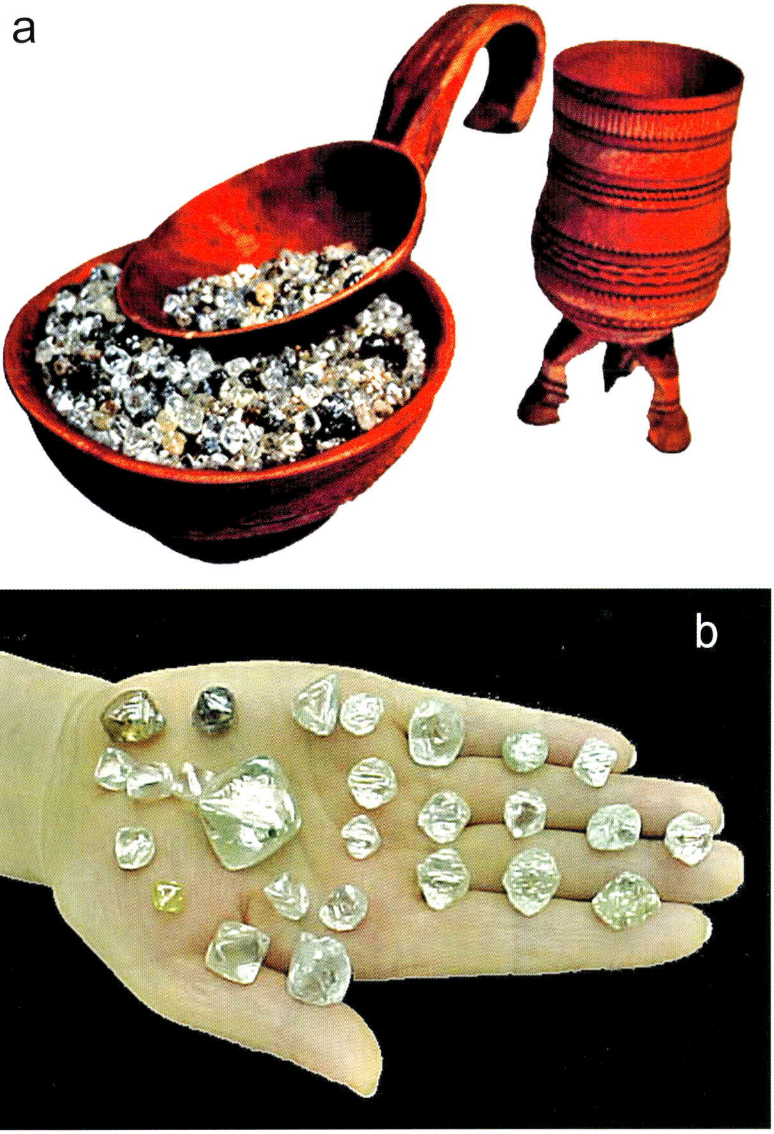

Diamonds – "From Russia with Love." a) Charon (national Yakutian goblet) full of diamonds. b) A beautiful array of spectacular diamonds on a gorgeous hand; the smallest diamond is 8 carats.

Chapter 1

Introduction to Kimberlites

Great advances have been made in the last few decades in the understanding of the processes and environments of diamond formation (e.g., Sobolev, 1976; Haggerty, 1986; Gurney, 1989; Pearson et al., 1995). Much of the knowledge on the genesis of diamond has accumulated from the study of mineral inclusions, particularly their geochemistry and age dating, the examination of the internal stratigraphy of diamonds, and the relationships of diamonds to their host rocks. Various aspects regarding these beautiful and unique gem stones have now been generally agreed upon: 1) Virtually all diamonds have their origins in mafic and ultramafic mantle rocks, rather than from the kimberlites or lamproites in which they occur; this statement has the caveat that some coated/fibrous diamonds may have grown metastably in the kimberlite melt; 2) Diamonds are considered xenocrysts and are older than their kimberlite hosts; 3) Diamonds can reside in the mantle for long periods of time before they are entrained by ascending magmas, which are intruded to the upper crust and extruded onto the Earth's surface. It will become increasingly obvious as one proceeds through this book that the origin(s) of diamond is still uncertain, as is the nature of the source fluids from which they grew. Nevertheless, the presence of diamond in mantle xenoliths indicates that favorable conditions for growth have been attained within their mantle host rocks.

Several lines of evidence have suggested that natural diamond has grown directly from carbonaceous fluids or melts rich in volatiles, rather than by solid-state transformation of graphite. But is diamond only a metamorphic mineral? Does it have such unusual properties by which it can grow to perfect crystals, commonly free of inclusions? In most current hypotheses, fluids rich in dissolved carbon complexes, water, and sulfides are believed to be involved in the diamond crystallization process, a C-O-H-N-S fluid, similar to that suggested by Haggerty (1986). The nature of these C-bearing fluids, however, is still unknown. Some believe that these fluids are rich in CO_2 and CH_4, the result of mantle degassing or liberated from subduction zones, and hence, recycling of crustal carbon (e.g., Sobolev and Sobolev, 1980). Oxygen activity (fugacity) is obviously an important factor in diamond genesis, since the speciation of carbon is controlled by the redox conditions of both the fluid and the host-rock environment (e.g., Palyanov et al., 1998)

The origin of diamond is always referenced to the experimentally determined graphite-diamond stability curve of Kennedy and Kennedy (1976), as shown in Figures 1.a & 1.b. Thereby, growth of diamond is expected to take place in the lithospheric mantle, which has pressure, temperature, and oxygen fugacity conditions compatible with the diamond stability field (e.g., Navon, 1999, and references therein). This diamond-

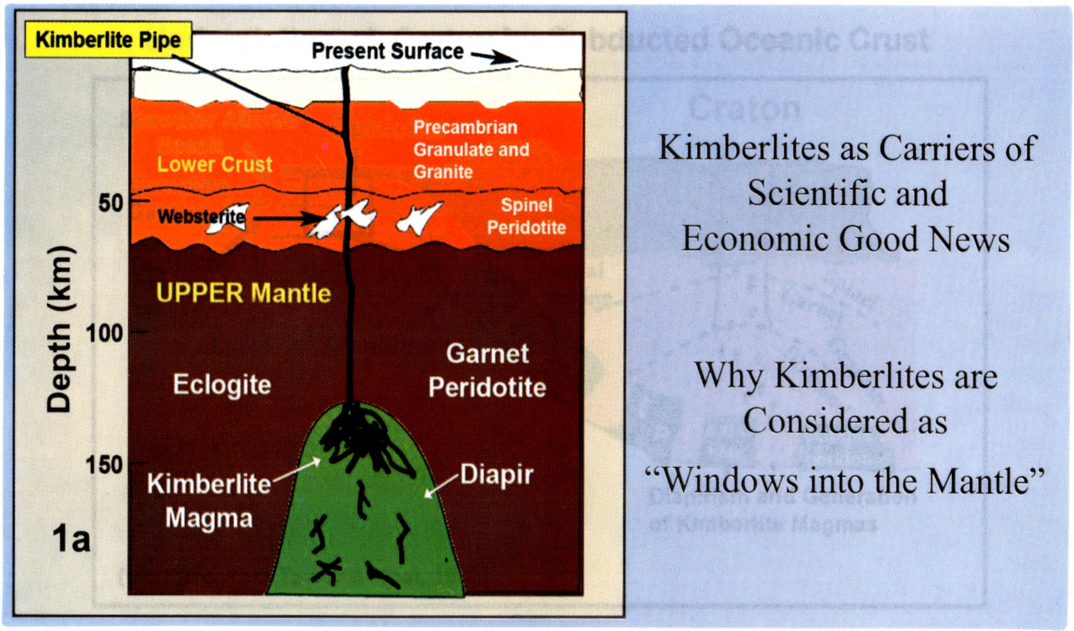

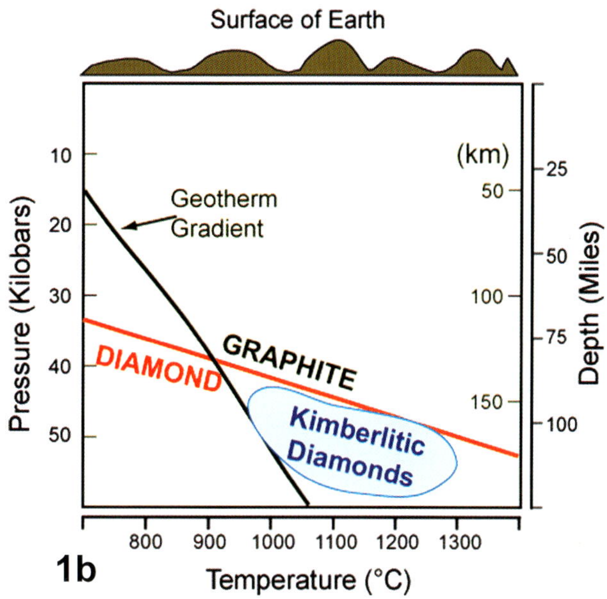

Figure 1a. Schematic cross-section of the earth illustrating the strange type of volcanic activity of the emplacement of a kimberlite. Almost all other types of "normal" volcanoes come from rather shallow depths (5-10 km), stopping to form magma chambers just below the surface.

Figure 1b. Pressure-Temperature profile downward from the surface of the earth. Note the region at high pressures above the graphite/diamond stability curve. In reality, diamonds can originate at any depth below this.

graphite stability curve reflects the relatively low cratonic geothermal gradient necessary for the growth and preservation of diamond in the mantle. However, it has become increasingly obvious from diamond-inclusion data that diamonds have also grown in deeper portions of the mantle, well within the asthenosphere.

The possibility of diamond growing and/or residing outside its stability field can not be discounted. Bundy (1963) predicted the existence of a metastable region containing graphite and diamond, which would account for the presence of graphite inclusions encountered in diamond. In addition, conditions have been documented that would appear to have diamond formation of entirely outside the diamond stability field. New types of primary deposits have been identified during the past two decades that include ultramafic lamprophyres, minettes, and picrites, whose parental melts may have been generated at P-T conditions outside the diamond stability field (Janse, 1994). Diamonds have also been recovered from metamorphic rocks formed in continental collision zones, and in such cases may have distinctly different growth mechanisms, compared to diamonds of deep-seated origin (e.g., Sobolev and Shatsky, 1990).

Diamond formation in the upper mantle of the Earth is based on the attainment of conditions that favor the nucleation and growth of diamonds. This involves the proper P-T environment, favorable fO_2, and a medium that is conducive to crystal growth. It is entirely possible that these conditions can be met by more than one scenario of events: in silicate-melt systems; in metasomatic fluids; and with metasomatic fluid interactions with primary minerals, such as in late-stage partial-melting episodes (Spetsius and Taylor, 2000).

The formation of diamonds in upper-mantle rocks involved a long and complicated, multi-stage process that occurred at different times throughout the growth and development of continental cratons. The numerous evidences for the protoliths of diamondiferous eclogites possessing distinct crustal signatures has led to the theory that many mantle eclogites are direct products of subduction of ancient oceanic crust beneath continental cratons. This is well illustrated in Figure 1.2. The multiple crustal signatures are reviewed in detail in Taylor and Anand (2004), reprinted at the end of this book.

The secondary nature of eclogitic diamonds is generally agreed upon, but the primary nature of peridotitic diamonds is still debated (Taylor et al., 2005). Likewise, the nature of the processes that formed the diamonds, be they from silicate melts, sulfide melts, or metasomatic fluids, remains a major quandary for mantle scientists. Studies of showing evidence for growth, as well as with different diamond generations in mantle xenoliths may reveal clues to the mystery of diamond formation. However, in spite of a huge number of published articles on diamonds with condiderable amonts of new data, it has been only during the last two decades that the issue of diamond genesis has been seriously addressed. This has been made possible, largely due to definitive stable-isotope studies coupled with high-pressure experimentation.

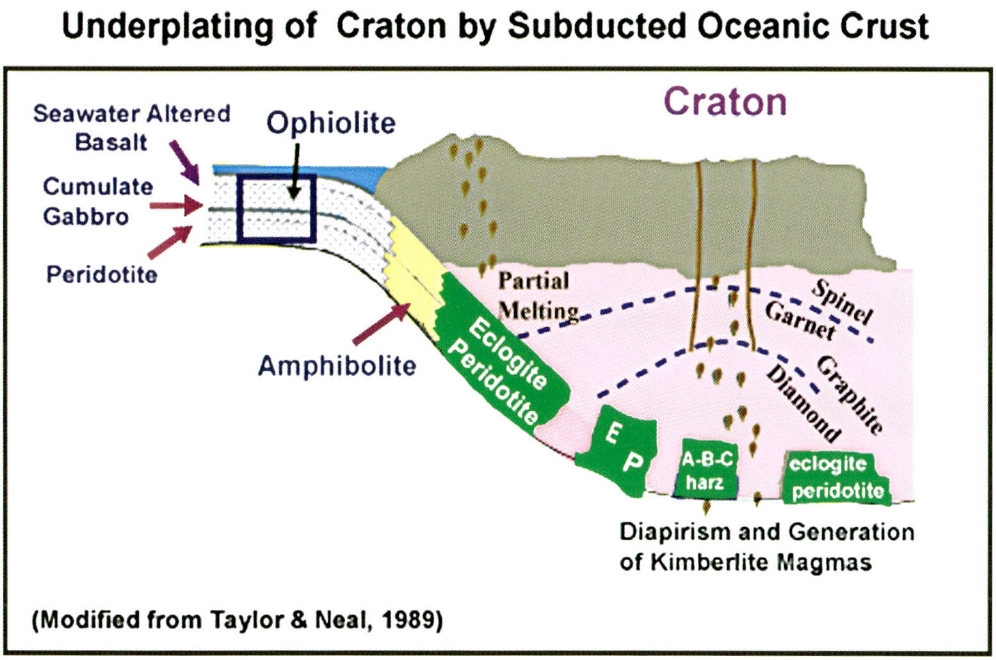

Figure 1.2. Schematic cross-section of the Earth illustrating the subduction of oceanic crust to beneath a continental craton, where the mafic upper portion of the crustal slab is metamorphosed into eclogite and the ultramafic portion into mantle peridotite.

One of the problematic aspects of diamond genesis is the ratio of diamonds of eclogite (E-type) to those of peridotite parageneses (P-type=U-type of Sobolev, 1978) in different kimberlite pipes, with a reported predominance of P-type diamonds. Because the number of diamonds with visible inclusions of silicate and oxide minerals is considerably <10 % of the total population of crystals recovered, simple examination of diamonds with inclusions (P- versus E-type) from any kimberlite pipe may introduce a bias. From our somewhat limited sampling of diamondiferous eclogites (~300), it appears that diamonds in most eclogites are prone to be crystal-clear, uncolored, and inclusion-free. Furthermore, the diamond concentrations in eclogites are commonly pronounced. One 50 gm Udachnaya eclogite contained >70 macrodiamonds, some up to several carats (Anand et al., 2003). More recently, an 8.8 kg diamondiferous eclogite xenolith, containing thousands of diamonds has been discovered in the Udachnaya pipe (Taylor et al., 2005). Although diamondiferous peridotites are rare by comparison, it would appear that various evidences are mounting to indicate that the number of eclogitic diamonds (E-type) is far more numerous than 'biased' mineral-inclusion statistics would indicate. In the cases of the Udachnaya and Mir pipes, the amount of E-type diamonds may be >20 %, rather than the <5 % that the inclusions indicate. The bottom line is that some method of classifying the inclusion-free diamonds would seem necessary, perhaps based on trace-element chemistry and FTIR spectra. Investigation of the physical and chemical properties of diamonds and their mineral inclusions from diamondiferous eclogite and peridotite xenoliths presents the closest relationships of diamonds to their original settings in the mantle.

A paramount subject to be addressed is the determination of the ages of diamonds. Although Sm/Nd determinations by silicate inclusions in most cases give old ages for peridotite diamonds and a fairly wide spread for crystals with eclogite inclusions, relatively younger ages are being reported each year. But most important is the question of what is being dated. The aggregation of garnet and pyroxene from *hundreds* of diamonds is necessary in order to obtain large enough masses with which to perform Rb-Sr and Sm-Nd age analyses (Navon, 1999). Taylor et al. (2003) have also addressed this issue of the synchronous nature of the mineral inclusions with the time of diamond formation, emphasizing that many, if not most, mineral inclusions in diamonds are protogenetic – formed before their encapsulation by diamonds. Although not a new concept, this is one that is most definitely probable. High-precision Re-Os dating of well-characterized, single-sulfide inclusions, particularly *in-situ*, would seem to be the most promising inclusion-dating method at present (Pearson & Shirey, 1998). But, here again, one is only dating the inclusion, not the diamond itself.

Summary: It is apparent that the kimberlitic magma acted mainly, if not entirely, as the "transporter" of the xenoliths from the upper mantle, and thereby, mechanically crushed most xenoliths, freeing their diamonds. Therefore, the grade of a kimberlite is a function of the nature of the diamondiferous xenoliths originally picked up by the kimberlitic magma, the efficiency of the magma in disaggregating the xenoliths, and the oxidative properties of the ascending kimberlitic magma. The oxidation state of the kimberlite can influence the resorption of diamonds to various extents, changing their morphology, and in severe cases, completely "dissolving" the diamonds. Accordingly, it can be said, somewhat metaphorically, that "*kimberlites are responsible for the handling and safety of diamonds during their transportation to the surface.*"

Inasmuch as diamonds in kimberlites are xenocrysts from disintegrated mafic and ultramafic mantle rocks, it is to the study of these rocks in which the diamonds originally formed that we now turn. In so doing, we hope to be able to determine and better understand the chemical and physical conditions that prevailed during diamond genesis.

1.1 Geography of Yakutian Diamond Locales

Diamonds occur mostly in rocks of upper mantle origin, including kimberlite, lamproite, and other related alkaline rocks. In addition, they have also been found as pre-solar grains in meteorites (Fukunaga et al., 1987), as shock-related metamorphic minerals in impactite rocks (Lipschutz and Anders, 1961), and in high-pressure metamorphic terrains (Rozen et al., 1973). Diamonds in these host rocks are chiefly discrete crystals. But polycrystalline aggregates of diamond, such as carbonado and framesite, have also been found. Carbonado is a porous, black, irregularly shaped, multi-crystal aggregate, typically observed in alluvial deposits. Its relationship to a possible primary source in the mantle remains unclear (De Corte et al., 1998). Framesite is a nonporous, polycrystalline diamond aggregate found in kimberlite. It has recently become recognized as a potential recorder of secondary processes in the lithosphere (Burgess et al., 1998; Jacob et al., 2000). Macrodiamond (>1 mm) pseudomorphs have been found in orogenic lherzolite (Pearson et al., 1989) and metamorphic eclogites (Shumilova, 2002).

Kimberlites, lamproites, and related rocks originate as volatile-rich, ultramafic, and ultra-potassic mantle melts, which are intruded into the crust as pipes, dikes, or sills (e.g., Dawson, 1980). The high-volatile content of these melts results in their rapid ascent through the mantle, facilitated by crack propagation during magma decompression and fluid liberation (e.g., Eggler, 1989; Mitchell, 1995). Emplacement of diamond-bearing diatremes has been episodic over most of the history of the Earth (e.g., Richardson et al., 1984). The oldest documented diamondiferous intrusions are the Argyle lamproite in Australia and the Premier kimberlite in South Africa, both with ages around 1180 Ma (Kirkley et al., 1992). The youngest diatremes discovered to date are the lamproites of the Ellendale field in Western Australia, with ages around 20-25 Ma (Jaques et al., 1986). Kimberlites and related alkaline rock are restricted to stable, continental regions or cratons, with low-geothermal gradients (Gurney, 1989). As observed worldwide, most diamondiferous pipes occur on Archean cratons (Janse, 1984); an exception is the Argyle lamproite, which is located in a mobile belt adjacent to the East Kimberley craton of north-west Australia.

The Siberian craton occupies an area of about 4 x 10^6 km^2, mostly overlain by 1-8 km (4 km average) of Phanerozoic sedimentary cover. The main structural blocks and tectonic zones, as well as known kimberlite pipes, are illustrated in Figure 1.3. The Siberian craton is believed to have formed from the collision and amalgamation (accretion) of micro-continents (i.e., sialic masses) of various ages and geneses (Rosen et al., 2002). In the zones of collision, traces of tectonic compression and over-thrusting can be observed; accretion of terrains occurred in several stages. Generally, larger super-terrains or tectonic provinces appeared before they consolidated into the craton structure.

Delineated by two orogenic fold belts, the eastern Siberian Platform is penetrated by a system of deep fractures that has been associated with kimberlitic activity during

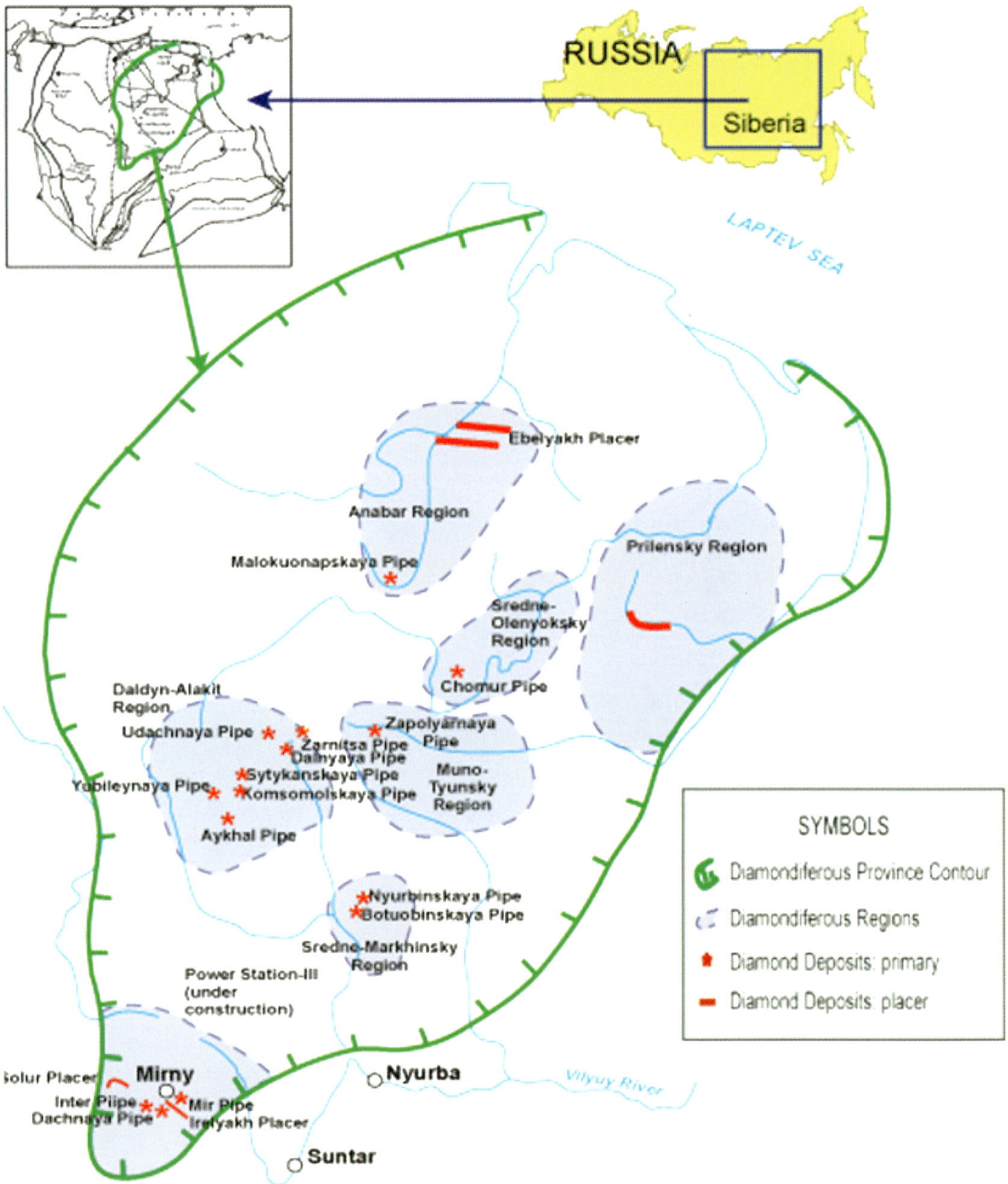

Figure1.3. Siberian platform in the upper left, with the Yakutian Diamond field expanded to show the locations of the various diamond deposits.

periods of fault reactivation within the basement (Brakhfogel, 1984). Basement rocks are exposed in the Anabar and Aldan Shields, located in the north and south, respectively, of the eastern Siberian Platform (Afanasiev et al., 1995), which is believed to have been stable since the early Proterozoic (~2.0 Ga) (Mityukhin, 1995). Over seven hundred kimberlite diatremes and dikes occupy the eastern Siberian Platform, within an area known as the Yakutian kimberlite province (Fig. 1.3). These kimberlite occurrences extend from the Viluy River in the south to the lower reaches of the Olenek and Kotui Rivers in the north, covering an area more than 1100 x 800 km^2. Geochronological studies have shown that most kimberlite emplacement occurred during the Paleozoic between 230 and 380 Ma (Brakhfogel, 1984), and that the upper mantle below the Siberian Platform has been stable since 2.8 Ga (Rosen et al., 2002).

The first kimberlite pipe ever discovered on the Siberian platform is the Zarnitsa. It was discovered on the 21st of August in 1954 by Larisa Popugaeva, a geologist at the Institute of Geological Investigations (St. Petersburg), who used pyrope as an indicator mineral to locate kimberlites. The discovery of this Zarnitsa pipe; [translation: "lightning"], was the real beginning of successful findings of many kimberlite pipes and subsequent mining in the territory of the Yakutian kimberlite province. The Zarnitsa pipe consists predominantly of gray and blue kimberlitic breccias at its upper levels. Until recently, this pipe was considered non-economic, but during the last 8-10 years, it has supplied kimberlitic ores that are processed together with kimberlites from the Udachnaya pipe. It should be noted that diamonds of significant sizes (>10-20 cts) are often recovered from these two pipes.

A number of kimberlite fields have been recognized within the eastern Siberian Platform. Among them, the Daldyn-Alakit and Malo-Botuibinsk fields represent the major diamond-bearing regions within the Yakutian kimberlite province (Fig. 1.3). Of the 700 kimberlites that have been discovered within the province, only ~20 are presently of economic value. Short descriptions of many of these pipes are provided below. These are the pipes from which the xenoliths and diamonds pictured herein were taken.

Discovered in 1955, the Mir kimberlite is well known as the first economic-grade pipe with good-quality diamonds. The pipe is a steeply dipping, intrusive body that is elongated along a north-west direction, with surface dimension of 400 x 320 m (Fig. 1.4a & b). It contains three varieties of facies of kimberlitic breccia. Open-pit mining of this pipe was started in 1956 and ended in 2002, at a depth of about 370 meters. At present, the pipe is being developed for underground mining, with the permafrost posing an engineering challenge.

Another economic pipe in this field is the International pipe. It consists of porphyritic and, more abundantly, kimberlitic breccias (Fig. 1.5). This pipe is unique in its low content of mantle xenoliths and ilmenite in its concentrate, and in high contents of perfect gem-quality diamonds. Of special note is the fact that this is the only Yakutian pipe with water-blue diamonds. Similar to the Mir pipe, the open-pit operation at the

Figure 1.4a & b. The Mir Pipe, the first diamond mine in Yakutia, at early and later stages of its development. Underground mining has now begun.

Figure 1.5. The International Pipe, near Mirny, which has been continued with underground mining.

Figure 1.6. The Aikhal Kimberlite Field, some 100 km SW of Udachnaya.

International pipe ceased at 280 m. Importantly, this is the first Yakutian pipe with underground mining, in operation since 2002.

The Daldyn-Alakit region covers a large area in the central part of the Yakutian kimberlite province in which diatremes are controlled by flexures and deep NE-trending faults. Over 100 kimberlites occur in this region, which can be subdivided into the Alakit-Markhinsky and, towards the southwest, the Daldyn kimberlite fields. Among these kimberlite occurrences, Aikhal, Sytykanskaya, and Udachnaya are the three pipes well known for their economic grades of diamonds, and have been mined during the past three decades. The Aikhal kimberlite, located in the eastern part of the Alakitsky field, is composed essentially of two intersecting bodies of porphyritic kimberlite and kimberlitic breccia (Fig. 1.6). In plan view, it forms a narrow, elongated diatreme concordant with a deep NE-trending fault (Kharkiv et al., 1997). The Sytykanskaya diatreme contains two kimberlite varieties, porphyritic and autolithic kimberlites, within a NE-trending elongated body. The Udachnaya pipe is made up of two steeply dipping intrusive bodies (Fig. 1.7a & b), both containing porphyritic kimberlites and kimberlitic breccias, which were emplaced into sedimentary rocks of Ordovician age (Zinchuck et al., 1990). The two adjacent bodies (western and eastern) are readily discernible by their contrasting degrees of weathering, and have yielded ages of 361 and 367 Ma, respectively (Kinny et al., 1997). the East-pipe is remarkably fresh containging plentiful olivine, whereas the West-pipe contains no real olivine, only weathered relicts.

One of the largest pipes in the Alakitsky field is the Yubileynaya pipe (Fig. 1.8), which has been mined for more than a decade. This pipe is located near the middle of the Alakitsky field, where it is surrounded by Ordovician-Silurian sedimentary rocks. The Yubileynaya pipe is composed of different kimberlite varieties, including porphyric kimberlite, auto-lithic kimberlitic breccia, and kimbelitic-tuff breccia, which were intruded as separate phases during kimberlite emplacement (Kharkiv et al., 1997). A U-Pb age for this pipe has been determined to be 344 Ma (Davis et al, 1980).

Two kimberlite pipes, Botuobinskaya and Nurbinskaya, are known within the newly discovered Nakyn field (Cherny et al., 1998). The Botuobinskaya pipe comprises kimberlite-tuff breccias, auto-lithic breccias, and porphyritic kimberlites. These kimberlites are highly altered, resulting in the formation of carbonates in their groundmass at the upper levels of the pipes and intensive chloritization of phlogopite down to a depth of about 200 meters. They contain no olivine relics, extremely low contents of chromites and picroilmenites, rare pyrope, but abundant almandine garnets, indicative of eclogitic affinities. The Nurbinskaya has yielded the largest number and array of diamondiferous xenoliths ever discovered in any single pipe, including several peridotitic xenoliths. The oxygen isotopes in garnets from these numerous xenoliths shows the most diverse array of values ever seen for mantle xenoliths (Spetsius et al., 2006). The vast majority of these eclogitic xenoliths have $\delta^{18}O$ values significantly above those typical of the mantle, with some of the highest values yet reported for eclogites.

Figure 1.7 a & b. The large Udachnaya Pit with its East and West separate Kimberlite pipes at two stages of its development.

Figure 1.8. The Yubileynaya Pipe in early stages of its development, just northwest of Aikhal.

Chapter 2
DIAMONDS

CHAPTER 2

DIAMONDS

2.1. Diamond Characteristics

Kimberlites are characterized by wide variations in content, morphology, and physical properties of their enclosed diamond crystals (e.g., Orlov, 1977, 1984; Bokil et al., 1986; Harris, 1987, 1992; Gurney, 1990, 1995). It has been amply demonstrated that each kimberlite pipe is characterized by definite morphological varieties of diamonds (e.g., Bobrievich et al., 1959; Bartoshinsky, 1977; Harris et al., 1986, 1987, 1992). These characteristics also include the internal structure of the diamonds as revealed by catholuminescence (CL) of polished plates (Beskrovanov, 1986; Bulanova, 1995) or laser-luminescence tomography of integral crystals (Spetsius, 1994). These regularities were established and repeatedly verified during investigations of representative collections of diamonds from several Yakutian pipes, including Mir, Udachnaya, and Zarnitsa. As detailed by Spetsius (1995), about 50% of the optically clear diamonds from each of these pipes exhibit zonal or sectoral structure caused by the heterogeneous distribution of defects in the diamonds structure. The proportion of such diamonds in Udachnaya is greater than 50%, and somewhat lower in the Mir and Zarnitsa pipes.

Diamonds from these Yakutian pipes also differ in cathodoluminescence of their central domains. The simplest explanation for these distinct diamond characteristics being present within each pipe probably is that different regions of the mantle were sampled and transported by the ascending kimberlite magma. In addition, there are more differences in diamond features between different kimberlite fields than among individual pipes of any separate field.

In a discussion of diamond characteristics pertaining to their genesis, Gurney (1990) noted that these properties could be divided into two categories: 1) primary growth features that characterize the environment and conditions of the diamond crystallization; and 2) secondary physical features that have been superimposed on the diamonds by subsequent processes of resorption and deformation. The paucity of resorbed diamond crystal surfaces in xenoliths of mantle rocks led Robinson (1979) to conclude that this characteristic feature of diamonds in xenoliths was the result of the isolation of the diamonds by the enclosing xenolith, thereby protecting them from being resorbed by the kimberlite melt. The kimberlitic melt can definitely destroy an entire diamond that is not so protected, especially small crystals. This is further substantiated by the definite relationship between the octahedron versus dodecahedron (I.E., secondary) morphology of diamonds as a function of their size, such as occurs in several pipes in Yakutia and in the majority of southern African pipes (Robinson et al., 1984, 1989; Gurney, 1990). The number of dodecahedral crystals varies from one deposit to another and even in different intruding kimberlitic facies (Bartoshinsky, 1961; Gurney, 1990), and is thought to be due largely to the length of time the diamonds were exposed to the resorbing kimberlite medium during transport.

A secondary characteristic of diamonds in different kimberlites is the natural resorption and corrosion of the crystals. The origin of these features depends upon the physical/chemical conditions of the kimberlite magma (i.e., oxygen fugacity) and its intrusive activity. For example, with a given crystal sieve size, the corrosion of diamonds from the Mir pipe is greater than for those from Udachnaya, as detailed in Table 2.1. These morphological features are also thought to be in part determined by the properties of the diamonds themselves (Spetsius, 1995).

It can be seen from Tables 2.1 that the extent of natural corrosion and distortion of diamonds is apparent most when the crystals were sheared, and the sheared planes have regained their characteristic growth features (i.e., "re-generated planes"). Crystals with such re-generated shears increase with depth in many of the Yakutian kimberlites.

Table 2.1. Corroded Features on Diamonds from the Mir and Udachnaya Pipes

Characteristic of corroded features	Size 1-2 mm				Size 1-2 mm			
	Mir		Udachnaya		Mir		Udachnaya	
	# Ds	%	# Ds	%	# Ds	%	# Ds	%
Integral crystals	445	**16.5**	305	**36.0**	460	**13.6**	930	**29.8**
Crystals with **regenerated shear shape**	2146	**79.8**	398	**47.0**	2360	**69.6**	1706	**54.7**
Crystals with **combined shears**	93	**3.5**	68	**8.0**	504	**14.9**	206	**6.6**
Crystals with **mechanical shears**	5	**0.2**	76	**9.0**	67	**2.0**	275	**8.8**
Total	**2689**		**847**		**3391**		**3117**	

Note: aggregates and polycrystalline intergrowths are excluded from these data. Combined shears signify that the diamonds' surfaces contain both natural corrosion and mechanical textures well.

2.2. Crystal Morphology

As a single crystal, diamonds occur in several forms, such as the most common, the octahedron, as well as the cube, cubo-octahedron, and dodecahedron (e.g., Harlow, 1998). Experimental studies have suggested that the morphology of natural diamonds may be a function of pressure and temperature. The classification schemes adopted for describing diamond morphologies are based on external form and surface features (Orlov, 1977; Robinson et al., 1989). Both of these parameters vary widely, and much of the information obtained from synthetic diamond-growth experiments has aided in resolving the specific growth conditions necessary for their development (e.g., Arima et al., 1993; Chepurov et al., 1997). Octahedra are supposedly the high-temperature growth form, whereas cubo-octahedra form at lower temperatures (Wilks and Wilks, 1991; Chepurov et al., 1997). Although the major morphological features are the results of primary growth, these can be significantly altered by subsequent deformation, resorption to produce rounded forms (e.g., dodecahedrons), and etching by corrosive agents (Orlov, 1977; Robinson et al., 1989; Mendelssohn and Milledge, 1995a, 1995b).

Investigating external morphology alone is of limited value for clarifying the processes of diamond genesis, and thus, may not provide accurate representation of the growth history of diamonds. It is the internal morphology of the diamond that can reveal the details of diamond formation, complicated by several growth and resorption stages, and possibly periods of plastic deformation (Varshavsky, 1968; Bulanova, 1995; Taylor et al., 2000).

2.2.1. External form - Natural diamonds exhibit a number of primary forms that can be the result of the conditions at the time of their crystallization (e.g., Moore, 1985; Chepurov et al., 1997). Sunagawa (1984) developed a classification scheme based on the styles of growth displayed by diamonds as follows: 1) single-crystal diamonds, including single, twinned, and aggregated crystals that have a regular crystalline lattice; 2) coated diamonds composed of a crystalline core overgrown by fibrous diamond; 3) cuboid diamonds with a radiating structure of fibrous diamond growth; and 4) polycrystalline aggregates, such as carbonado, framesite, and bort. Except for polycrystalline aggregates, which are usually not considered primary crystals, these growth styles incorporate both primary and secondary forms of diamond, including octahedra, cubes, cubo-octahedra, and dodecahedroids, as seen in the accompanying Plates. Other descriptive classifications have been described by Orlov (1977). These schemes are more flexible and include transitional forms that combine single-crystals and crystal intergrowths, thereby expanding the descriptive terms for the morphological diversities of diamond. In some cases, morphology is assigned to a diamond using its most dominant form shown by the crystal faces. When a diamond demonstrates little or no recognizable form, it is classified as 'irregular.'

The development of crystalline or fibrous diamonds is dependent on the degree of fluid/volatile super-saturation in the fluid source during their crystallization (Sunagawa, 1984). Under low-carbon super-saturation conditions, single-crystalline diamond favors tangential growth and regular atomic arrangement in the crystal lattice. In contrast, fibrous diamond grows during periods of enhanced fluid and carbon super-saturation, a condition that increases the rate of diamond precipitation, resulting in a structurally imperfect growth mode (Sunagawa, 1984).

Octahedron, cube, and cubo-octahedron are three primary or natural growth forms of diamond. The octahedron is the most common primary form of natural diamond that grows via a spiral-growth mechanism. Octahedra display faces corresponding to {111} growth planes, often with deviation away from the ideal isometric form, resulting in a variety of distorted habits. There are two varieties of natural cubes: 1) transparent, plane-faced crystals that display no internal structure (type II) and 2) semi-transparent crystals with complex internal zonation (type III) (Orlov, 1977). Type II cubes have hummocky faces that develop through non-faceted crystal growth, whereas type III cubes have fibrous internal structures. Cubo-octahedron is the primary morphology that is distinguished by the presence of {100} and {111} faces, as a result of a sectoral growth mode.

Contemporaneous growth of fibrous diamond along the faces, edges, and vertices of an octahedral core may produce a combination of all three primary forms. Pseudo-cubic diamonds, exhibiting cubic or cubo-octahedral forms, may evolve from the fine-scale combination of octahedral edges and vertices during growth along surfaces equivalent to the {100} plane (Orlov, 1977). Most rhombic or plane-faced dodecahedra are believed to be coated stones, shaped by fibrous diamond growth and/or resorption. Rounded-faced dodecahedroids (-oid = 'looks like') are considered, however, to be secondary forms caused by resorption. According to Otter and Gurney (1989), resorption is defined as any process that reduces the size of a diamond, and imparts a rounded morphology to the stone. The original growth structure of the resorbed diamond is retained, but is disguised by curved surfaces that truncate any internal layering. The effect of resorption on any crystal is a function of its primary morphology and the extent of the resorption process (Robinson et al., 1989). In the conversion of an octahedron to a dodecahedroid, resorption begins at crystal vertices and extends towards edges, ultimately developing curved-faced crystals that have lost 45% or more of their original mass. The resorption process gives rise to a series of transitional forms that Otter et al. (1994) classified with respect to the degree of resorption. Resorption also affects crystals with cubic faces. During resorption, cubes convert to curve-faced cuboids that ultimately give way to the dodecahedroid form.

Multiple crystal forms develop through twinning or aggregation of diamond during growth, but the conditions in which these forms are produced largely unknown. Twins and aggregates are composed of diamonds with either primary or secondary crystal forms, and are classified according to the form of their constituent crystals. Diamonds can display either inter-penetrating or contact twinning, but the most common one is the macle twin, which is a contact twin whose twin axis is perpendicular to the octahedral plane. The ideal macle is triangular and tabular in habit, with re-entrant or convex angles, often producing a "Star of David" form, which consists of a six-pointed star from 180° rotation about an axis through the octahedral plane. Multiple crystals of diamond may also form simple aggregates, in which the orientation of individual crystals is random, or complex aggregates with parallel growth axes of individual crystals,

2.2.2. Surface features - Diamonds possess a range of surface textures that reflect conditions both during and following growth, i.e., paragenetic history, of a diamond (e.g., Orlov, 1977). Robinson et al. (1989) identified the existence of over 40 individual textures, each of which is considered to develop under specific conditions and can possibly be used to constrain a chronological sequence of events for the formation of diamond. The surface features of diamonds are largely controlled by growth, deformation, resorption, and late-stage etching or corrosion. Crystal growth is known to produce only two surface textures (Robinson et al., 1989), both of which are ascribed to octahedral growth. Crystal growth is also recognized to have produced hummocky {100} surfaces as a result of non-faceted growth of cubic faces.

Surface textures fabricated by deformation imply that the diamond has at some point resided in a solid medium. A commonly observed deformation on diamonds is plastic deformation, which is caused by movement along glide planes parallel to octahedral cleavage, giving rise to lamination lines. Usually, the deformation feature is not obvious unless the affected diamond crystals have been significantly resorbed. Textures resulting from resorption, in general, are influenced by the underlying growth structure of affected crystals and hence, are expressed differently by octahedral, cubic, and dodecahedral surfaces (Robinson et al., 1989).

The remaining surface textures are the products of diamond etching or corrosion that takes place under a range of redox and temperature conditions. Etch features are controlled by the crystallography of the affected crystal, and have specific forms on diamond surfaces. In contrast, corrosion produces irregular surface textures. Some late-stage etch and corrosion textures are manifested by truncating features that develop through a series of growths and resorptions.

Diamonds can also eventually be destroyed by graphitization, dissolution, and oxidation, although only the latter is believed to be important in the mantle environment (e.g., Robinson et al., 1989). The most likely oxidizing agents affecting diamond are volatile species (e.g., CO_2, H_2O) carried by mantle fluids/melts or by kimberlite magmas.

2.2.3. Internal morphology - Whereas external morphology gives some indication of the final stages of diamond growth or destruction, investigation of the internal structure of a diamond provides information on its entire growth history, as so well displayed by several fine investigations (e.g., Varshavsky, 1968; Mendelssohn and Milledge, 1995a). Much of this work can be complemented and extended by recognizing the internal features present in diamond and distinguishing their evolutionary sequence. The influence of various processes during the development of diamond is distinctly visible in polished sections of diamonds through the use of cathodoluminescence (CL). Several Plates at the end of this Chapter illustrate the concept of CL observation of diamonds.

2.2.4. Cathodoluminescence – The use of electron excitation of diamonds to give off typically blue fluorescent colors has proven to be a valuable petrographic tool and has been utilized by many workers (e.g., Milledge et al., 1993; Bulanova 1995; Chinn et al., 1995; Taylor et al., 2000) to investigate the growth history of diamonds. The luminescent properties are revealed when a sample emits non-thermal radiation following excitation by some form of energy (Milledge 1993), particularly by high-energy electrons, thereby producing cathodoluminescence images. Most CL is produced 'near-surface', at the maximum depth of beam penetration in the sample. The intensity of the emitted spectrum is dependent on the depth at which the CL is generated, i.e., the initial beam energy and on the extent to which internal absorption takes place before emission at the surface.

The intensity of intrinsic CL is a reflection of the degree of structural defects in the diamond lattice and the extent to which any defect works to modify light emission. Impurities and radiation damage are both activators of extrinsic CL that contribute to the variations in luminescent properties of diamonds, or any other minerals. As with structural defects, the CL intensity generally increases with the concentration of activator impurities.

Natural diamond is rarely pure and thus, generally displays some degree of cathodoluminescence due to the presence of impurities, particularly nitrogen. The recognition that diamond may possess complex zonation based on its nitrogen distribution has initiated the application of CL to investigate diamond internal structure (e.g., Trautman et al., 1998; Taylor et al., 2000; Bulanova et al., 2002, 2005).

Different stages of diamond growth are usually accompanied by marked changes in zonation that can be readily identified by CL imaging of sectioned crystals (e.g., Bulanova, 1995). The variation in signal intensity as a function of different nitrogen defects and nitrogen-substitutional forms is presently not well understood; therefore, the application of CL to diamond internal morphology is largely qualitative. Luminescence centers may also form in response to radiation damage and deformation. Slip planes are visible under CL in diamonds that have undergone plastic deformation, whereas evidence of brittle deformation and subsequent annealing is usually observed as brecciated regions in affected crystals (Bulanova, 1995; Taylor and Anand, 2004).

The investigation of CL zones in diamonds is somewhat similar to examination of "growth rings in trees " in that these zones reflect the progressive growth and associated events of the diamond formation as recorded in successive rings, the so-called 'stratigraphy' of a diamond. A number of pioneering studies of such internal morphology have improved the recognition and understanding of diamond growth patterns (e.g., Bulanova and Milledge, 1995 and reference therein). According to Bulanova (1995), the internal morphology of diamond may be divided into three descriptive classes. Internally uniform diamonds essentially display homogeneous cathodoluminescence, and are believed to have undergone a single stage of growth. Simple zoned diamonds are characterized by a single mechanism of growth that generates a sequence of zones, distinguished by variations in luminescence intensity. In complex-zoned crystals, the CL patterns vary significantly between the core and rim, typically distinguishing multiple stages of growth. These changes may appear as a continuous progression or abrupt transitions caused by interrupted growth and resorption or deformation. Complex zonation is the most significant evidence for a variety of source conditions during diamond formation. Similarly for micro-diamonds, using cathodoluminescence imaging, Trautman et al. (1998) described three growth sequences: 1) homogeneous or simple growth, 2) changing growth habits, and 3) discontinuous growth involving crystal breakage and/or resorption. More unusual features, such as polycentric cores, are only rarely observed.

It should be noted that CL imaging is only truly representative as displayed in well-prepared diamond plates, thus making it difficult and work-intensive to study whole populations of diamonds. This problem can be overcome, to some extent, with laser crystal tomography that permits investigation of the internal morphology of whole crystals (Spetsius, 1995). A preliminary study of a representative collection of diamonds (~200) has suggested that approximately 50% of the diamonds from Mir, Udachnaya, Yubileynaya, and Zarnitsa are zonal or sectoral, but variations within different groups between pipes are occur.

It is generally expected that diamonds in an assemblage are heterogeneous, because of various combinations of morphology, nitrogen aggregation, and carbon isotopic compositions. However, diamond populations from different localities and hosts reveal some common denominators. For example, euhedral forms, including octahedra, cubes, and more rarely, cubo-octahedra are usually a substantial, if not dominant, form of diamonds within an assemblage. Indeed, the various external morphologies of diamonds are the most conspicuous evidence that would appear to reflect changes in the P-T-X conditions within the mantle during diamond formation.

As we have discussed above, the observable properties of diamonds support a model for periodic and continuous growth of octahedral forms in the mantle environment, although late-stage participation of mantle melts and fluids cannot be precluded, possibly including those associated with kimberlitic magmas. Inhomogeneities of nitrogen and carbon isotopic characteristics within and between growth zones of octahedral diamonds also testify to time-dependent changes in the source region (e.g., Taylor et al., 2000; Bulanova et al., 2002). Notably, these variations are represented by abrupt steps in compositional gradients that are most easily modeled on the premise that there are intermittent changes in source composition and/or growth conditions.

Cubic growth forms would appear to result from a restricted set of P-T-X conditions in the mantle. In contrast, "cuboids" are invariably nitrogen-rich and accompanied mainly by A-centered nitrogen-carbon defects. However, a small number of crystals accommodate these paired defects in combination with singly-substituted or more aggregated nitrogen. Under CL, cuboids luminesce poorly and commonly have simple or homogeneous internal structures. Such features may form concurrent with the development and ascent of mantle melts. The physico-chemical identities among diamonds would suggest preservation, rather than growth, as responsible for such morphological discrepancies. This suggests that the importance of ascending magmas as the source of resorption may have been over-emphasized for development of such morphologies

2.3. Impurities in Diamond Chemistry

The presence of impurities within the atomic structure of diamond is manifest in many of its chemical and physical properties (e.g., Davies, 1984). Despite its relative purity, natural diamond contains a variety of contaminants which can be grouped into: 1) impurities that occupy interstitial sites or substitute for carbon in the diamond lattice (trace elements), and 2) impurities such as fluids, melts, or mineral inclusions enclosed within the diamond body (Meyer, 1985). Both types of impurities provide information on growth processes and compositional variations in the diamond source regions.

Despite the enormous bond strengths of the diamond structure that provide little potential for atomic substitution, impurities are common, with around 57 elements believed to substitute into natural diamond (e.g., Bibby, 1982; Davies, 1984). Of these elements known to exist in diamond, nitrogen and hydrogen are the only significant trace elements to substitute for carbon in the crystal structure. The role of these impurities in the genesis of diamond is not well understood at present. Impurities are believed to be incorporated by diamond during growth, but for many chemical species, their behavior and structural state in the crystal structure have not been adequately characterized. In fact, much of the trace-element chemistry reported for diamond is probably present in the form of micro-inclusions of silicate and sulfide minerals within the diamond-cell network. It is possible that variations in the diamond source regions can be determined by the trace-element signatures of diamond. It would seem desirable, despite the present difficulties of interpretation, to collect information on the trace-element contents of diamonds for relative comparisons. These data may prove valuable for the possible identification of "conflict" (i.e., rebel, blood) diamonds from central and western Africa. A recent development at Macquarie University has been the development of standards and a laser-ablation technique for in-situ measurement of the trace elements in diamonds. The preliminary results from such studies of a world-wide array of diamonds has demonstrated good possibilities for "finger-printing" conflict (rebel, blood) diamonds (Diamond Finger Printing Workshop, Geel, Belgium, 10/2007).

Nitrogen - Nitrogen is the major substitutional impurity found in diamond (Sellschop et al., 1980). Two classes of substitution, Types I and II, are distinguished by the presence or absence of nitrogen in the crystal structure (e.g., Harris, 1987). For nitrogen-bearing specimens, both the nitrogen content and defect nitrogen have proven useful in defining discrete populations with similar genetic affinities (Davies et al., 1980; Taylor et al., 1990; 1996). For this reason, diamond is typically described in terms of the "defect type" created by substitutional nitrogen.

The accompanying reprint by Taylor and Anand (2004) in the latter portion of this book contains figures and discussion on nitrogen substitution in diamond. Nitrogen-bearing (Type I) diamonds typically accommodate between 10 and 5000 atomic ppm of nitrogen, and contain defects that form part of a progressive aggregation sequence (Davies, 1976; Harris, 1987). Nitrogen is believed to be incorporated into the diamond structure during growth, with the substitution of single nitrogen atoms for individual carbon atoms. Type Ib diamonds are dominated by singly-substituted (C-center) defects that are present in synthetic, cubo-octahedral diamond. In nature, C-center defects are common in diamonds derived from ultra-high pressure metamorphic rocks (Sobolev et al., 1994). These defects are mostly restricted to fibrous overgrowths (coats) and crystals with cubic habit (e.g., Taylor et al., 1996). The relative scarcity of natural type Ib diamond in deposits derived from kimberlites implies a low-activation energy required to convert singly-substituted C-defects into aggregated nitrogen defects. Nitrogen aggregation is thought to be a kinetic process, amplified at higher temperatures. Thus, the preservation of type Ib point defects in any diamond requires a cool source region and/or a short-residence interval prior to surface eruption.

With extended resident time in the mantle and/or at higher temperatures than those of type Ib, the singly-substituted nitrogens become aggregated and develop into type Ia paired defects (Evans and Qi, 1982). Nitrogen-pair defects (A-defects), i.e., two nitrogen atoms on adjacent lattice sites, define type IaA diamonds. Further aggregation results in four nitrogen atoms arranged in a tetrahedral coordination with a carbon-site vacancy in the center (B-defects), thereby producing type IaB diamonds. The presence of nitrogen in diamond gives rise to characteristic absorption spectra in the infrared over the 1400 to 1000 cm-1 region. Types C, A, and B-centers generate absorption peaks at wavelengths specific to each defect, so that the degree of nitrogen aggregation at a locale within a diamond can be determined (e.g., Mendelssohn and Milledge, 1995a).

2.4. Stable Isotopes in Diamonds

2.4.1. Carbon Isotopes - Stable carbon isotopes represent an additional method of resolving chemical differences within zones of a diamond, as well as between diamonds (e.g., Boyd et al., 1987, 1994; Javoy et al., 1984; Deines, 2002). The isotopic compositions of carbon reflect the $^{13}C/^{12}C$ ratio, which is expressed relative to PDB ($\delta^{13}C$ = 0‰). The global distribution of $\delta^{13}C$ values for diamond ranges from +5‰ to -36.4‰ and peaks between -5 and -6‰ (e.g., Deines, 2002).

The carbon isotopic ratios of diamond broadly reflect the nature of the possible carbon source from which the diamonds were derived. These isotopic compositions have been shown to correlate strongly with mineral paragenesis (e.g., Sobolev et al., 1979; Deines et al., 1993, 2002). Peridotitic diamonds display a restricted range of $\delta^{13}C$ values, between -1 and -10‰, whereas eclogitic diamonds have compositions that virtually span the entire known range of carbon isotopic compositions for diamond, from 0 to -36‰.

Several hypotheses have been put forth to explain the range and distribution of these carbon isotopic compositions, especially with reference to the restricted range of values displayed by diamonds from individual pipes. Primordial isotopic heterogeneity in the mantle, analogous to that observed in iron meteorites, is one process by which discrepancies in $\delta^{13}C$ composition may take place, provided that significant homogenization has not ocurred since the Earth's accretion. Homogenization of the carbon reservoir in the mantle may account for the relatively restricted range of $\delta^{13}C$ values attained from peridotitic diamond; however, this model finds difficulty in the explanation of isotopically light crystals. Several authors have invoked Rayleigh fractionation as an alternative explanation (e.g., Deines, 1980; Galimov, 1991), but fractionation of a homogeneous carbon reservoir alone is unlikely to account for the known distribution of isotopic compositions.

Cartigny et al. (1998) suggested that high-temperature fractionation of mantle fluids might account for the complete $\delta^{13}C$ range of diamonds, given that these fluids evolved separately in peridotitic and eclogitic environments. The loss of CO_2 from eclogitic sources is thought to leave a $\delta^{13}C$ depleted residue that may crystallize diamonds with a broad $\delta^{13}C$ range. However, experimental studies have suggested that peridotitic diamonds were once in a wide range of mantle conditions, in contradiction with their restricted range of $\delta^{13}C$ values. Other researchers believe that eclogitic diamonds inherit some of their carbon isotopic characteristics from crustal carbonaceous materials that display strong enrichment in ^{12}C, possibly from biogenic sources. Metamorphism or partial melting of subducted, ancient oceanic crust with crustal carbon, combined with a primitive carbon source in the mantle, are considered capable of generating the observed range of $\delta^{13}C$ values for diamonds (Sobolev and Sobolev, 1980). But, much research remains before the definitive answer can be attained.

The C-isotopic compositions of diamonds have been explained by a variety of processes: 1) intra-mantle isotopic fractionation during fluid transport and diamond growth (Deines, 1980; Galimov, 1991), 2) primordial mantle heterogeneities (e.g., Deines et al., 1987), and 3) introduction of heterogeneous subducted carbon (e.g., Sobolev et al., 1979; Kirkley et al., 1991; Navon, 1999). Carbon isotopic studies alone are not able to resolve the source of such C isotopic variability, and the relative importance of these competing processes. Thus, current approaches are integrated studies of C-isotopic composition with N-abundance, N-aggregation, and N isotopic composition, along with age determinations on the same diamonds.

2.4.2.. Nitrogen Isotopes - The application of nitrogen isotopic analyses to diamonds is well established by the pioneering investications of Boyd et al., (1992), Cartigny et al., (1998), Deines et al., (1993, 2001), and Javoy et al., (1984). The data set for N is less extensive than that for C, chiefly due to the lower abundance of N (<10 ppm to 1000 ppm) that requires low analytical blanks. The N-isotopic compositions of diamonds show more isotopic heterogeneity and do not display any clear paragenetic distinction (P- vs E-type), as observed with the distribution of C isotopes.

Except for the unusually large N isotopic range ($\delta^{15}N$) seen in P-type diamonds from Pipe no. 50 in China (Cartigny et al., 1997) and the 'low $\delta^{13}C$ group' of Boyd et al., (1992), P-type and E-type diamonds from many localities have an overlapping N isotopic composition, ranging from –12‰ to +4‰ $\delta^{15}N$ (Cartigny et al., 1998b). This range is identical to that of the modern terrestrial depleted upper mantle, as measured in ocean ridge basalts (Javoy et al., 1986; Marty and Zimmerman, 1999). Close examination of the N isotopic values of diamond allows a crude paragenetic distinction to be made. For instance, E-type diamonds are light in C isotopic composition, but contain isotopically heavy N, extending towards $\delta^{15}N$ values characteristic of sediments and S-type granites (+5‰ to +10‰).

2.4.3. C-N Isotopic Systematics - The combined systematics of $\delta^{13}C$-$\delta^{15}N$ and N_{total} abundances in P-type and E-type diamonds provide the strongest constraints on the abundance and ultimate sources of carbon and nitrogen in the mantle. Cartigny and coworkers (2001) estimated the C/N ratio of a diamond growth-medium to be around 200-500 (mantle C = 400 ppm; mantle N = 1-2 ppm) or as low as 10, depending on the assumptions used for the partitioning behavior of N. The former values would put continental-lithospheric mantle C/N ratios in line with estimates for the modern depleted upper mantle from MORB analyses (Cartigny et al., 2001). P-type diamonds, based on their mantle-like C isotopic composition, would appear to have predominantly mantle N. Collectively, when including the diamonds from Pipe 50, China, the entire P-type diamond population shows a large range in $\delta^{15}N$ from –25‰ to +6‰, which has been ascribed to primordial mantle heterogeneity, an idea inherited from the large isotopic range seen in meteorites (Cartigny et al., 1997). To explain the diamonds with non-mantle-like C and N isotopes characterized by low-$\delta^{13}C$ and high-$\delta^{15}N$, Cartigny and coworkers (1998, 1999, 2001) argued for a fractionation process in the diamond-forming fluid. They reasoned that mixing between mantle and sedimentary end-members could not reproduce key features of their data. Specifically, recycling of crustal material is not consistent with the occurrence of E-type diamonds with negative $\delta^{15}N$ or a lack of concave-upward data arrays in the $\delta^{13}C$-$\delta^{15}N$ plot that would be expected by the high C/N of a crustal fluid. Other features that are inconsistent with this model include the drop in N content of a diamond when $\delta^{15}N$ increases and a positive correlation of N content with $\delta^{13}C$ seen in some eclogitic diamond suites. Indeed, detailed C and N isotopes on synthetic diamonds demonstrate the unexpectedly large variations that can occur at a given P-T, starting with a homogeneous melt system (Reutsky et al., 2002).

Navon (1999) objectively argued against recycling models and proposed a model in which the C-N isotope systematics of E-type diamonds might be explained as a mixture of mantle and subducted components. This model overcomes the objections of Cartigny et al. (1997, 1998) to subduction scenarios by using mantle and sedimentary-mixing end-members of variable compositions, variable C/N ratios of the diamond-forming fluids, and poorly known N-isotopic fractionations during diamond growth. However, these N fractionations may be determined by an in-situ, high-spatial-resolution technique (Taylor et al., 2000; Hauri et al., 1999, 2002; Bulanova et al., 2002). Indeed, the large range for N -isotopic composition of P-type diamonds (Cartigny et al., 1997) and the lack of data on N isotopes in Archean crustal rocks require variable end-members to be considered. The future resolution of the subduction-fractionation debate may well be addressed by detailed *in-situ*, high-spatial resolution C-N isotopic studies on typically complex-zoned diamonds, using a combination of CL, FTIR, and SIMS analyses. Such studies could possibly demonstrate large co-variations in coupled C-N isotopic compositions that correlate with diamond growth zones, these generally do not corroborate thermodynamic fractionation during diamond growth models (e.g., Fitzsimons et al., 1999; Bulanova et al., 2002). Furthermore, they suggest decoupling of the C and N isotopic compositions and point toward temporal changes in fluid composition during diamond growth (Fitzsimons et al., 1999; Cartigny et al., 2001), perhaps, inherited from isotopic heterogeneities in the continental lithosphere.

2.5. Volatiles and Fluids in Diamonds

The first indication of the presence of fluid inclusions within diamonds was reported by Chrenko et al. (1967), who used infrared spectroscopy to identify the presence of water and carbonate in the fibrous coating of an octahedral diamond. Subsequent studies have demonstrated that fibrous diamonds, with their porous nature, contain high densities of fluids trapped during diamond growth. Hence, this type of diamond offers an insight into the chemistry of mantle fluids from which fibrous diamonds grow (Navon et al., 1988). Fibrous diamonds commonly occur either as cubic crystals, where the entire diamond is fibrous throughout, or as coatings on octahedral diamonds. Fibrous diamonds are widespread and have been reported from almost every pipe worldwide. Although generally forming less than 1 wt.% of mine production, fibrous diamonds can represent up to 8 wt.% of production at Jwaneng, Botswana (Harris, 1992). Fluid inclusions have also been found in "cloudy" monocrystalline diamonds (Izraeli et al., 2001).

Fibrous coats have been ascribed by Sunagawa (1984) to supersaturation of the growth medium with respect to carbon, resulting in rapid cubic growth around pre-existing crystalline diamond nuclei. Navon et al. (1988) found micro-inclusions rich in water and other volatiles in cubic and coated diamonds from Zaire and Botswana, and interpreted these inclusions to represent volatile-rich fluids from the upper mantle. The bulk compositions of the micro-inclusions resemble a potassic melt and do not correlate with either a peridotitic or eclogitic paragenesis. Navon et al. (1988) suggested that the growth medium contained residual volatile-rich fluids, originating from alkaline melts, which crystallize at depth. Ascent of a later kimberlitic or related magma was required to transport these diamonds to the Earth's surface. It should be pointed out that fluids of similar compositions influence partial melting in eclogites (Spetsius and Taylor, 2002), thereby allowing a possible genetic link to both fluids. With a similar model, Boyd et al. (1994) suggested that coated diamonds form in spontaneous events associated with periods of mantle metasomatism. Cartigny et al. (1999) determined that the young-age nature of the fibrous coating on many diamonds is in keeping with their possible derivation from kimberlitic fluids.

Volatile-rich fluids could be released during the pre-kimberlite development of asthenosphere-derived kimberlitic magmas that have failed to erupt. Evolved fluids may initially be saturated with respect to carbon or become so during interaction with a reduced lithospheric environment. Infiltration of these carbon-saturated fluids into regions of diamondiferous lithosphere and consequent interaction with pre-existing diamond nuclei possibly induce the growth of new and fibrous diamond.

Analyses of the fluid inclusions within fibrous diamonds from several worldwide locales show that the composition of fluids in diamond populations from different pipes is uniform but varies within individual crystals. The fluids range in composition from carbonatitic to hydrous end-members. The carbonatitic one is rich in carbonate, CaO, FeO, MgO, and P_2O_5, and the hydrous fluid is rich in SiO_2 and Al_2O_3 (Schrauder and Navon, 1994). K_2O contents are high, and Mg# is low in all inclusions. Fluid inclusions from cloudy diamonds contain much higher Cl contents and are classified as brines that are distinct from the other fluids of fibrous diamonds (Izraeli et al., 2001). Pal'yanov et al. (2005, 2007) have performed numerous experimental studies directing addressing the nature of the mantle-fluids during diamond genesis. Tomlinsin et al. (2006) have reported fluids from diamonds that they feel characterize mantle fluids. Recent studies by Klein-BenDavid et al. (2006, 2007) on fibrous diamonds have elucidated considerable on the possible nature of the diamond-forming fluids. The convergence of diamond observations with the experimental data are rapidly closing the gap in knowledge ofthe fluids from which diamonds originate in the mantle.

It is noteworthy that beside the main volatile fluids rich in hydrogen, noble-gas and halogen components are reported in coated diamonds from Africa and Canada (Burgess et al., 1989, 1992). Recent investigation of coated diamonds from the Aikhal kimberlites, showed that only low abundances of halogens (Cl, Br and I) are present in fibrous coats of these diamonds (Burgess et al., 2002). Abundant sub-micron fluid inclusions have an isotopic signature similar to that of upper mantle mid-ocean ridge basalt (MORB). Based on halogen concentrations, the presence of three distinct Ar components, and a high $^{40}Ar/^{39}Ar$, it is most probable that the coatings on these diamonds were grown just prior to or during kimberlite eruption. . Indeed, the recent findings of primary alkali chlorides and carbonates in the extremely fresh portion of the Udachnaya-East pipe by Kamenetsky et al. (2004), Zedgenizov et al. (2004), and Maas et al. (2005) have provided definite evidence for the nature of the diamond-forming fluids.

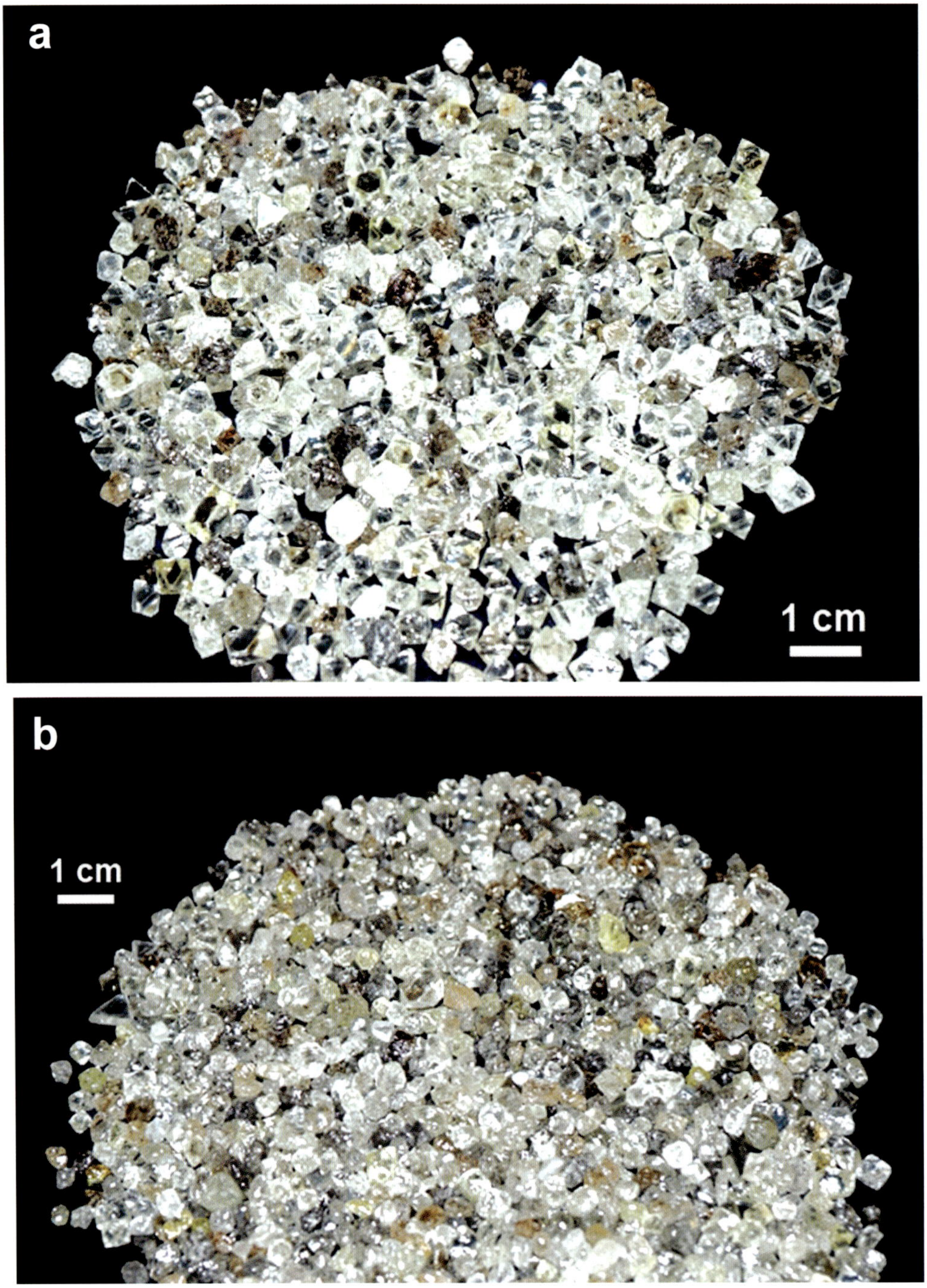

Plate 2. 1. Diamond Production from the well-known kimberlitic regions of the Yakutian Province. a) Malo-Botuobinsky region; b) Alakitsky region (see map, figure 1.3). These are the average diamond populations, unsorted, but sieved to size. Notice the high proportion of clear stones and wide distribution of octahedral crystals.

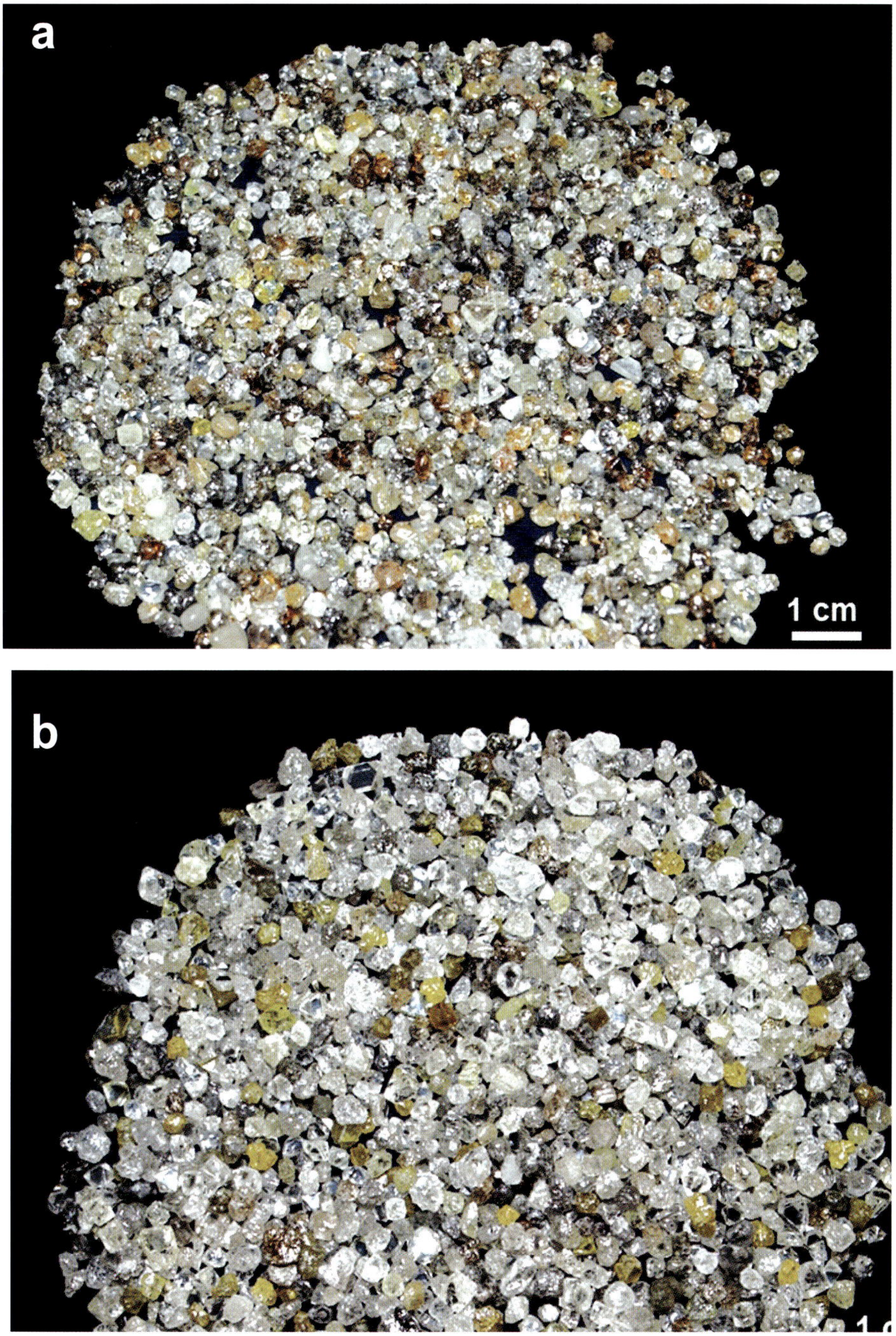

Plate 2. 2. Diamond Production from newly developed kimberlitic regions of the Yakutian Province. a) Anabarsky region; b) Nakynsky field (Nyurbinskaya Pipe). In comparison with the Malo-Botuobinsky and Alakitsky region in Plate 2.1, there is a larger distribution of rounded and dodecahedral crystals, as well yellow cubes and grey aggregates from these locales.

Plate 2. 3. Colored Diamonds of Various Sizes (10 - 50 carots) from the International Pipe, near Mirny.

Plate 2.4. Different Diamond Crystal Morphologies. a) Clear octahedron in kimberlitic rock from the Udachnaya pipe; b) Combined growth form of water-clear diamond, containing a small garnet inclusion; c) Rounded octahedron, due to the stepped growth pattern on the {111} faces; notice the 100 μm pyrope inclusion; Yubileynaya pipe; d) Dodecahedron, largely due to the resorption of the diamond rounding the crystal edges, Mir Pipe.

Plate 2. 5. Different Diamond Crystal Morphologies. a). Yellow cubic diamond from the Udachnaya pipe; such cubic diamonds are generally rare from Yakutian pipes; b). Diamonds intergrown based on a spinel twin on the [111].

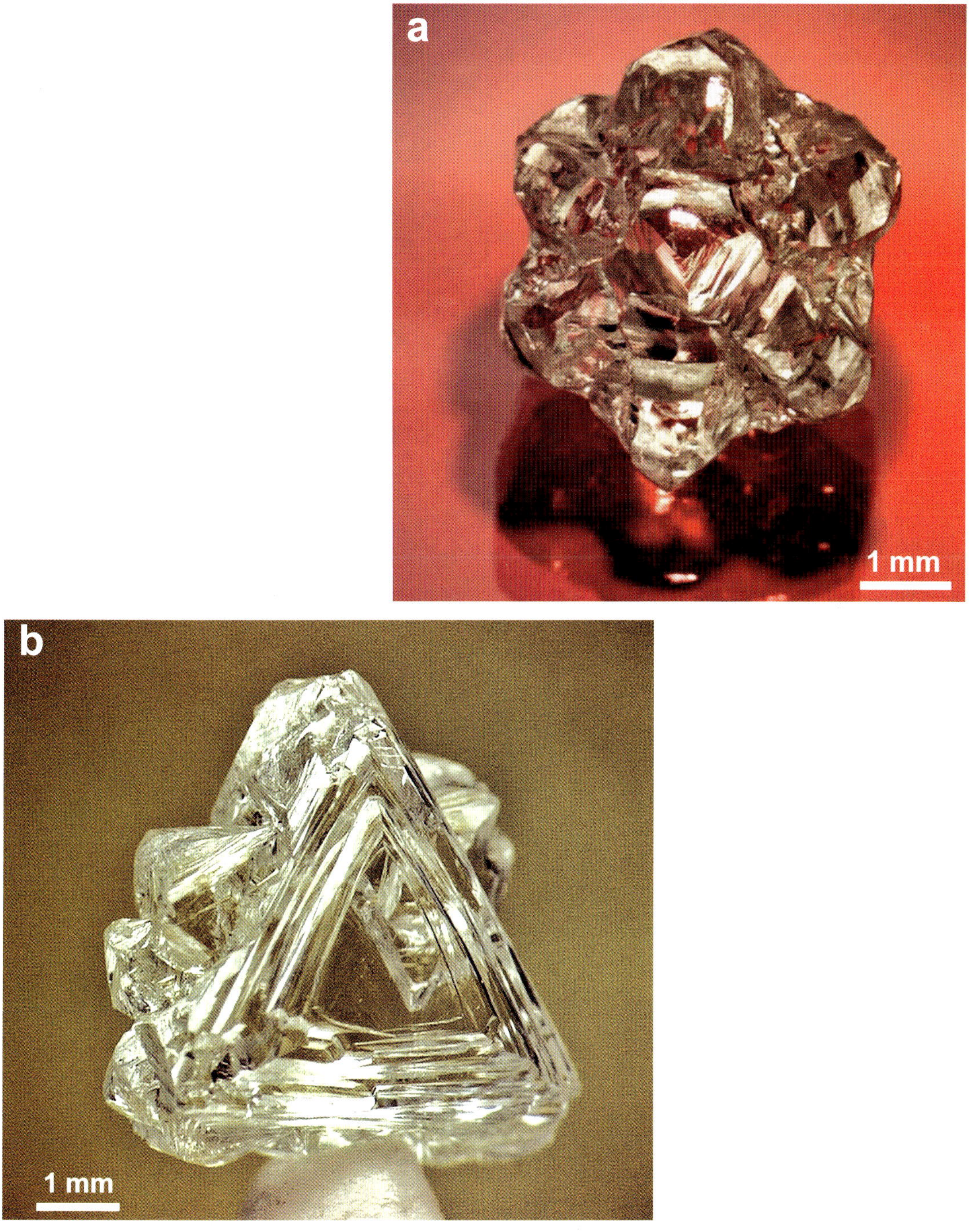

Plate 2. 6. Rare and Unusual Diamond Forms from Yakutian Kimberlites.
a) Transparent, slightly yellow rose-like twinned intergrowths, weight 1.3 carats;
b) Intergrowths of 5 small octahedral crystals twinned about the [111], demonstrating multiple stages of growth.

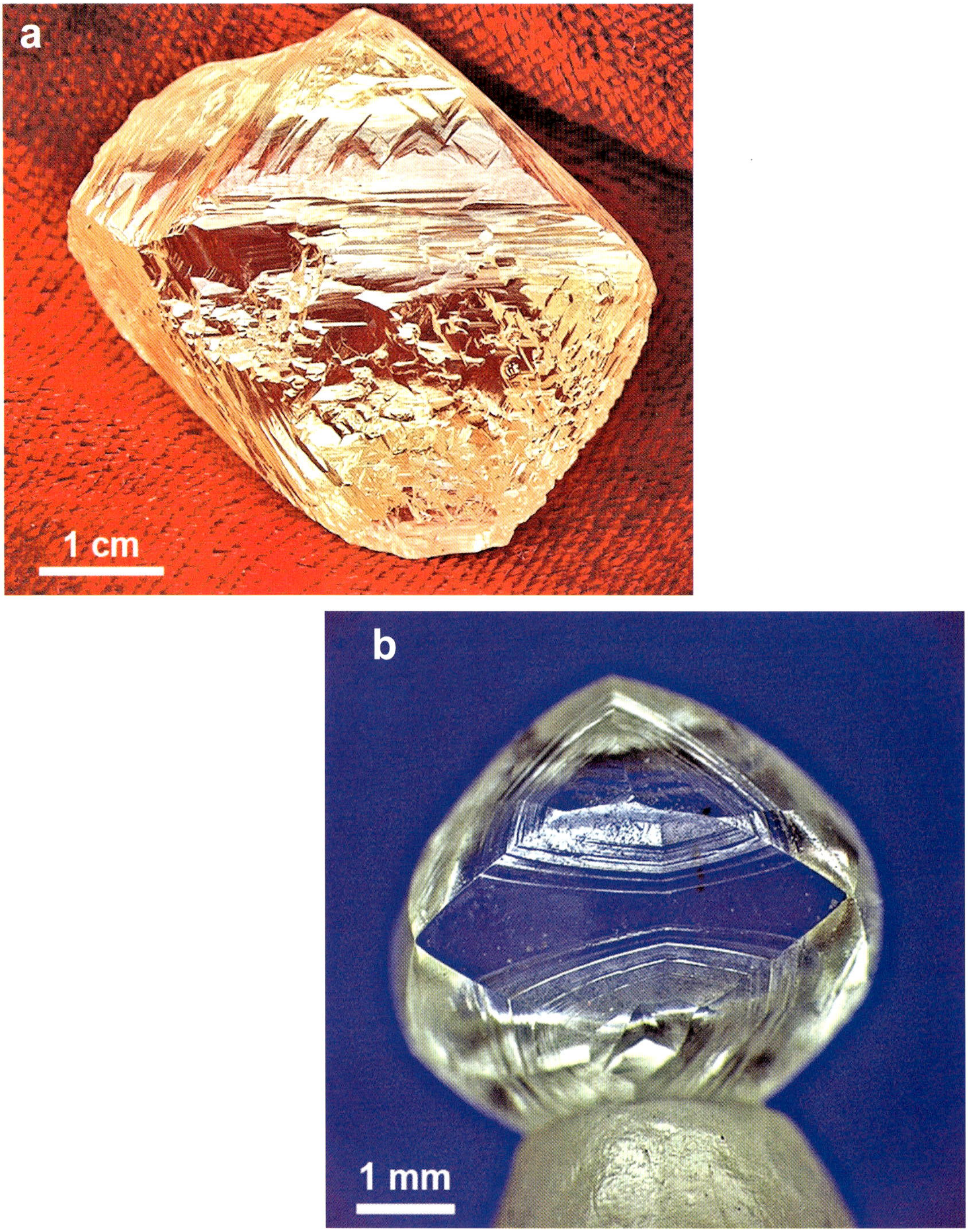

Plate 2. 7. Diamond Surface Features Resulting from Resorption and Corrosion. a) Corrosion sculpture on translucent yellowish octahedral form crystals from the Yubileynaya pipe, weight of crystal 125.8 carats; b) Transparent octahedral crystal with smooth regenerated surface from the Anabar placer; weight 0.7 carats.

Plate 2. 8. Examples of Etching and Resorption on Nurbinskaya Diamonds. a) An octahedral crystal with evidence of strong etching; notice the two parallel channels that have resulted from the corrosion; 1.8 carat; b) An octahedral crystal with strong etching features on its edges, deeply cut and corroded; 3.1 carat.

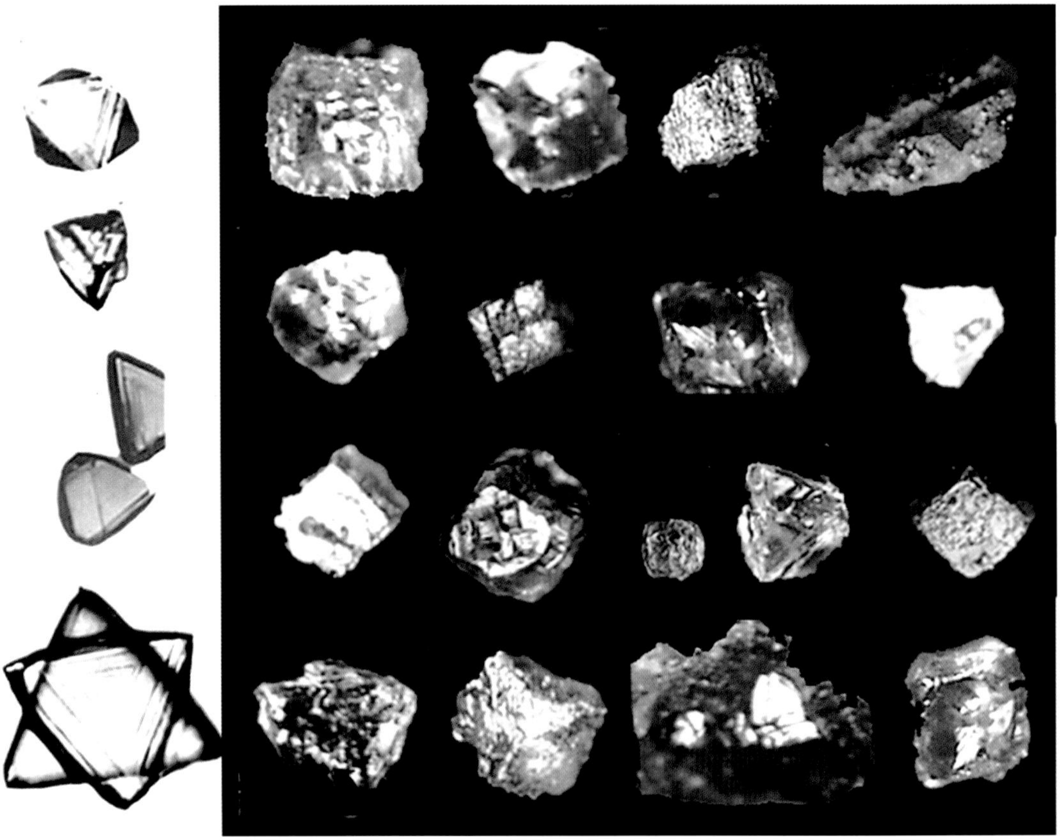

Plate 2. 9. Diamond Morphologies from Udachnaya. (black-white photo) Notice that many diamonds are of cubic habit, and some crystals have irregular forms. On the left side of the plate are crystals from a Gt-orthopyroxenite, mostly with octahedral forms; a rare "Star of David" is composed of a macle, a pair of octahedral diamonds twinned about the [111], perpendicular to this (111) faces

Plate 2. 10. Complex diamond intergrowths from the aikhal Pipe. a). Several octahedral diamonds; b). Intergrowth of diamonds that have been rounded by resorption, approximately 100 carats.

Plate 2.11. Rare and Unusual Diamond Forms from Yakutian Kimberlites. a) Transparent circle twin, Solur mine. b) Translucent irregular intergrowth of octahedral crystals displaying strong resorption.

Plate 2.12. Rare and Unusual Diamond Forms from Yakutian Kimberlites, both set in Kimberlite.

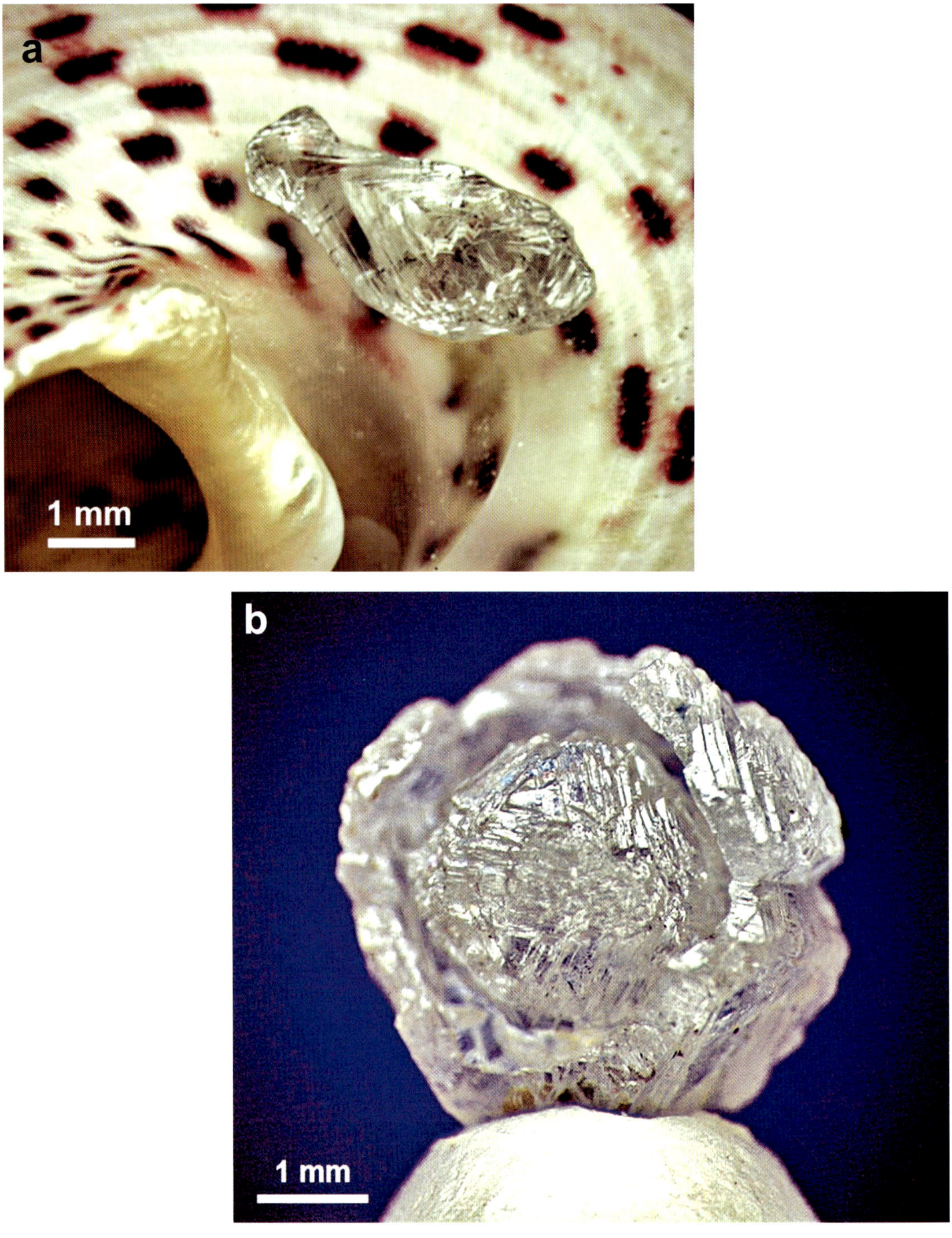

Plate 2.13. Corroded Diamonds. a) Strongly resorbed transparent crystal fish-like form with traces of octahedral faces; b) Rounded crystal with corroded and regenerated faces, from the Anabar placer.

Plate 2.14. Primary and Secondary Features of Udachnaya Diamonds. a) Intact crystal of perfect flat-faced octahedron with some rounding on edges; b) Broken and corroded crystal of step-faced octahedron with partially regenerated surface; c) Crystal with combination of forms but with strong evidence for resorption and corrosion. Late-stage corrosion is particularly apparent from the presence of elongated channels from etching.

Plate 2.15. This 100 carat gray-black diamond from Udachnaya is but a small part of a huge (4000-5000 CT) diamond that was detected by the x-ray radiation in the production cycle. Unfortunately, the crushing process had broken this suspected huge diamond into numerous pieces all gathered at the same time, some as large as 6-700 carats. This multi-crystal aggregate contains "veins" of clear single crystal diamond as well.

Plate 2. 16. Cubic Diamonds with Zoned and Complicated Internal Morphologies. Much of the mass of each stone represents a fibrous-growth mode, typically indicative of rapid growth, culminating in the yellow outer rim. a) From Udachnaya; note the dull, almost granular aspect to the inner portions of these diamonds; the tree-ring design in b) represents a discontinuity (hiatus) in its growth.

Plate 2. 17. Cubic Diamonds with Zoned and Complicated Internal Morphologies. Much of the mass of each stone represents a fibrous-growth mode, typically indicative of rapid growth, culminating in the yellow outer rim. From Yubileynaya; the fibrous-growth zones are well displayed showing changes in the diamond growth over time; this type of diamond will probably contain numerous melt and fluid inclusions, such as described by Navon et al. (1999); different growths zones are present in inner part of crystals, culminating in the yellow over-growth a) and b).

Plate 2.18. Rare and unusual diamonds forms from Yakutian kimberlites.
a) Transparent circle twin, Solur mine. b) Translucent irregular intergrowth of octahedral crystals with the strong resorption, Yubileynaya pipe.

Cathodoluminescence

Cathodoluminescence Imaging of Diamonds

- Illustrates Growth Patterns & Diamond stratigraphy.
- Reflection of Nitrogen Abundances & Aggregation Status
- Shows Episodes of Crystallization, Resorption, changes in growth modes, even Plastic Deformation.

Nitrogen Aggregation in Diamond

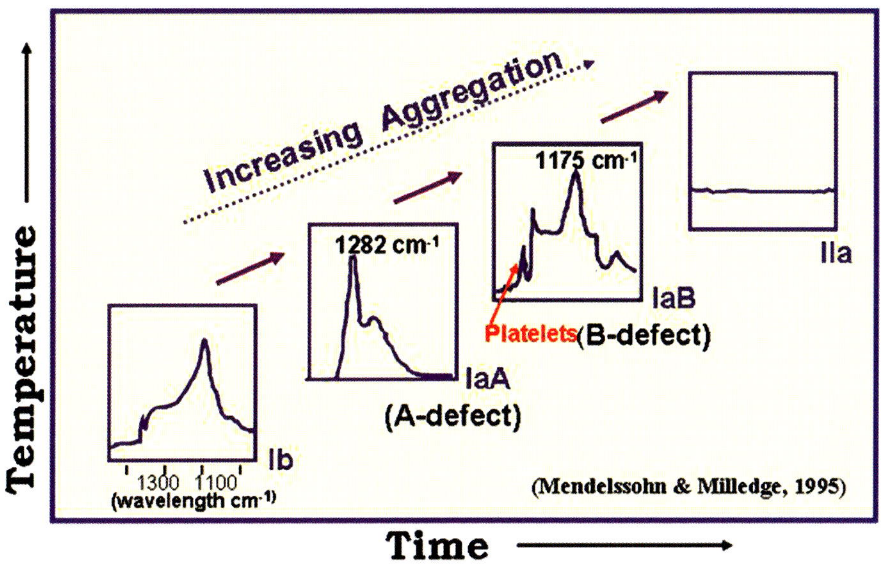

1b = Single N, as a point defect, into the diamond structure. Most synthetic diamonds

1aA = Pair of N on adjacent sites, A defects, in structure.

1aB = Four N arranged tetrahedrally about a vacancy in structure. Development of "platelets" as well

IaB in yellow color = Three N and a vacancy (N3 centers).

II = No N substitution in structure.

Plate 2.19. Principles of Cathodoluminesence in Diamonds

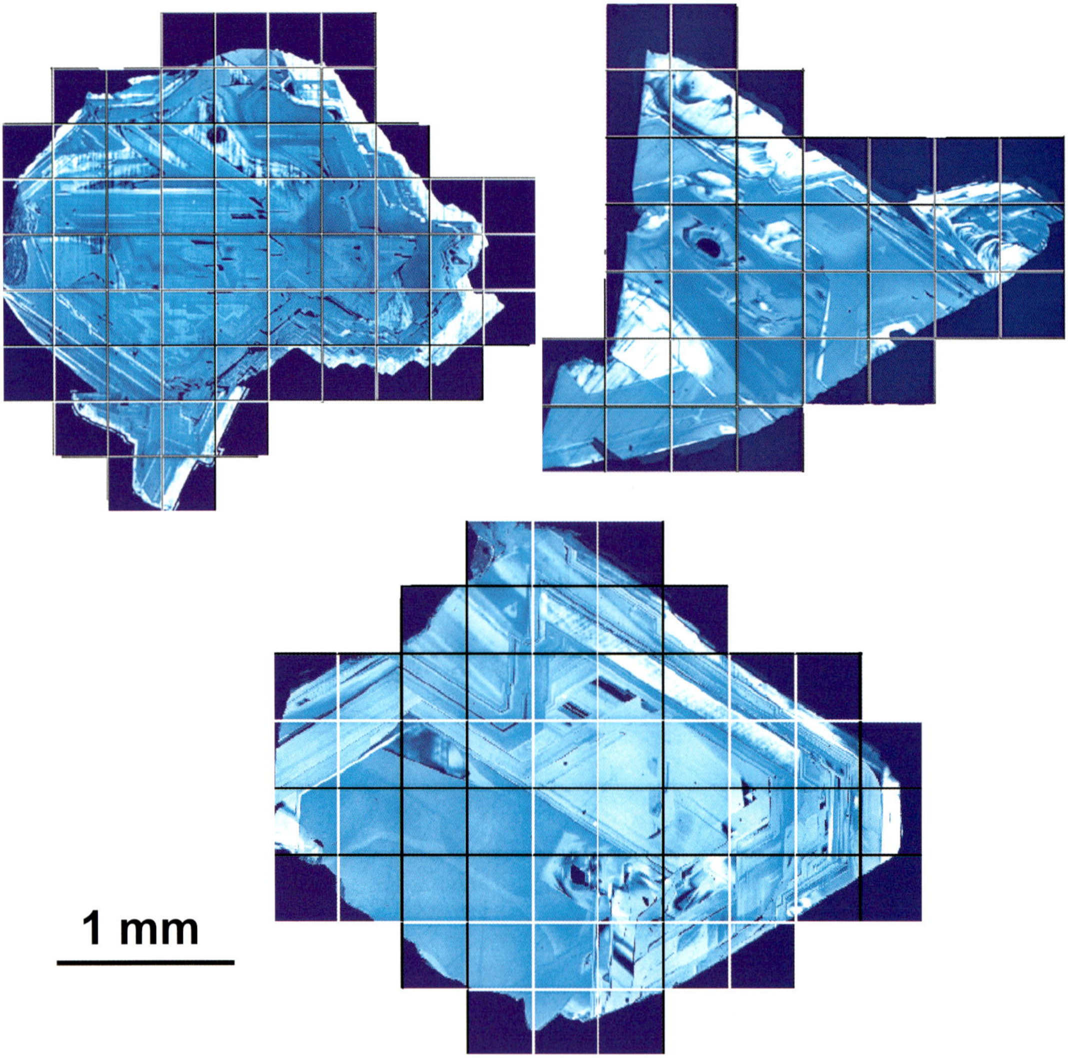

Plate 2.20. Cathodoluminescence (CL) of Eclogitic Diamonds from Udachnaya. The various zones in the diamonds, called stratigraphy by diamond scientists, are akin to the growth features of tree rings. They reflect different degrees of nitrogen aggregation and/or total nitrogen content. These zones can be used to unravel the "life history" of a diamond, as to its initial growth mode (e.g., cubic, octahedral), cessation of growth with possible resorption, changes in growth mode, etching episodes, brittle deformation, incorporation of mineral inclusions, and even plastic deformation.

By polishing diamonds to reveal from two up to as many as six mineral inclusions on the same surface is that their relative ages can be determined. In this manner, it is possible to track the nature of the chemical and mineralogical environment of the diamond during the various time-intervals of its growth-life.

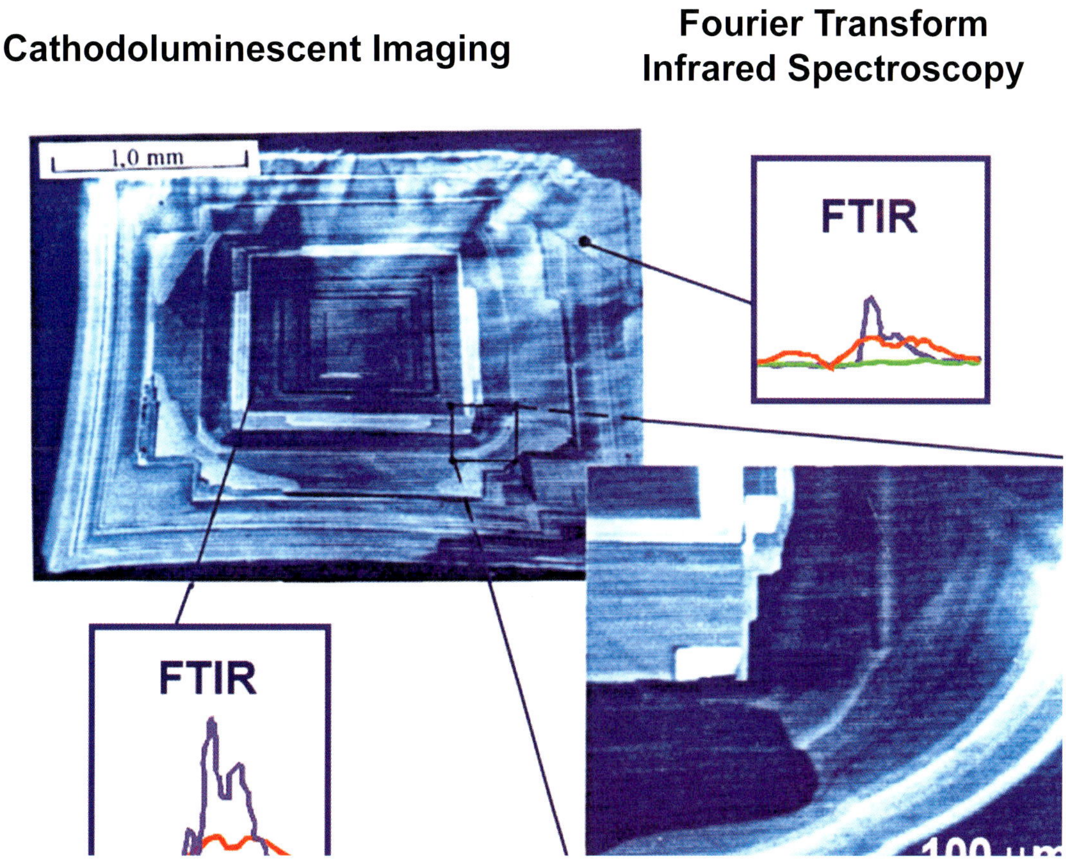

Plate 2.21. CL Imaging of a Diamond Cut approximately Parallel to the (100) face. It can be readily seen that the diamond had periods of no growth accompanied by resorption (insert), followed by renewed growth. FTIR analyses of the two main portions of the crystal, namely its core and rim, reveal entirely different nitrogen aggregation states. Such apparently large differences in nitrogen aggregation are not well understood, although they are largely attributed to differences in annealing temperatures and times for the zones (Taylor et al., 1990).

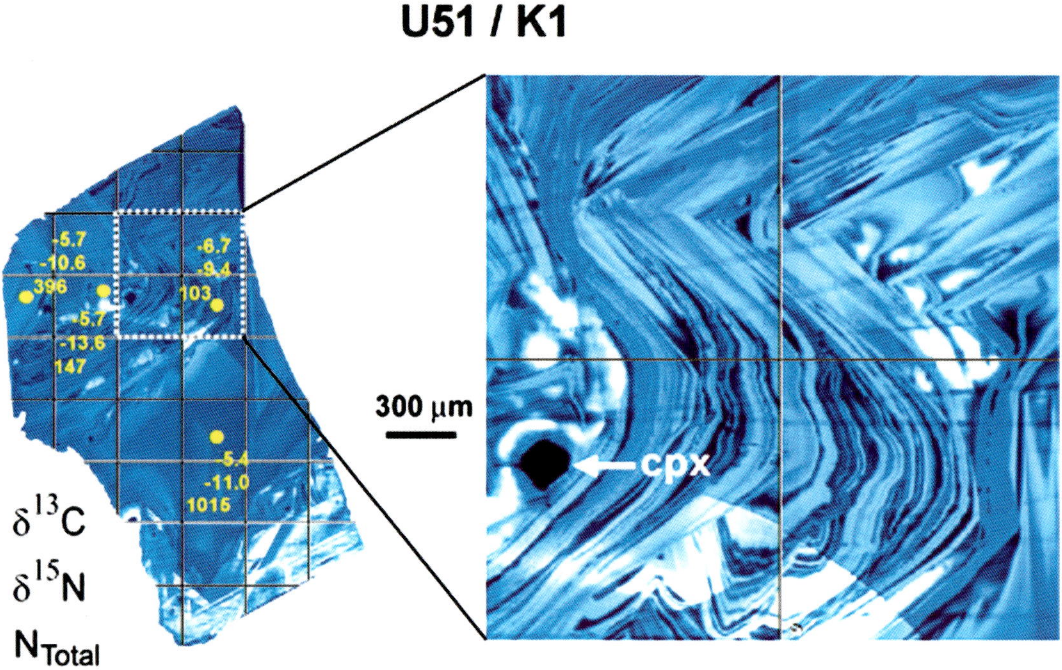

Plate 2.22 CL Imaging of an Unusually Complex Diamond from an Eclogite collected in the Udachnaya Pipe. This diamond has been polished to reveal a clinopyroxene at its surface. SIMS analyses of the different zones in this diamond were performed for δ13C, δ15N, and total Nt. Notice the large variation in Nt over a few hundred microns (modified from Taylor et al., 2000).

In the case that a diamond grew as a constant rate always in the same chemical environment, it would have stratigraphy quite similar to tree rings. However, this diamond has obviously undergone a hyper-eventful life with all sorts of intervening events, followed by extreme plastic deformation. How all this strange behavior of the diamond occurs is largely unknown at present.

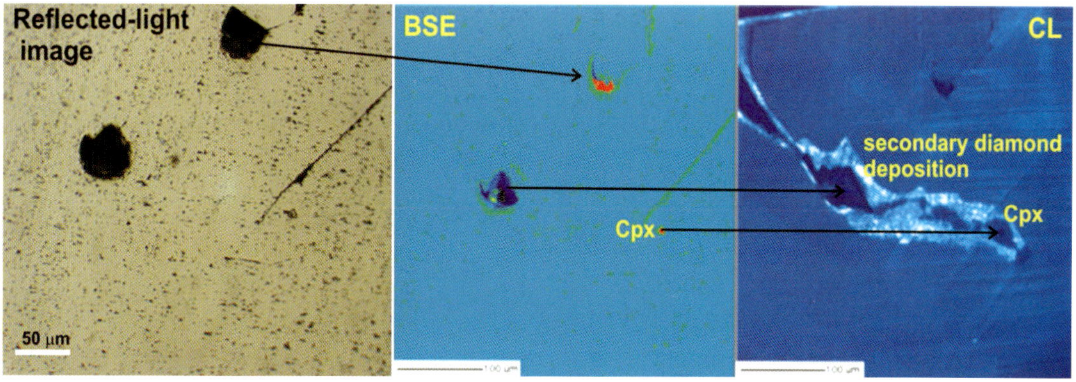

Plate 2.23. llustration of the Use of CL for Finding Otherwise Invisible Cracks Along which Later Metasomatism May have Occurred. The three pictures are of the exact same areas of a polished diamond surface. The left photo taken in reflected light shows only the two inclusions, the same as shown by transmitted light photographs. The middle photo is a Back-Scattered Electron (BSE) image of the spot revealing only the same two clinopyroxene inclusions. However, the CL image on the right reveals the previous presence of a large crack, at least partially healed by secondary diamond growth.

The presence of such optically invisible cracks may have been avenues for "Open-System" metasomatism, which could have easily modified the inclusion, including complete alteration to some unexpected phases, such as amphibole or chlorite. Caution must always be practiced when interpreting the mineralogy and chemistry of diamond inclusions because of the possibilities of such possible open-system cracks.

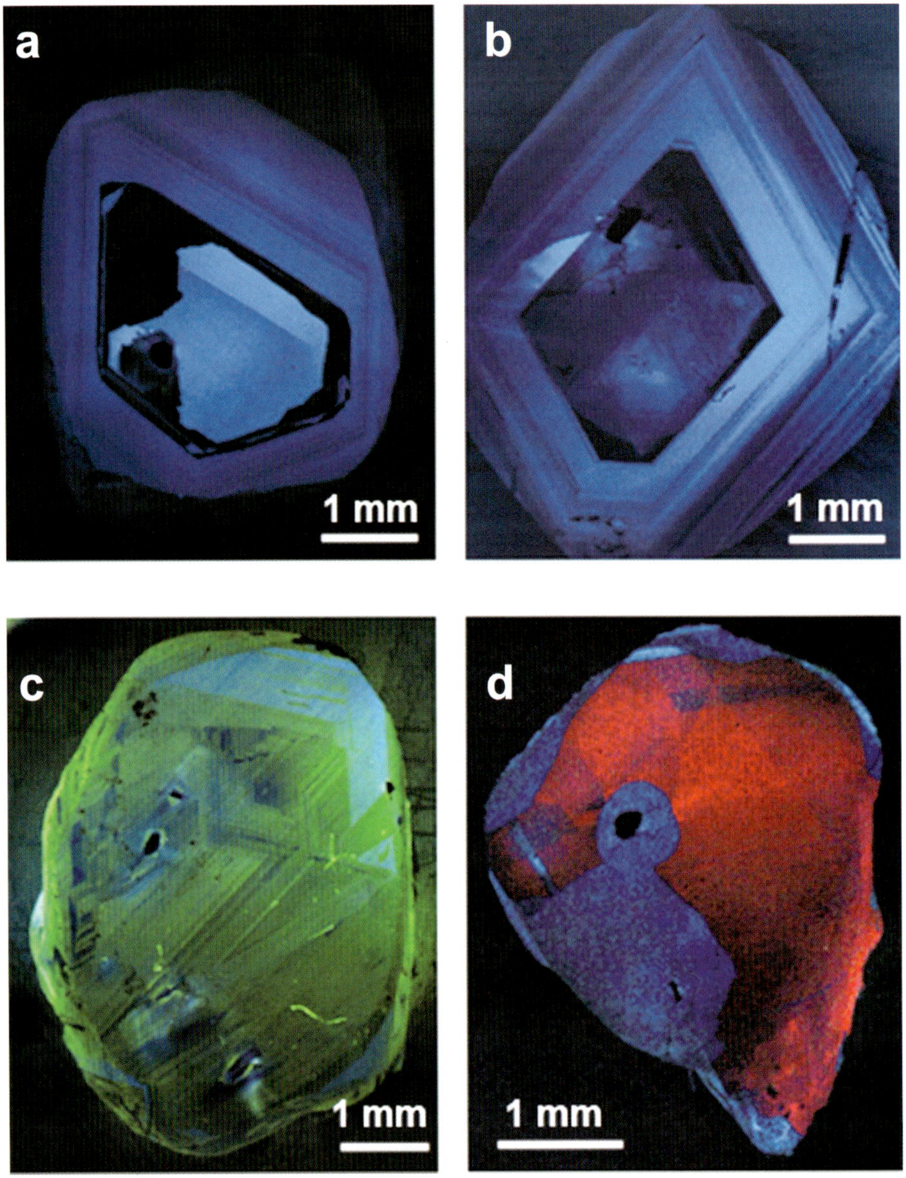

Plate 2.24. Cathodoluminescent Images of Diamonds Showing Complicated Internal Morphology and Zonal Growth of Crystals. a) Complicated and interrupted growth of diamond is obvious in central part of crystal; Yubilena; b) Different zones, showing interrupted diamond growth; Udachnaya; c) Dodecahedral crystal that most probable was formed as the result of resorption; Yubilena; d) Rounded and partly broken crystal showing a high variation in nitrogen content exhibiting secondary re-growth; Mir.

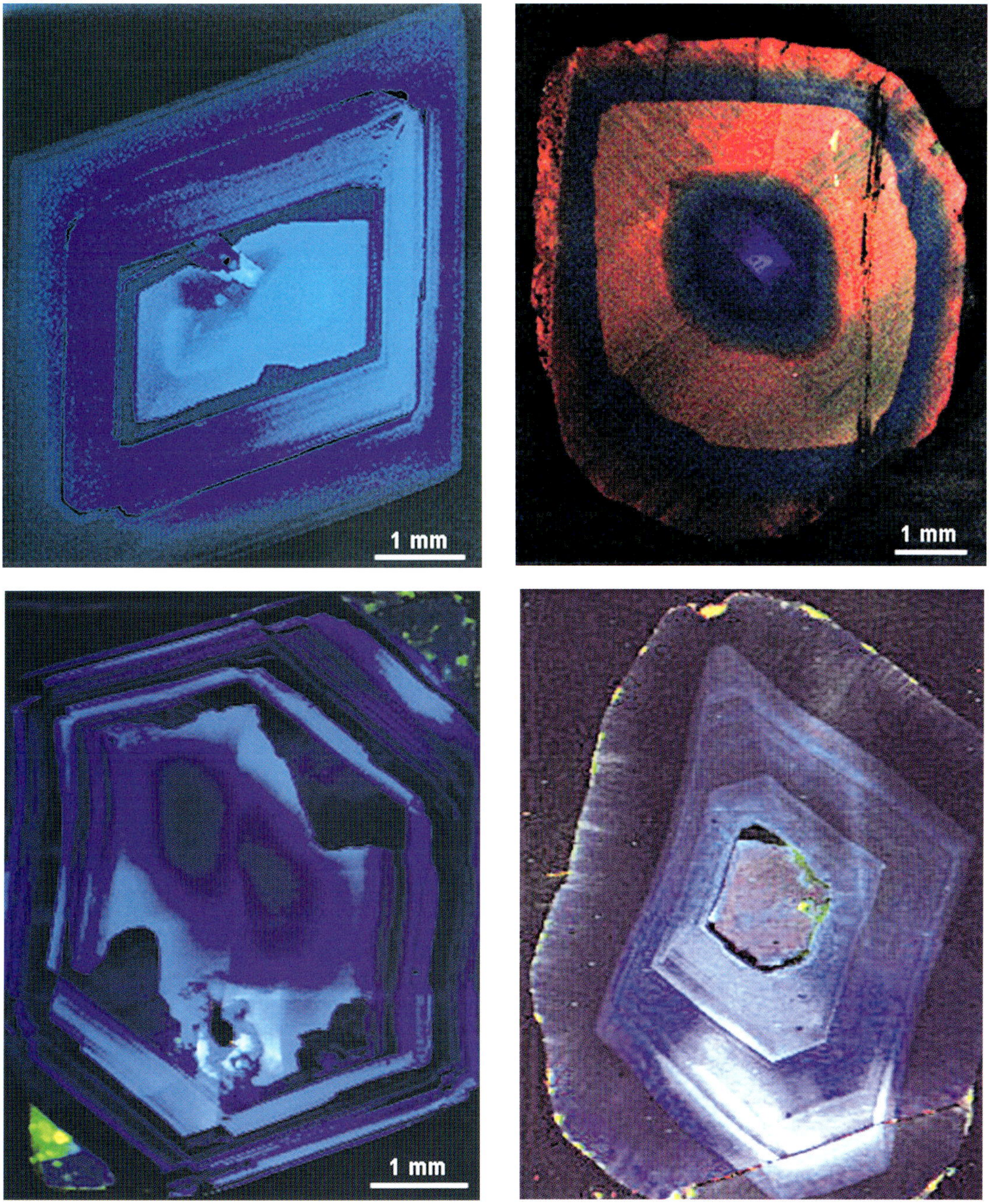

Plate 2. 25. Cathodoluminescent Images of Multi-Stage Diamond Growth, Resorption, and Continued growth; Mir.

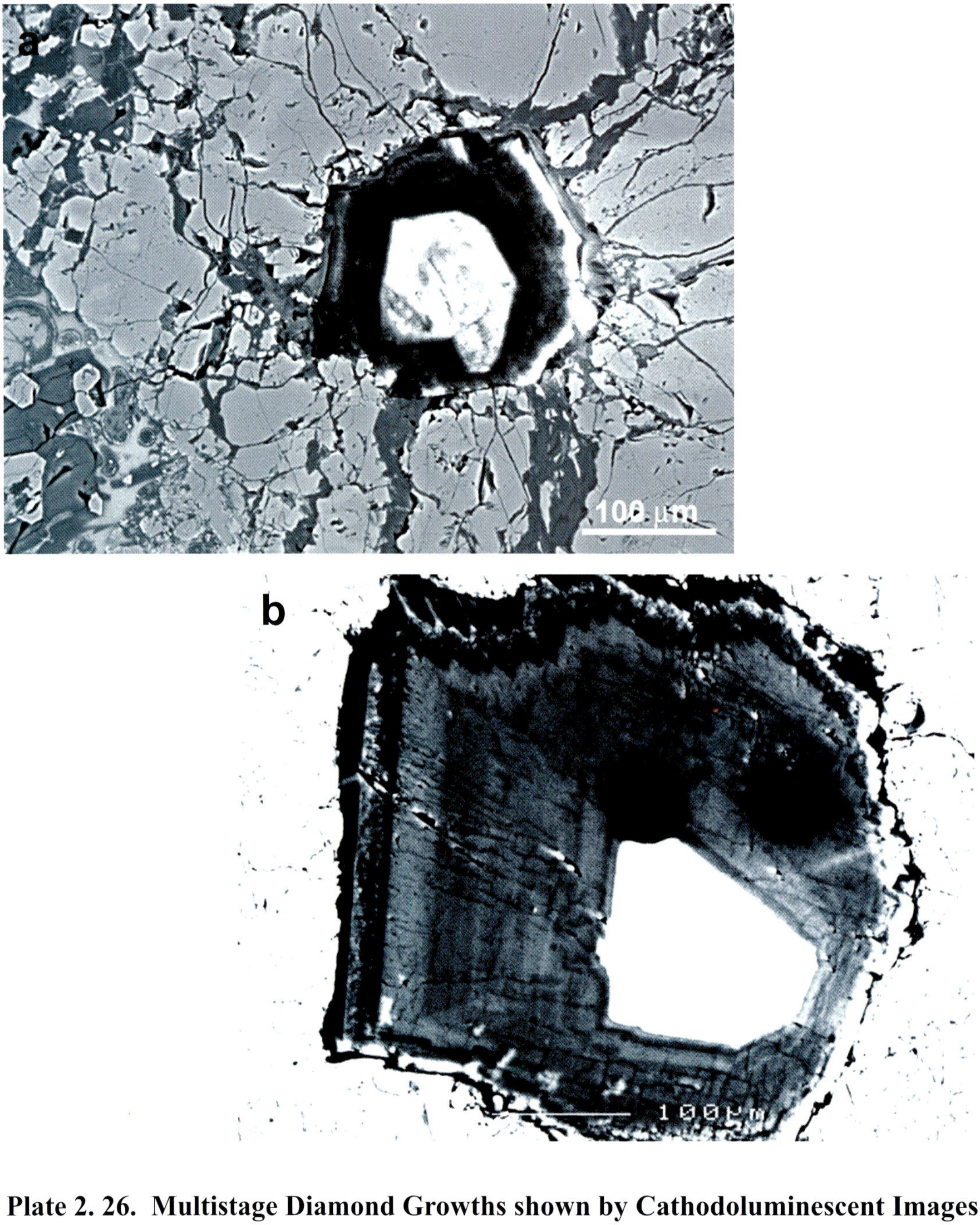

Plate 2. 26. Multistage Diamond Growths shown by Cathodoluminescent Images of Udachnayan Diamonds in Gt-Pyroxenite and Eclogite Xenoliths (BSE-Images+ CL). a) Octahedral crystal in clinopyroxene, two growth zones with interrupted rim can be seen; note the selvage of alteration between the diamond and the pyroxene; b) Octahedron with at least three obvious growth zones, in eclogite; two round sulfide inclusions are seen close to the surface.

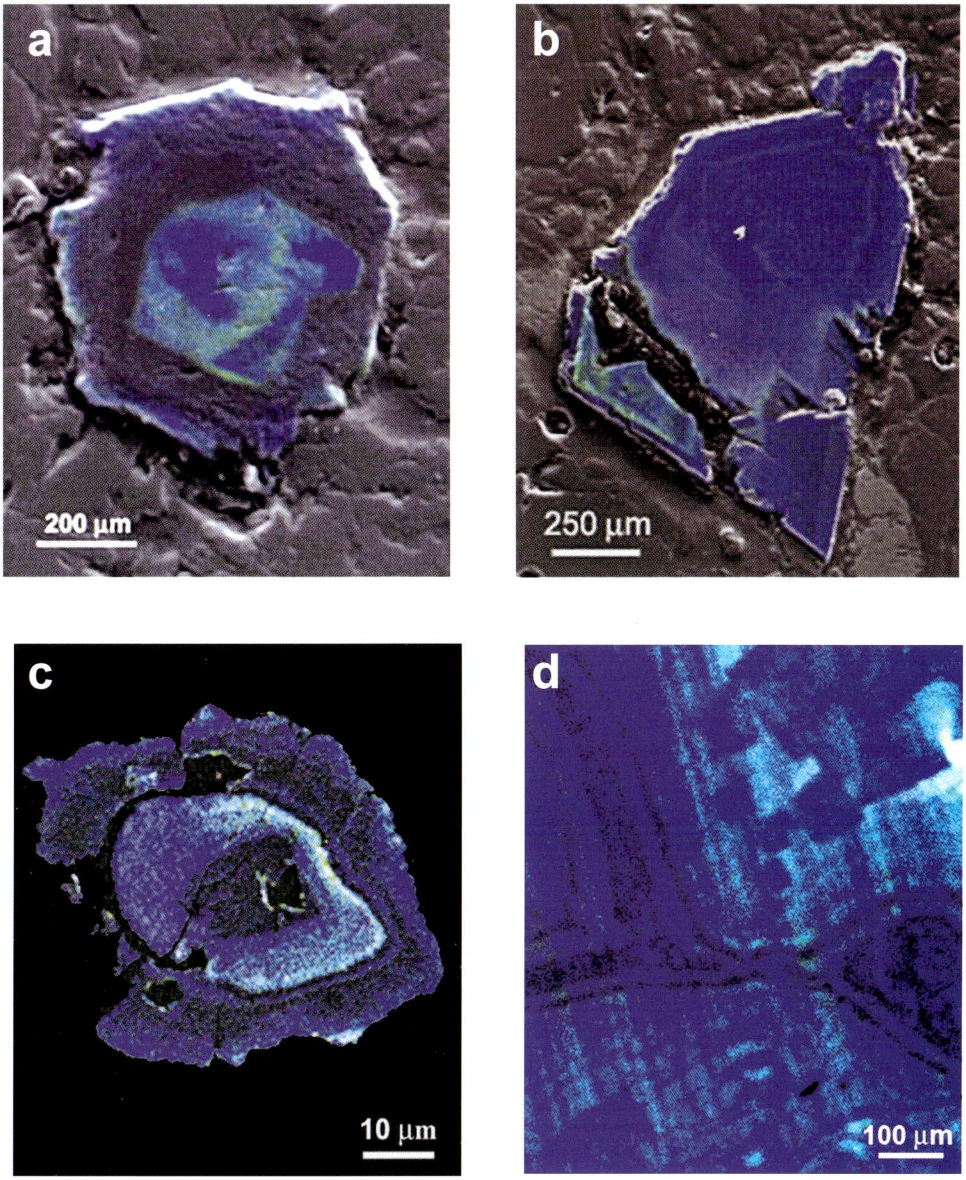

Plate 2. 27. Cathodoluminescent Images of Diamonds in Eclogite Xenoliths from Udachnaya. Note the complicated internal morphologies and growth zones. a) Two growth zones are obvious; notice resorption of crystal rim; b) Slightly zoned growth in octahedral diamond; new center of crystallization is obvious at the top; c) Three growth zones are obvious in this crystal with a cubic center, interrupted growth between intermediate and outer zones, which shows traces of resorption, and interrupted growth and resorption also present in the outer zones; d) Complicated growth with two growth centers in a single crystal.

CHAPTER 3

MINERAL INCLUSIONS IN DIAMONDS

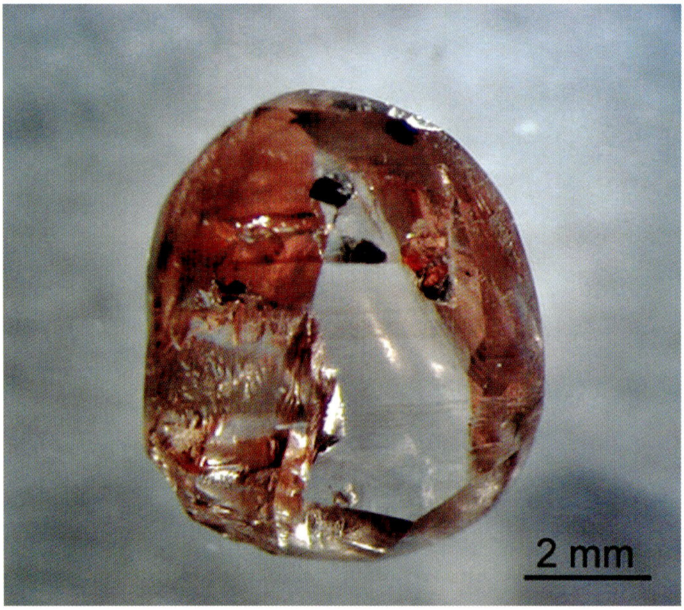

Chapter 3

Mineral Inclusions in Diamonds

Meyer (1987) documented over 20 minerals that have been proposed as protogenetic (formed before diamond growth) or syngenetic (formed with diamond) on the basis of modern analytical methods (X-ray or electron microprobe). Additionally, 10 minerals have been proposed to be of epigenetic origin (formed after diamond growth). Recently, Taylor et al. (2003) suggested, based on trace-element characteristics of harzburgitic garnet inclusions, that some inclusions that are assumed to be syngenetic could in fact be protogenetic. Syngenetic minerals are high pressure-temperature phases that crystallize simultaneously with the host diamond, and have the same genesis that can be used to constrain the conditions necessary for diamond growth. Inasmuch as there are no known eclogitic or peridotitic melt from which both silicate minerals and diamonds can grow, Taylor et al. (2003) doubted entirely the gross assumption of the syngenesis of diamonds and their mineral inclusions. Epigenetic minerals are typically associated with overgrowth of fibrous diamond or alteration of pre-existing inclusions through fracture/crack systems. Diamond overgrowths have a porous nature due to their fibrous structure, and commonly contain a wide variety of epigenetic minerals resulting from fluid interaction either in the mantle or within the transporting host magma (e.g., Navon et al., 1988). The chemical distinction between proto- and syngenetic-diamond inclusions and the mineral phases that crystallize within kimberlitic or lamproitic melts preclude any genetic link between the two.

A principal outcome from the study of mineral inclusions in diamonds, regardless of their genesis relative to that of the diamond, has been the establishment of two paragenetic associations affirming that two chemically distinct environments are responsible for diamond growth in the mantle (e.g.,Harris, 1968; Meyer, 1987). Diamonds of the peridotitic (P-type) paragenesis contain an assemblage of olivine, ortho- and clinopyroxene, garnet, chromite, etc. Eclogitic (E-type) diamonds commonly contain omphacitic clinopyroxene, garnet, and sulfide. The relative proportions of each genetic type vary, and P- and E-type diamonds almost always co-exist in the same kimberlite pipe. The distribution of "syngenetic" inclusions in many well-studied kimberlite pipes of Yakutia is shown in Table 3.1. The well-known data of Meyer (1985) and others (e.g., Gurney, 1990 and references therein) for inclusions in diamonds from kimberlites of South Africa are similar to those from Yakutia, in terms of the presence of P- and E-type diamonds and the compositions of mineral inclusions.

Examples of the most common inclusions in diamonds are shown in several Plates at the end of this Chapter. Mineral inclusions from a particular diamond population can differ not only in their compositions but also in quantity and morphology. The origin of most diamonds in the mantle has been verified by the geothermobarometry of mineral

Table 3.1. Classification of Mineral inclusions in Diamonds. (modified from Meyer, 1987)

	Post 1950[1]			
	Protogenetic and/or Syngenetic			**Epigenetic**
Pre 1950	**Peridotite (P) Ultramafic (U)**	**Eclogitic (E)**	**lower mantle phases[2]**	
Quartz	**Forsterite**	**Omphacite**	**TAPP[5]**	Serpentine
Diamond	**Enstatite**	**Pyrope-almandine**	**$CaSiO_3$**	Calcite
Pyrite	**Diopside**	**Kyanite**	**$MgSiO_3$**	Graphite
Hematite	**Cr-pyrope**	**Sanidine**	**Ferropericlase**	Hematite
Ilmenite	**Cr-spinel**	**Coesite (quartz)**	**Majorite Gt**	Kaolinite
Garnet	**Mg-ilmenite**	**Rutile**	**Stishovite**	Acmite
Graphite	**Sulfides**	**Ruby**		Richterite
Diopside	**Zircon**	**Ilmenite**		Perovskite
Olivine	**Diamond**	**Chromite**		Mn-ilmenite
Calcite	**Native iron**	**Sulfides**		Spinel
Chromite		**Diamond**		Xenotime
Zircon				Goethite
Biotite				
Phlogopite[3]				
Biotite[3]				
Muscovite[3]				
Amphibole[3]				
Magnetite[3]				
Apatite[3]				
Gas(CO_2)/ Liquid[4]				
Topaz[4]				
Plant remains[4]				
Gold[4]				

[1]Minerals identified positively by X-ray diffraction and electron microprobe.
[2] From Stachel 2001.
[3]Uncertain paragenesis.
[4] Probably misidentified.
[5] Tetragonal almandine pyrope phase.

inclusions (e.g., Griffin et al., 1992). Co-existing polymineralic inclusions in diamond are rare; however, individual inclusions at different places within a given diamond are "assumed" to be in equilibrium, an assumption inconsistent with several observations (Sobolev et al., 1988; Taylor et al., 2000). Estimations from such 'co-existing' pairs suggest a temperature range for diamond formation of 900-1400 °C and 40-60 kbar, i.e. depths of 150-200 km are suitable for diamond growth (Meyer, 1985).

Mineral inclusions can be invaluable as "recorders" of the growth conditions of their host diamonds. Sulfide inclusions are abundant in diamonds from all kimberlite and lamproite pipes (e.g., Bulanova et al., 1992; Deines at al., 1998) and present us with a vast array of possibilities for study. It cannot be emphasized enough that the advent of micro-analytical instrumentation has opened up new realms for scientific study of mineral inclusions. Consequently, there have been numerous significant studies of inclusions in diamonds, among them the discovery of new phases such as ferropericlase and MgSi-perovskite, and the majorite component in garnets. These rare minerals are found in diamonds from Brazil, South Africa, and Canada (e.g., Hart et al., 1999; Kaminsky et al., 2002). The findings of such unique inclusions have led to the possibility that such inclusions, as well their host diamonds, could have been derived from the deepest portions of the mantle, perhaps near the core-mantle boundary. In spite of the fact that the inclusions in diamonds from Yakutia are well studied, such deep-mantle inclusions have not been reported to date from Yakutian kimberlites. However, majoritic components in garnet are now well established from Yakutia (Sobolev et al., 2001). This may be indicative of the different evolutionary histories of the mantle beneath different cratons. It is possible that the continental roots under South Africa and other cratons were thicker than those beneath the Siberian craton, and some diamonds have encapsulated inclusions from deeper mantle levels. Such a concept was supported by Griffin et al. (1999), based on the distribution of garnets and other macrocrysts in kimberlites from different cratons.

3.1. Geochemistry of Mineral Inclusions.

The geochemistry of diamond inclusions has been discussed in review articles by Meyer (1987), Harris (1992), Gurney (1989), and Kirkley (1991), and was summarized and discussed in the context of newer SIMS trace-element data on inclusions by Pearson et al. (2003). Meyer (1987) pointed out the importance of inclusions in diamonds for study of the mantle petrology. First, inclusions present us with the necessary data with which to relate the diamonds to their mantle source lithologies. Secondly, inclusions could represent pristine, geochemically unaltered samples that have not been subjected to the chemical re-equilibration and alteration that affects the minerals in xenoliths and megacrysts.

In the P-type diamonds, six minerals typically occur as mineral inclusions (Table 3.1), whereas in E-type diamonds, only five minerals are usually found (Gurney, 1989). The P-type inclusions are subdivided into harzburgitic and lherzolitic on the basis of Ca and Cr contents in garnets, i.e., harzburgitic garnets have much higher Cr and lower Ca contents than lherzolitic garnets (e.g. Sobolev et al., 1973; Gurney et al., 1973). The P-type versus E-type distinction for garnet and clinopyroxene is evident on major-element plots (Meyer, 1987). P-type garnet inclusions generally have lower FeO, CaO, and higher MgO compared to E-type garnet inclusions. P-type clinopyroxenes have considerably lower Na_2O, Al_2O_3, and FeO, and much higher MgO and Cr_2O_3, compared to E-type clinopyroxene.

Sulfides that consist dominantly of Fe and S, with lesser Cu, Ni and Co, have been classified into P versus E genesis. P-type sulfides have higher Ni contents, typically >13 wt.%, versus E-type sulfides with <10 wt.% (Yefimova et al., 1983; Deines and Harris, 1995). It is possible to have some overlap of Ni contents for sulfides, but the genetic distinctions can be clarified using Re-Os systematics (Pearson et al., 1999). E-type sulfide inclusions have low Os and high Re, resulting in the high Re/Os compared to P-type sulfides, which have lower Re and much higher Os (Pearson et al., 1999).

While the major-element compositions of inclusions have been used to identify the genetic types of the inclusions (Peridotite/Ultramafic versus Eclogite), it is their trace-element contents that contain the most information concerning their formation in the mantle and the complex metasomatic processes to which they have been subjected. The Sm-Nd isotopic analyses (Richardson et al., 1984) and the first SIMS trace-element analyses on harzburgitic garnets from the Finsch and Kimberley pipes (Shimizu and Richardson, 1987) confirmed their light rare-earth-element (LREE) enrichment, coupled with their large depletion in major-element chemistry. **This was recognized as a signature of ancient mantle metasomatism before incorporation of the garnets into diamonds**, and is a feature of nearly all harzburgitic garnets analyzed for Sm-Nd isotopic ages (e.g., Pearson et al., 1995; Pearson and Shirey, 1999). Taylor et al. (2003) insisted that the presence of complex sinusoidal rare-earth-element patterns for all harzburgitic garnets worldwide is proof of an initial partial-melting event early in the history of the garnets, followed by subsequent metasomatism. **This is evidence that harzburgitic garnets definitely pre-existed prior to diamond encapsulation – harzburgitic-garnet inclusions are NOT syngenetic or synchronous with the formation of their host diamonds.**

Subsequent data on diamond inclusions from Udachnaya (Pearson and Milledge, 1999) and other pipes (e.g., Stachel and Harris, 1997; Stachel et al., 1998, 1999) have shown a whole range of multi-stage (e.g., sinusoidal) REE patterns. These trace-element complexities have implicated the following processes in formation of the harzburgitic garnets: 1) Initial depletion by polybaric partial melting; followed by 2) enrichment, particularly of MREEs and LREEs, by mantle metasomatism; and 3) late-stage LREE depletion, possibly due to re-equilibration of the garnet with clinopyroxene (Stachel and Harris, 1997; Stachel et al., 1998) or another partial melting event

(Taylor et al., 2003). **All of these processes occurred prior to diamond encapsulation of the garnets** (Pearson, 2003; Taylor et al., 2003). Similar trace-element patterns between harzburgitic garnets from xenoliths (diamondiferous and non-diamondiferous) and harzburgitic garnet inclusions (Stachel et al., 1998) suggest that host rocks reflect the same processes as seen in the inclusions – i.e., a complex petrogenesis.

Lherzolitic garnets show much less complexity in their REE patterns. They are LREE depleted and could have formed in equilibrium with kimberlitic/lamproitic or carbonatitic fluids (Stachel and Harris, 1997; Stachel et al., 1998). The lherzolitic garnets are proposed to be distinctly different than harzburgitic garnets, and are thought to be associated with metasomatic enrichment of the lithosphere, which introduced additional Ca, Fe, and Si (Stachel et al., 1998).

Eclogitic garnet and clinopyroxene inclusions from Udachnaya (Ireland et al., 1994; Taylor et al., 1996) Mwadui (Stachel et al., 1999), and Kankan (Stachel et al., 2000) are characterized by more 'regular' REE patterns that reflect typical equilibrium distribution of the REEs between the two minerals. As is the case with P-type paragenesis, similar trace-element patterns between garnets and clinopyroxene from eclogite xenoliths and those occurring as inclusions in diamond (Taylor et al., 1996) show that at least some host xenoliths reflect the same processes as seen in their corresponding inclusions. Since the mineral trace-element patterns are 'regular' and eclogites approximately consist of ~50% each of garnet and clinopyroxene, meaningful bulk-rock, trace-element patterns can be constructed for these rock-forming minerals (Ireland et al., 1994; Stachel et al., 1999; Taylor et al., 1996). These reconstructed whole-rock compositions show strong LREE depletions that are interpreted by Ireland et al. (1994) as the possible result of subduction of basalt residues, but after partial-melt extraction in the garnet stability field to form the tonalite-trondhjemite-granodiorite (TTG) magmas.

Taylor et al. (2000) have performed X-ray tomography on diamondiferous eclogites from Udachnaya, effectively producing Computer Assisted Tomography scans of the rocks, similar to medical CAT scans. Using these maps for planning dissections, they were able to successfully extract some 30 macro-diamonds from one 50 g eclogite, 74 diamonds from another (Anand et al., 2003). **Examination of the inclusions in the diamonds from a single xenolith revealed that they were very different from each other and from the host eclogitic minerals. In fact, the chemical differences between multiple inclusions of clinopyroxene within single diamonds were large as well.** This first-ever diamondiferous eclogite dissection showed that the diamonds from one eclogite xenolith can vary tremendously. This questions possible over-interpretation from the use of garnets and clinopyroxenes from hundreds of diamonds from one pipe for determining the ages of diamonds (e.g., Richardson et al., 1984, 1993). Similar diamondiferous eclogite xenolith studies by Anand et al. (2002, 2003) have further verified such strange findings and warn that the use of diamond inclusions for various scientific purposes must proceed cautiously.

3.2. Age of Mineral Inclusions and Host Diamonds

The ages of diamond inclusions, assumed to be the ages of the diamond formation, have been summarized by Harris (1992), Pearson and Shirey (1999), and Navon (1999). Analyses are typically performed on inclusions, assumed to be syngenetic, of both E-type and P-type parageneses in diamonds, using the following isotopic systems: Sm-Nd in garnets and clinopyroxenes, Re-Os in sulfides, Ar-Ar in clinopyroxenes, and Pb-Pb in sulfides.

In reviewing the literature, the reader is cautioned that some ages have been determined from a mixture of inclusion suites, and this may have caused apparent age discrepancies. For example, Sm-Nd isotope studies of garnet or other silicates typically require composites of inclusions from hundreds of diamonds and may provide a statistical average for many diamonds (Richardson et al., 1984, 1990), whereas Re-Os dating can be performed on single sulfide grains (Pearson et al., 1998).

Until the application of the Re-Os isotopic system to whole-rock peridotites, the Archean and Proterozoic model and isochron ages, obtained from Pb-Pb and Sm-Nd isotopic studies on diamond inclusions, were the chief constraints on the antiquity of the continental lithospheric mantle (e.g., Kramers, 1979; Richardson et al., 1984). Early work, most of which was on diamonds from the Kaapvaal-Zimbabwe craton, had suggested a simple difference from Meso-Archean P-type (harzburgitic) diamonds and Proterozoic E-type diamonds (Richardson, 1986; Richardson et al., 1984; Richardson et al., 1993). More recent Re-Os work on single sulfide inclusions from the Kaapvaal-Zimbabwe craton underscored the importance of a Neo-Archean E-type diamond suite (Shirey et al., 2001) and has resolved Proterozoic diamond populations (Pearson et al., 1998).

The accuracy of age-determinations of diamonds, based on their inclusions, has engendered many debates in the literature (Navon, 1999; Richardson, 1989; Spetsius et al., 2002; Taylor et al., 2000, 2003). The main issue is to what extent the diamond ages of included minerals are averages of many ages, and to what extent younger diamonds can encapsulate older, protogenetic inclusions. It has been shown that some diamonds occur in fracture systems in their host xenoliths, thereby post-dating the rock itself (e.g., Taylor et al., 2000). Most data suggest that diamonds can be Archean and Proterozoic in age, although some are as young as the Phanerozoic age of most kimberlites (Pearson et al., 1998). Evidence for Proterozoic and Archean diamonds includes the occurrence of Proterozoic kimberlites such as Premier, which carry abundant xenocrystic diamonds (Navon, 1999), and Re-Os ages of the host peridotites that agree with the diamond ages (Shirey et al., 2001). In addition, Type Ia diamonds show advanced aggregation which would appear to require billions of years of residence at high-mantle temperatures (Richardson and Harris, 1997). Furthermore, Nd and Sr isotopic differences between a) garnets from kimberlitic mineral concentrates and b) garnets encapsulated in diamond suggest encapsulation for billion-year timescales (Pearson and Shirey, 1999; Richardson, 1989; Richardson et al., 1984).

The most reliable dates for mineral inclusions are probably those done on individual minerals in-situ, now only possible for Re-Os dating on sulfide inclusions. Limited Ar/Ar dates obtained from single clinopyroxene inclusions do exist, but are not as precise as Re-Os ages. The results and data compilation by Pearson (1999, 2000, 2003) on inclusions in diamonds from different deposits have been used to address possible discrepancies in diamond age-dating These data indicate that the age of diamond formation varies predominantly between 2.0-3.3 Ga. These estimations of timing are similar for diamonds - both from the Siberian and South African cratons, and suggest the similarity of timing for diamond formation in different areas, as well as the probability of multistage formation of diamonds in mantle rocks.

Diamond Inclusions are Time Capsules from Long Ago and Far Away.

Long-Ago – Udachnaya Kimberlite: 340 Ma
Diamonds: 3,000 Million years old.

Far-Away – Only samples from >120 km, providing a "window" into the mantle".

Pressure Capsules – contain exotic phases from great depths

Pristine Samples – Diamonds have kept their inclusions in perfect isolation – i.e., a closed chemical system after encapsulation

Plate 3. 1. Diamonds with Mineral Inclusions. a) Inclusion of purple garnet of harzburgitic composition with an intergrowth of Cr-spinel in transparent planar octahedron from the Udachnaya kimberlite. b) The same inclusion with high magnification.

Plate 3. 2. Diamonds with Mineral Inclusions. a) Inclusions of pyropic garnet in octahedral crystal from the Aykhal; a little inclusion of olivine can be seen to the left of the pyrope; b) The same garnet inclusion at high magnification, small black spots are probably sulfides; octahederal form of the garnet crystal suggests syngenetic growth of inclusion with diamond. Taylor et al. (2003c) have addressed the validity of this assumption.

Plate 3. 3. Rare Form of a Cr-Pyrope Inclusion in Diamond. a) Elongated pyrope inclusion in an octahedral partly rounded diamond from the Udachnaya pipe. b) The same crystal at high magnification. Note that the elongated form of the garnet, not of the inherited diamond morphology, is indicative of a protogenetic origin for the inclusion (Taylor et al., 2003c).

Plate 3. 4. Rare Form of Cr-Pyrope Inclusions in Diamonds. a) Elongated pyrope inclusion in round diamond with corroded and resorbed surface from the Zarnitsa pipe. b) The same crystal at high magnification. Note that the elongated form of the garnet, not of an inherited diamond morphology, is indicative of a protogenetic origin for the inclusion (Taylor et al., 2003c).

Plate 3. 5. Diamond with Clinopyroxene Inclusion. a) Transparent smooth-faced planar octahedron from the Udachnaya pipe with inclusions of chrome-diopside and a smaller opaque chromite (left). b) The same inclusion of chrome-diopside at high magnification (rotated CW 90 °).

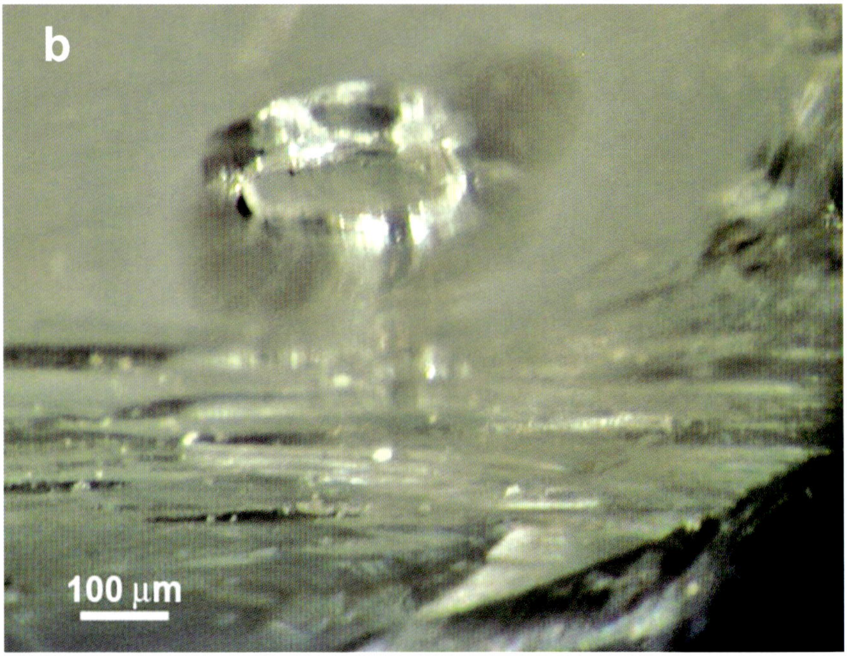

Plate 3. 6. Diamond with a Clinopyroxene Inclusions. a) Combined octahedral form from the Aykhal pipe with three clinopyroxene inclusions and small garnet; Notice the presence of an opaque sulfide inclusion, as well. These inclusions are probably of eclogite paragenesis. b) One of the clinopyroxene inclusions at high magnification.

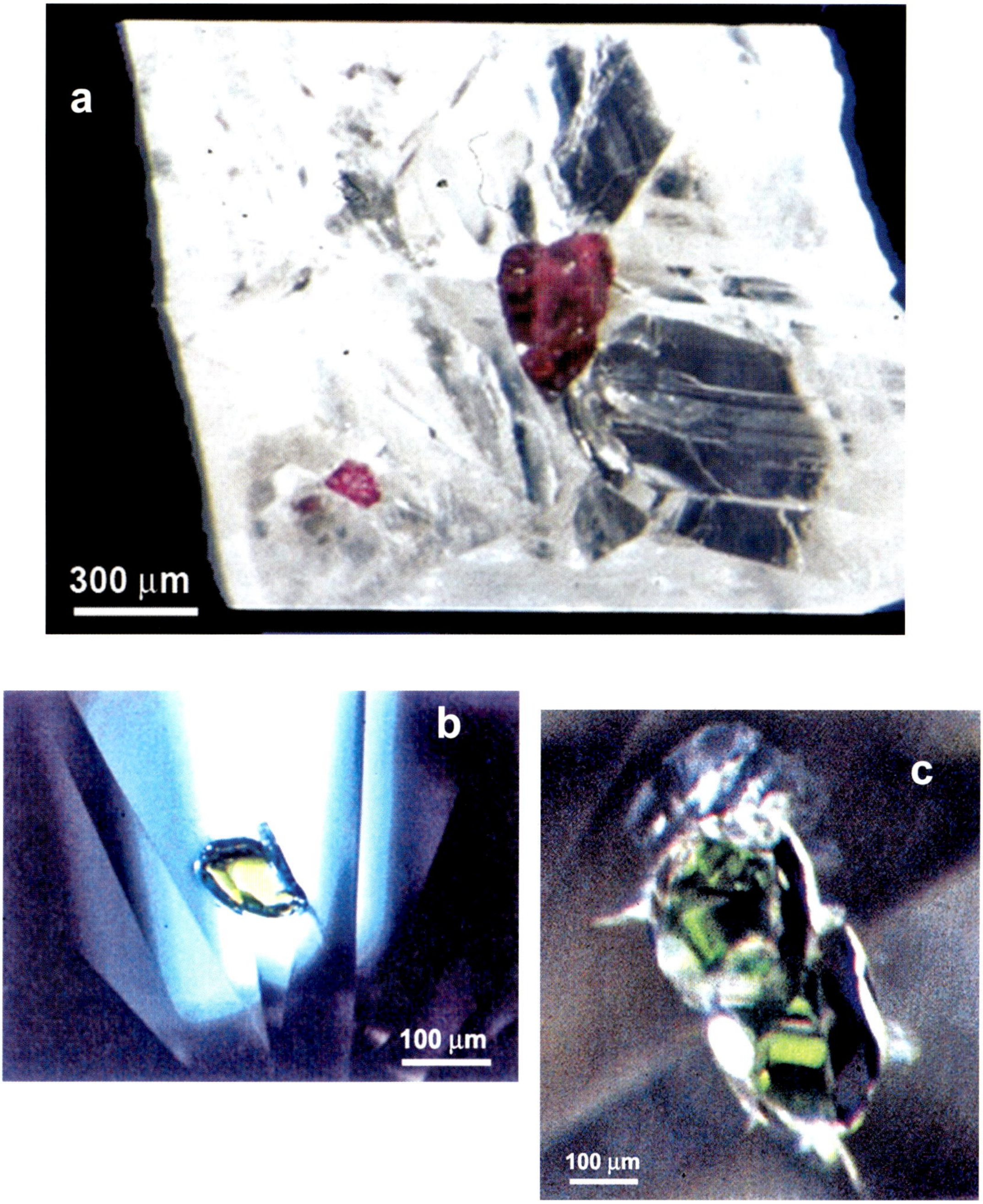

Plate 3.7. Mineral Inclusions in Diamonds. a) Garnet inclusion in a crushed diamon. b) + c) Chrome diopside crystals in P-type diamonds. Photo taken from Sobolev (1977).

Plate 3. 8. Mineral Inclusions in a diamond from the Yubilanya Pipe. a) Clear octahedron with rounded edges, with a purple harzburgitic garnet inclusion in the center; note the smaller clinopyroxene to upper-left of the pyrope; b) Close-up of the pyrope; c) Close-up of the clinopyroxene.

Plate 3. 9. Mineral Inclusions in Diamonds. a) Near-perfect clear octahedron containing sulfide inclusions. b) Polished diamond exposing purple harzburgitic garnets and olivines. Note the cubo-octahedral morphologies to the garnets, supposedly an indicator of syngenetic growth of the diamond and the garnet, a long-held assumption recently questioned by Taylor et al. (2003c).

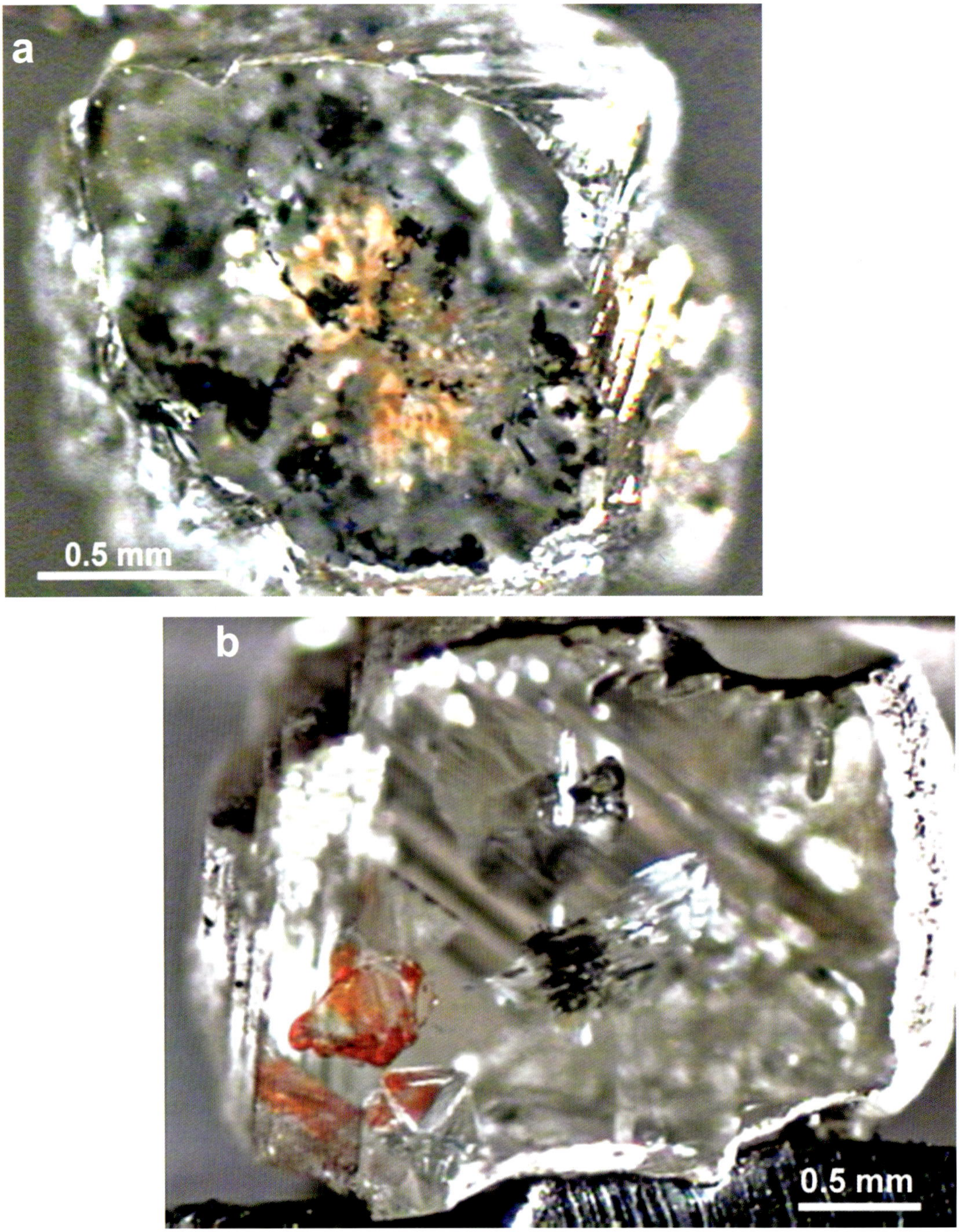

Plate 3. 10. Mineral Inclusions in Diamonds. Both of these diamonds have had "windows" polished into them revealing the fine details of the numerous inclusions, mostly sulfides.

Plate 3. 11. Sulfide Inclusions in Diamonds. a) Accumulation of sulfides in the the center of a planar, colorless, flat-faced octahedron, from the Mir pipe. b) Sulfide and chromite inclusions on a polished surface of an octahedral diamond, from the Botuobinskaya pipe; notice that in both cases, the presence of black cracks around some inclusions are filled by the same sulfide minerals, as a secondary process.

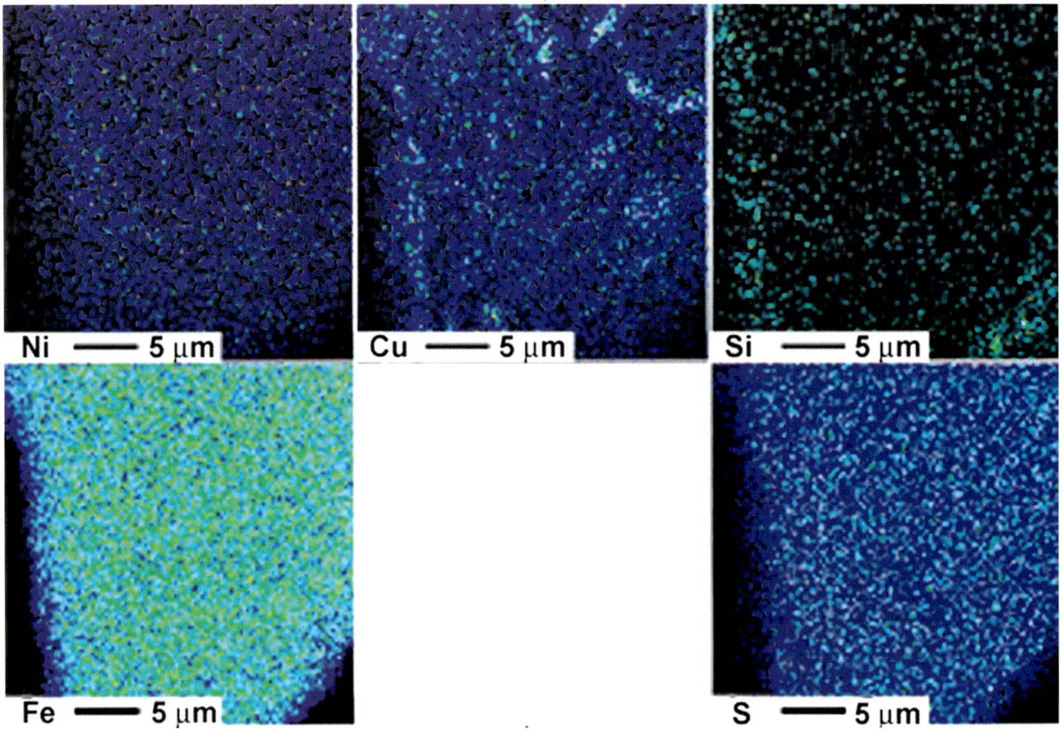

Plate 3. 12. Sulfide Inclusion in a Diamond. X-ray image maps of pyrrhotite in an octahedral crystal from the Botuobinskaya pipe. Such inhomogeneities of Cu, Ni, and Fe are typical for virtually all sulfide inclusions. The sulfide phase that was originally incorporated into the diamond was a homogeneous mono-sulfide solid solution of Fe-Ni-Cu-S. However, due to the large differences in thermal expansion of this phase and the diamond, the diamond cracked around the inclusion. This permitted the Mss to exsolve into pyrrhotite, pentlandite, and chalcopyrite. Notice particularly the inhomogeneous distribution of Cu (chalcopyrite) in the inclusion.

Plate 3.13. Sulfide inclusions in Diamonds. a) rounded dodecahedron, due to resoption, with cracks on the right with weathered products and black sulfide inclusions in the center. b) at higher magnification.

Plate 3.14. Sulfide inclusions in diamonds. Notice the numerous cracks in the diamonds along which secondary minerals have migrated.

Plate 3.15. Sulfides in Highly Corroded and Resorbed Diamonds.

Plate 3.16. Mineral Inclusions in Diamonds. a) Garnet and sulfide in complex E-type diamond; b) Chromite in clear stepped octahedron; c) Harzburgitic pyropes.

CHAPTER 4

DIAMOND HOST ROCKS

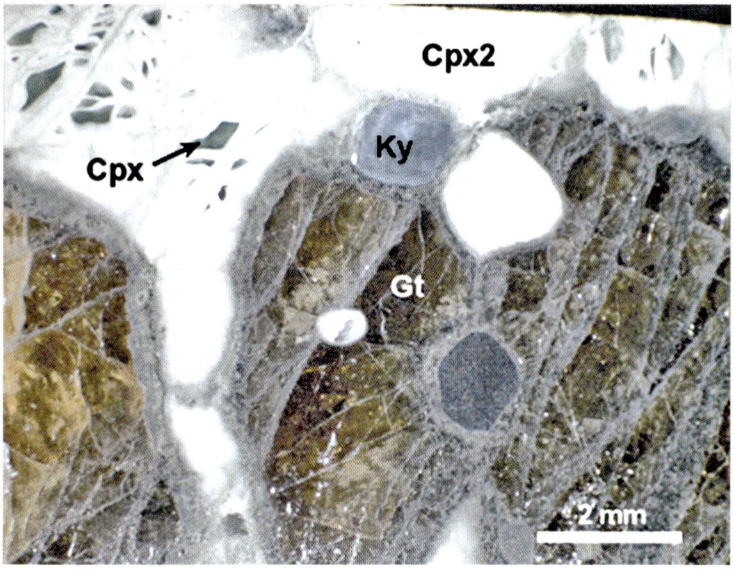

Chapter 4
Diamond Host Rocks

Decades ago, diamonds were believed to have originated from the same melt that resulted in the kimberlite itself. As time went on and more investigations were performed into the petrologic nature of kimberlites, it was discovered that many diamonds actually occurred in xenoliths carried from depths by the kimberlite. In fact, the support for this has become so massive during the past 20 - 30 years or so (e.g., Meyer, 1985) that today it is generally accepted that kimberlites are simply the carriers of the diamonds. These diamonds originated in eclogitic, peridotitic, and other ultramafic rocks within the mantle long before being “picked up” by the kimberlite magma on its ascent from the mantle to the surface. The individual diamonds occurring in the kimberlite are considered to simply be disaggregated portions of xenoliths, similar to the common xenocrysts of garnet and ilmenite also found in the kimberlite matrix. More recently (e.g., Boyd et al., 1987, 1994; Akagi & Masuda, 1988), it has been argued that some portions of many diamonds (e.g., the outer, frosted, fibrous layers) are actually grown in the kimberlitic magma and that some micro-diamonds are possibly of this same origin (McCallum et al., 1994; Pattison & Levinson, 1995). In this chapter, we will address the genesis of diamonds within kimberlite versus those in mantle xenoliths. In addition, the oxidation state of the kimberlite will be addressed, as this affects diamond resorption and preservation processes as evidenced by diamond morphologies.

4.1. Diamond Origin(s)

The two contrasting hypotheses to explain the origin and presence of diamonds in kimberlite pipes are the magmatic versus xenocrystic theories. The former theory assumes crystallization of diamond in the deep-seated kimberlite magma, just prior to eruption. There have also been supporters of diamond formation at shallow depths (i.e., 40 kbars) in the rising kimberlite (e.g., Trofimov, 1966) or even in the near-surface conditions (Mikheenko et al., 1970). We will not further address these lower - pressure origins, particularly since these authors seem to have erroneously interpreted the significance of diamond inclusions, which most likely have had a deep-seated origin.

The hypothesis of diamond crystallization from kimberlitic magma was originally supported by several investigations (e.g., Williams, 1932; Bobrievich et al., 1959; Harris, 1979; Vaganov, 1996). Various observations were used as evidences to support the magmatic hypothesis for the genesis of diamond: a) differences in the crystal morphology of diamonds from different pipes; b) common occurrences of zoned diamonds; c) the presence of "diamond in diamond", et cetera. However, we do not feel that these 'evidences' are unique to this kimberlitic magma theory and are equally explainable as resulting from a prior existence of diamonds in mantle xenoliths. The xenocrystic theory is detailed in the review presented below.

The hypothesis of the xenocrystic nature of diamonds (xeno = foreign; cryst = crystal) in kimberlites is not recent, having been proposed in 1899 by Bonny, after the discovery of diamondiferous eclogites as xenoliths in kimberlites in South Africa. Indeed, it was he who first reasoned that the kimberlite was merely the carrier of the xenoliths. The numerous discoveries of diamondiferous eclogites and peridotites in kimberlite pipes worldwide further supported this theory (e.g., Sobolev, 1974; Robinson, 1979; Gurney, 1989, 1990; Jaques et al., 1990). This xenocrystic/xenolithic origin, based upon abundant evidence, continues to be the most widely accepted theory.

The origin of diamond in mantle rocks finds direct evidence in the presence of silicate mineral inclusions in the diamonds from all diamondiferous kimberlite pipes (e.g., Sobolev, 1968, 1974; Meyer and Boyd, 1968). These inclusions appear to have originated from two major suites of rocks, ultramafic peridotites (P/U) and mafic eclogites (E), and their presence permits the distinction of P- and E-type diamond genesis, respectively. [It should be noted that P-type is used throughout this book as this is most widely accepted, whereas the U-type for ultramafic by Sobolev (1977) is probably more correct overall]. Other corroborating evidence for the xenocrystic nature of diamonds is the presence of graphite pseudomorphs after diamond, which was discovered in pyroxenites from the delaminated massif Beni-Bousera (Slodkevich, 1982; Pearson et al., 1995). The xenocrystic presence of diamonds is additionally supported by the discovery of diamonds in Australian lamproites, where numerous diamondiferous mafic xenoliths were recovered. Indeed, the diamond inclusions there are mostly representative of an eclogite paragenesis as well (e.g., Sobolev, 1987; Jaques et al., 1989). Perhaps the most convincing evidence, however, in favor of xenocrystic nature of diamonds occurs with the dating of diamonds using their mineral inclusions (Kramers, 1979; Richardson et al., 1984, 1993). The diamonds are reported to be significantly older than the enclosing kimberlite -- for example, 3 Ga versus 90 Ma, respectively. In addition, strong evidence for old-age diamonds versus young-age kimberlites has been obtained from the Re-Os isotopic data from individual sulfide inclusions in diamonds from the Koffiefontein and Udachnaya kimberlites (Pearson et al., 1998, 1999). Indeed, three sulfide inclusions from the core and intermediate zones of one Yakutian diamond possess 3.1 to 3.5 ±0.3 Ga for Re-Os model ages, indicating that the growth of this single diamond possibly occurred over an extended time period, perhaps tens of millions of years (Pearson et al., 1999).

4.2. Kimberlite Characteristics

In general, potassic representatives of intrusive, ultramafic rocks having high contents of H_20 and CO_2 are classified as kimberlites (Bobrievich et al., 1959; Mitchell, 1986). Mineralogically, kimberlites contain phenocrysts and xenocrysts of olivine and additional minerals such as garnet, phlogopite, and ilmenite, all set within a fine-grained groundmass consisting of olivine, serpentine, calcite, dolomite, various micas, perovskite, etc. Chemically, kimberlites are high in K_2O and Mg#, low in $^{87}Sr/^{86}Sr$, and enriched in LREEs. In fact, kimberlites have distinct similarities to the KREEP (K, REE, & P) signatures found in lunar rocks, and have been referred to as terrestrial KREEP.

Throughout the 20th century, considerable research was directed at effectively discerning diamondiferous kimberlite from a more-common barren kimberlite. Probably the most reliable indicator mineral for the possible presence of diamonds is the occurrence of high Cr, low Ca garnets (Sobolev et al., 1973, 1974; Gurney and Switzer, 1973). The Cr and Ca contents of garnet were further used to distinguish different types of garnet, especially the G-9 and G-10 garnets, indicative of potential diamond-bearing kimberlites (Dawson, 1980). It is generally agreed that such garnets belong to garnet peridotite/harzburgite rocks, and provide additional proof of the *absence of a syngenetic origin* for diamonds and the kimberlites in which they may occur.

Kimberlites show wide variations in mineral compositions between pipes, but also within a single pipe (e.g., Bobrievich et al., 1959; Mitchell, 1986). As an example of the heterogeneous nature of a kimberlite, we use the Udachnaya pipe in the center of the Siberian Platform. Based on an extensive sampling on four levels of exploration of this multiphase pipe, the kimberlite chemistry has revealed a systematic behavior of major-and trace-element contents in the pipe as a function of depth, to over 500 m (Spetsius et al., 1996; Zinchuck et al., 1993). This change in the chemistry of the kimberlite was accompanied by change in the content of secondary minerals such as serpentine, dolomite/calcite, phlogopite, etc. Therefore, the change in kimberlite composition may simply reflect the intensity of post-magmatic processes, along with the types of sedimentary and other crustal rocks into which the kimberlite was intruded. Indeed, the similarity of composition between the main stage of intrusion and subsequent intrusions within the pipe testifies to the similarity of the initial melt composition for each stage. This indicates the relatively small degree of differentiation that occurred in a given kimberlite magma source at depth. Such constancy of melt chemistry is apparent in more than ten multi-phase kimberlite pipes investigated in Yakutia (Spetsius, 1987; Zinchuck et al., 1993).

In contrast, variations in the contents and quality of diamonds within a pipe can vary considerably (e.g., Bobrievich et al., 1959; Harris, 1987; Gurney, 1990). Detailed examination of representative collections of diamonds from different exploration levels of the Udachnaya two pipes showed considerable variations in diamond content and quality. However, it was noted that there is a distinct similarity in the eastern and western pipes, making up the overall Udachnaya pipe. The diamond populations from these two separate, albeit close, bodies of the Udachnaya pipe are nearly identical, based on their morphology, "resistance to shock," and X-ray- and photo-induced luminescence. Such close similarities in diamond populations have also been noted between different intruding phases of several other Yakutian kimberlites, such as at Mir, Sytykanskaya, Yubileynaya, etc., as well as in the deposits in the Arkhangelsk region of western Russia.

There exists a constancy of diamond distribution by morphology, quality, and other properties in different phases of kimberlite pipes. That is, there is a lack of relationship between diamond characteristics and chemical and mineralogical features of their host kimberlites. This is additional evidence for the xenocrystic nature of diamonds in kimberlites. An important practical conclusion is that changes in processing characteristics of the kimberlite ores are defined predominantly by variation in the petrographic, mineralogic, and physical/mechanical properties of the kimberlite itself, not by the contents and properties of the diamonds. These variations appear in the vertical zoning typical for all kimberlite pipes, and they define changes in the contents of the original xenolithic, diamond-bearing materials with depth (Spetsius, 1996).

On the basis of kimberlite compositions and diamond contents (i.e., grade) of Yakutian pipes, it is possible to conclude that: a) substantial changes in mineralogy and chemistry of kimberlites occur with depth; b) the melt composition of kimberlite remains relatively constant throughout different stages of intrusion, such that multi-stage kimberlite pipes are characterized by compositions indicating little differentiation of the kimberlitic magma source at depth; and c) there is an obvious regularity in diamond grade between different kimberlite intrusive stages. All these lead to the conclusion that the grade of a given stage of a kimberlite does not appear to correlate with the petrochemical characteristics of the rock.

4.3. Mantle Host Xenoliths for Diamonds

The original source of diamonds found in kimberlites is considered to be mainly, if not entirely, in ultramafic and mafic rocks, e.g., peridotites and eclogites, from the mantle. These rocks differ significantly in their petrographic and chemical constituents, thus, their origins, and can be grouped into the peridotite-pyroxenite and eclogite series, respectively. Although the abundances of these xenoliths vary between kimberlites, the garnet-bearing peridotite-type xenoliths are more abundant by almost an order of magnitude in kimberlites of the Yakutian Province. In this section, we will address the petrologic nature of peridotite-pyroxenite and eclogite suites in light of the different portions of the mantle sampled by kimberlites and the inherent heterogeneity of this sampling process

4.3.1. Peridotite-Pyroxenite Series - Lherzolites, harzburgites, wehrlites, ortho- and clinopyroxenites, and even dunites make up the members of the ultramafic series of xenoliths in kimberlites. Both garnet and spinel varieties occur, garnet being a ubiquitous phase in diamondiferous xenoliths. In addition, these peridotites can contain ilmenite and phlogopite (1-30 vol%). More than half of all deep-seated xenoliths in kimberlites at the Mir, Udachnaya, Sytykanskaya, and non-diamondiferous Obnazhennaya pipes are of lherzolite association. The proportions of other ultramafic rocks, however, are not well documented. Pyroxenites are abundant in Obnazhennaya and Mir pipes, with porphyritic pyroxenites occurring only in the Mir pipe. Ilmenite-bearing xenoliths are abundant at Sytykanskaya.

Whole-rock compositions of the peridotite-pyroxenite xenoliths indicate a general increase in Al_2O_3, FeO, and Cr_2O_3 contents from dunites to peridotites, reflecting a differentiation trend (Spetsius, 1995). The most distinguishing factor in separating the dunite-lherzolite members is Mg# (100xMg/(Mg+Fe)), which varies within the range of 81-90, and can be as low as 76 for deformed garnet Iherzolites. Lower ranges of Mg# are observed in wehrlites (74.8-81.9), while the lowest values are characteristic of pyroxenite-suite xenoliths, ranging from 60 to 82. Additionally, these xenoliths have high contents of CaO (up to 12.5 %) and Al_2O_3 (up to 19.2 %), but are low in MgO (down to 13.9 %). Xenoliths of the peridotite series from Udachnaya and Sytykanskaya have higher Cr# (100xCr/(Cr+Al)), compared to similar xenoliths from the Mir and Obnazhennaya pipes. It is apparent from the statistics that the largest population of xenoliths from diamondiferous pipes is harzburgite-lherzolite. This observation is strengthened by the occurrence of websterites and clinopyroxenites as the major xenolith types in non-diamondiferous Obnazhennaya pipe (Taylor et al., 2003a). In general, it can be concluded that chemical variations among ultramafic xenoliths are specific for each pipe, reflecting the petrological nature of the mantle sampled by individual kimberlites.

4.3.2. Eclogite Series - Eclogite-series xenoliths constitute 2-10 % of the entire mantle-xenolith population in the Yakutian kimberlites. Despite being in the minority, eclogite xenoliths are particularly important in shedding light on a better understanding of the overall petrology of the upper mantle. This is due to their following characteristics: a) the tendency for most eclogites to be relatively fresh and usually less weathered than the peridotites, probably due to their lack of easily altered olivine; b) the observation that eclogite xenoliths, although less abundant, contain greater quantities of diamonds in comparison to ultramafic xenoliths (e.g., Taylor et al, 2000; Anand et al., 2003); and c) the relative pristinity of eclogites, relative to peridotites, allowing for absolute age dating, particularly by Re-Os systematics. [Unfortunately, most Sm-

Nd ages are those of kimberlite eruption (Snyder et al., 1997)].

Coleman (1965) originally introduced a classification for all different origins of eclogites that was based solely on garnet compositions. Subsequently, Taylor and Neal (1989) utilized both the garnet and omphacite compositions and devised the following three-fold classification for **mantle eclogites: Group A** (high Mg# gt; low-Jd cpx), **Group B** (intermediate Mg#, high-Fe gt; moderate-Jd cpx), and **Group C** (low Mg#, high-Ca gt; high-Jd cpx). In Russian literature, these groups of eclogites are often described as **"Magnesian, Ferroan,** and **High-alumina** varieties", respectively (e.g., Bobrievich et al., 1959; Sobolev, 1977). Mantle eclogites from worldwide localities, as well as many diamondiferous eclogite garnets from Udachnaya, are illustrated in

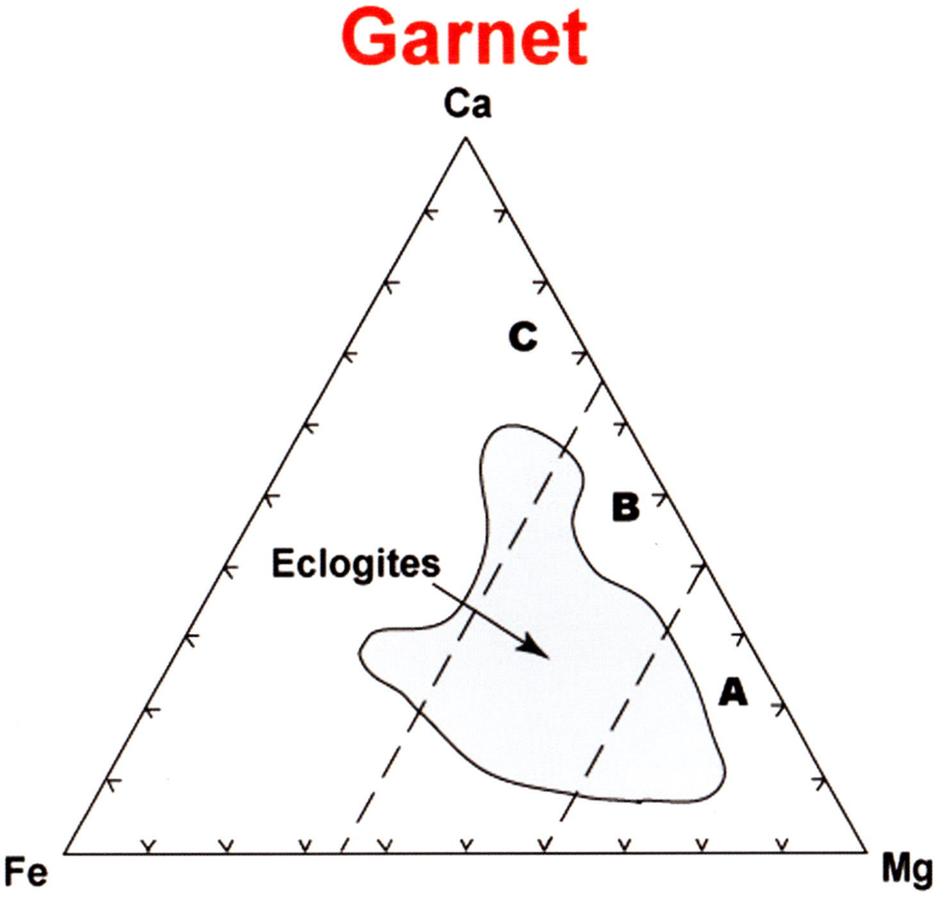

Figure 4.1. _ Plot of Mg (pyrope), Fe (almandine), and Ca (grossularite) of garnets in eclogite xenoliths from Udachnaya, depicting the three-fold (A, B, C) classification of eclogites, according to Taylor and Neal (1989).

Figure 4.1. As documented by several studies (e.g., Taylor and Neal, 1989; Snyder et al., 1997; and Taylor et al., 2003), many diamondiferous eclogites of Groups B and C, including those in Yakutian pipes, had protoliths that were portions of subducted oceanic crust. Recent studies on the Group A eclogites, especially from Obnazhennaya, have also concluded that Group A eclogites are most likely to be the result of subduction of ancient oceanic crust (Taylor et al., 2003). Diamond-bearing eclogites occur in all Groups, although most are within Group B and C eclogites. Garnet websterites, being an intermediate link between peridotite and eclogite series, also contain diamonds (Ponomarenko et al., 1980), and may have an ultimate crustal origin as well.

Group A (magnesian) eclogites are bimineralic, coarse-grained rocks having mosaic or cataclastic, and rarely poikilitic texture. They are composed of sub-equal proportions of orange garnet and pale-green, low-jadeite omphacite. Accessory minerals include rutile, sulfides, ilmenite, and occasionally a few olivine or orthopyroxene grains [although *sensu stricto*, olivine-orthopyroxene bearing eclogites are not true eclogites]. Group A eclogites have elevated Mg# (53-74), with 13-21 wt.% MgO, <12 wt.% total iron ($FeO+Fe_2O_3$), and are low in Na and K. Group A eclogites are distinguished from Group B/C eclogites by their elevated contents of Cr_2O_3, i.e., up to 1 wt% in whole-rock and >0.6-0.8 w.% in garnet. The higher Cr contents of Group A eclogites overlap, to some extent, with those of peridotites.

Garnet websterites, sometimes described as enstatite eclogites (Marakushev, 1973; Vladimirov et al., 1976) or combined within the Group A eclogites (Taylor and Neal, 1989; Taylor et al., 2003), are chemically similar to the high-Mg# eclogites. They are medium-grained rocks, with orange-red garnet (30-40%), grass-green, low-jadeite, high-Mg clinopyroxene, and pale-green orthopyroxene. The orthopyroxene, typically enstatite with high Mg#, can vary from 3 to 50 modal %.

Group B (ferroan) eclogites are coarse-grained rocks exhibiting mosaic texture. They contain orange-red garnet (about 40 modal %, 1-10 mm in size) and dark-green clinopyroxene, along with rutile, sulfides, and ilmenite as accessories. These eclogites have low Mg# (59-66%) and typically high contents of total iron ($FeO+Fe_2O_3$) = 10-14.4 wt.%, sometimes with elevated TiO_2. Neal et al. (1990) and Snyder et al. (1997) reported that the Group B ecolgites can have $\delta^{18}O$ values outside of the mantle region In addition, extremely high εNd values up to +250 have been observed. See discussion by Taylor and Anand (2004) reprinted at the end of this book.

Group C (high-alumina) eclogites are coarse-grained rocks typically containing kyanite (up to 30%) as an additional phase to the high-Ca garnets and high-jadeite clinopyroxenes (up to 11 wt.% Na_2O). Wide variations of the abundances of garnet (20-80%) and clinopyroxene (20-60%) are typical of this Group. Coesite and corundum are common, with rare rutile, sulfides, and ilmenite. A particular eclogite in this Group with abundant grossular (gros >50%) garnet, clinopyroxene (py), and diasthene (kyanite) (di) is called **"grospydite"**, first described by Bobrievich et al. (1960) and reported from the Zagadachnaya pipe (Sobolev et al., 1968; Sobolev, 1974). Similar to Group B eclogites, the Group C eclogites possess evidence for crustal origins, including +Eu anomalies in some garnet and omphacite, as well as non-mantle oxygen isotopes.

Abundances of eclogite xenoliths in the kimberlite pipes of Yakutia are quite variable, being 6-8 % at Udachnaya and <3 % at Mir. Grospydites, strictly with >50% grossular component in the garnet, and other Group C eclogites occur abundantly at Zagadochnaya, Dolgozhdannaya, and Zarnitsa (Bobrievich et al., 1959). However, only one such eclogite has been reported from the Mir pipe (Sobolev et al., 1974), and virtually none at Obnazhennaya (Ukhanov et al., 1988). At Udachnaya, all three Groups of eclogites are present, with a relatively higher abundance of jadeite-bearing Group B and Group C eclogites. This led Spetsius (1992) to suggest that the lithosphere underlying the central region of the Siberian Platform (Daldyn-Alakit region) is more aluminous than in other regions of the craton.

4.4. Upper Mantle Heterogeneity.

Kimberlites have sampled mantle rocks from depths of 100-300 km (e.g., Sobolev et al., 1975; Ringwood, 1982; Dawson, 1980; Gurney and Harte, 1980). The wide variety of deep-seated xenoliths in kimberlites, both worldwide and within a given craton, suggests significant vertical heterogeneity of the upper mantle. In contrast, lateral heterogeneity in the mantle is not well understood, largely because of the difficulties in comparing xenoliths from the exact same horizons, i.e., identical P-T conditions (Dawson, 1983; Kharkiv, 1995a). However, an attempt was made by Spetsius and Serenco (1990) in comparing the abundances and types of xenoliths from different segments of the Siberian Platform. As shown in Table 4, the types and abundances of xenoliths from the Obnazhennaya pipe (1000 km from the Mir pipe) are distinctly different than those from the Daldyn-Alakit region (e.g., Udachnaya, Zagadachnaya, and Aykhal), where highly aluminous Group B and C eclogites are present. However, it must be pointed out that the P-T conditions as revealed by the Obnazhennaya xenoliths are considerably less than for the other pipes (Taylor et al., 2003).

As discussed in Chapter 1, lateral heterogeneity in the mantle may be manifest in variations in different abundances of diamond morphologies across the Yakutian kimberlite province. For instance, cubic diamonds have been observed in most of the kimberlite pipes in the Daldyn-Alakit region, and are related to Group C eclogite xenoliths. However, both cubic diamonds and even eclogites are generally absent in pipes of the southern Malo-Botuobinsky field, possible evidence reinforcing the presence of lateral heterogeneity in the Siberian craton/mantle.

Table 4. Abundances (%) of the Mantle Xenoliths in Kimberlite Pipes of Yakutia.

Type of xenoliths	Mir	Int	Bot	Ud	Zar	Dal	Syt	Yub	Obn
Dunites	2.5	18.1	+	1.2	4	2	4.2	3	+
Gt-dunites	3.7	-	-	5.4	2	6	1.4	10	-
Sp-harzburgites	1.7	5.3	+	1.6	4	4	2.3	3	+
Gt-harzburgites	1.6	-	+	4	+	2	4.4	11	-
Sp-lherzolites	7	5.3	+	9.1	6	6	15.3	7	11.7
Gt-lherzolites	51.7	58.9	+	21.3	50	35	41.9	35	39.8
Gt-lherzolites, sheared.	7	-	+	39.2	18	20	2.8	22	-
Sp-wehrlites	0.4	-	-	0.2	-	-	0.5	-	-
Gt-wehrlites	1.2	-	+	0.3	+	3	2.3	+	+
Pyroxenites	1	-	-	1.8	4	4	0.5		3.7
Gt-pyroxenites	11.3	9.8	+	6.4	2	4	1.4	0.5	23.7
Ilm-peridotites	4.5	-	-	2.5	2	5	5.6	0.5	1.6
Ilm-pyroxenites	1.2	-	-	-	-	2	+	-	1.1
Phl-peridotites	-	-	-	0.8	-	-	4.4	6	-
Gt-Ilm-peridotites	1.9	-	+	1.4	2	4	6.5	-	5.1
Gt-Ilm-pyroxenites	1.7	-	-	+	-	2	0.7	-	-
Eclogites	1.6	2.6	+	4.6	6	1	5.8	2	13.3
Ky-eclogites, grospydites	-	-	+	0.2	+	-	+	-	-
Cs-eclogites	-	-	-	+	-	-	+	-	-
D- Peridotites	+	-	+	-	-	-	-	-	-
D- Pyroxenites	+	-	-	+	-	-	-	-	-
D- Eclogites	+	-	+	+	-	-	+	+	-
Number of samples	514	136	40	550	50	350	430	300	375

Pipe abbreviations: **Mir**- Mir, **Int**-International, **Bot**-Botuobinskaya, **Ud**-Udachnaya, **Zar**- Zarnitsa, **Dal**-Dalnaya, **Syt**-Sytykanskaya, **Yub**-Yubileynaya, **Obn**-Obnazhennaya.
+ present as a single sample; - not present.

4.5. Metasomatic-Induced Partial Melting.

The metasomatic influences of kimberlites on crustal and upper-mantle xenoliths have been addressed in numerous studies. Generally, low-temperature alterations, such as carbonatization, serpentinization, zeolitization, etc. are predominant (e.g., Berg, 1977). Nevertheless, the effects of high-temperature modification of primary mantle rocks are widespread among mafic and ultramafic xenoliths from Yakutian and South African pipes (e.g., Dawson, 1980; Spetsius and Serenko, 1990; Fadili and Demaiffe, 1999), and a number of notable investigations have been carried out on this subject (e.g., Lappin, 1978; Menzies, 1990; Dawson, 1999).

Many of these high-temperature alterations have been initiated by infiltration of metasomatic fluids, causing various degrees of partial melting of the primary minerals. Williams (1932) first noted evidence of such metasomatic-induced melting in xenoliths from the mines at Jagersfontein and Roberts Victor. In a classic study, high-temperature alterations were described in kyanite eclogites from the Roberts Victor pipe (Switzer and Melson, 1969), and attributed to partial melting under the influence of rapidly rising kimberlitic magma. Similarly, eclogites with extensive partial-melt products have been observed in the Mir and Udachnaya pipes of Yakutia (Ponomarenko, 1977; Spetsius and Ponomarenko, 1979; Spetsius, 1980), as well as in South Africa (Chinner and Cornell, 1974; McCormick et al., 1994). A similar type of melt product has been reported as occurring with garnet megacrysts in the kimberlite of Fayette County, Pennsylvania; here, the garnet exhibits "kelyphitic rims" with quenched melt (i.e., glass), indicating incongruent melting induced by metasomatic fluids (Hunter and Taylor, 1982).

Worldwide, most eclogite xenoliths from kimberlites display evidence of **partial melting processes**, brought about by the introduction of K-Na-rich fluids (Sobolev et al., 1998). This is particularly well developed in eclogites from Udachnaya, Mir, and other Yakutian kimberlites. In their studies, Spetsius and Taylor (2002) and Misra et al. (2003) have detailed the nature and characteristics of the products of metasomatically-induced partial melting by a K-rich fluid. Typically, the crystallization products of such melting are observed as a "spongy" texture around primary omphacite (Cpx1), wherein the primary and secondary omphacite are all in optical continuity. They are composed of secondary clinopyroxene (with lesser Na_2O; Cpx2), spinel, glass, and feldspar. With partial melting of kyanite eclogites, corundum and mullite are also encountered. In addition, primary garnet always shows evidence for metasomatic-induced partial melting, occurring as rinds of secondary minerals enclosing primary garnet core, commonly known as "kelyphitic rims". These kelyphitic rims are composed of orthopyroxene, spinel, olivine, and glass.

The majority of the fluids that induced partial melting of the eclogites were probably derived from the kimberlitic melt. Alteration of the primary eclogitic minerals by these fluids may have begun upon entrapment of the xenoliths by the kimberlite, and continued even during upward movement of the kimberlite to depths of less than 30 km or so. There are definite spatial correlations between the presence of some diamonds and the partial-melt products, as well as zones of alteration in many diamondiferous eclogites (Spetsius, 1999; Spetsius and Taylor, 2002; Misra et al. 2003;Taylor et al., 1996, 1998, 2000). It is postulated that the partial-melting process was connected to resorption of some diamonds, and to the formation of late-stage fibrous diamond coatings and possibly some microdiamonds.

4.6. Mantle Xenolith Mineralogy

Peridotite and eclogite xenoliths are fundamentally different in both abundance and chemistry of their constituent minerals. Peridotites consist basically of four phases, including olivine, orthopyroxene, clinopyroxene, and garnet, with the possibility of spinel in a lower-pressure assemblage. Eclogites are mainly bimineralic rocks, consisting of jadeite-rich clinopyroxene (omphacite) and garnet of various compositions. Below, we will present more details on the crystal chemistry of these and additional minerals within the ultramafic and mafic rocks from the Siberian craton mantle.

4.6.1. Mineralogy of Xenoliths from Ultramafic Suite - Most of the minerals in ultramafic xenoliths display a range in composition. The exception is olivine, a typical mineral in ultramafic rocks, which is comparatively constant in composition, regardless of the chemistry of the xenoliths. As such, its major-element chemistry makes it the most non-informative ultramafic mineral, having Fo contents of 84-94 and in most samples, 91-93. Lower Fo values are observed in olivines of ilmenite peridotites. A comparison of olivine chemistry of coarse-grained and sheared Gt-peridotites reveals a slight difference with average Fo contents of 90.9 and 91.8, resp. The CaO contents of olivines are variable in different types of xenoliths, as well as between xenoliths in a given pipe, and vary from 0.03 to 1.10 wt%. Notably, high-Ca olivines have been observed in ultramafic xenoliths of Obnazhennaya. The Cr_2O_3 contents range from near detection limit up to 0.15 wt.%. Olivines are usually enriched in nickel, with 0.30-0.50 wt.% NiO.

Orthopyroxene occurs mostly as enstatite, with bronzite observed in some pyroxenites, garnet websterites, and particularly common in ilmenite-bearing rocks. Minor and trace elements in ultramafic orthopyroxenes include Ti, Al, Cr, and Ca. Most variable are TiO_2 contents, ranging from <0.03 up to 0.21 wt.%, whereas Cr_2O_3 ranges from <0.03 to 0.45 wt.%. High-Ti orthopyroxenes occur in ilmenite-bearing peridotites from the Mir pipe. CaO usually varies in xenoliths from different pipes, being higher in ultramafic orthopyroxenes from the Udachnaya pipe (up to 1.08 wt.%). Al_2O_3 contents range from 0.3 to 3.65 wt.%, but most orthopyroxenes have <2.0 wt.% Al_2O_3. Peridotites with high-Al orthopyroxenes are present from Udachnaya (average 1.08 wt.% Al_2O_3), Mir (0.68 wt.% Al_2O_3), Sytykanskaya (0.97 wt.% Al_2O_3), and Obnazhennaya (0.87 wt.% Al_2O_3).

Most clinopyroxene in ultramafic xenoliths from Yakutian kimberlites is diopside-rich, frequently with substantial amounts of Na_2O and Cr_2O_3. They are augitic pyroxenes, with 44-50 Ca# {100x[Ca/(Ca+Mg)]}. Ca# is particularly useful in distinguishing between equigranular and deformed (sheared) peridotites. For instance, in clinopyroxenes in equigranular peridotites from Udachnaya, clinopyroxenes have Ca# >45, whereas those in the deformed peridotites are characterized by lower Ca#, usually <40% (Spetsius and Serenco, 1990). A similar division was also observed in the same type of peridotites from Mir.

It is typical for ultramafic clinopyroxenes to have low Na_2O (0.53-3.27 wt.%) and alumina contents (0.39-8.00 wt.% Al_2O_3); the alumina content is less in ilmenite-bearing rocks and in most diamondiferous peridotites. Generally, there are distinct compositional differences between ultramafic clinopyroxenes in diamondiferous pipes and those in non-productive pipes. In xenoliths from barren pipes, clinopyroxene has higher alumina, but low magnesian contents. With this contrasting chemistry, clinopyroxene is potentially useful as an indicator mineral to locate diamondiferous pipes. The Cr-content in clinopyroxenes varies from <0.03 to 3.0 wt.% Cr_2O_3, and shows only minor correlation with Mg#.

Garnet is common in mantle xenoliths of the peridotite-pyroxenite series, and it can vary significantly in composition between different rock types. In garnets, the Mg# ranges from 50 to 97, with lower values in ilmenite-bearing peridotites and the lowest Mg# in pyroxenites. The Al_2O_3 content is usually relatively constant, but can be as low as 14.3 wt.%, due to substitution by Cr, as in harzburgitic garnets. The CaO content of garnet can also vary. The characteristic feature of most ultramafic garnets is the elevated Cr-content, which can be as high as 15 wt.% Cr_2O_3 in the deformed and diamondiferous peridotites of the Udachnaya pipe (Logvinova et al. 2005). Until recently, the maximum content of Cr_2O_3 in pyrope was reported as 4.9 wt.%, occurring in wehrlitic garnets (Sobolev, 1977). Garnets with elevated Cr-contents are low in abundance in the ultramafic rocks from Obnazhennaya, a non-diamondiferous pipe, believed to have formed at less depth and pressure than most other Yakutian pipes (Taylor et al., 2003).

The most important chemistry of garnets, for the prediction of "diamondiferousness" (as they say in 'Russlish'), is the relationship of CaO and Cr_2O_3 (Sobolev, 1974; Kharkiv, 1995; Gurney, 1989). The Cr content in a concentrate of the high-Cr garnets (G10), which belong to the dunite-harzburgite paragenesis, has been used to define the diamond grade of a kimberlite. Such data mostly plot in the lherzolite field (Fig. 4.2). It has been repeatedly demonstrated through the years that Cr contents of

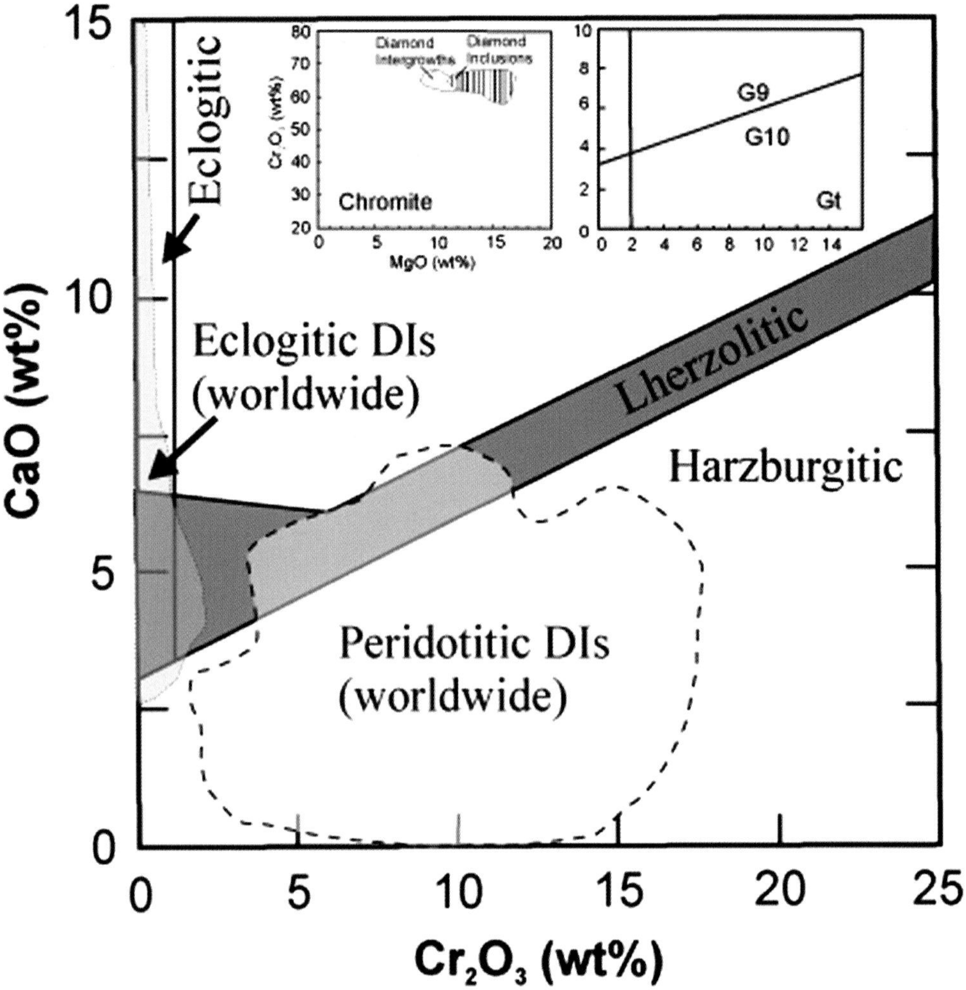

Figure 4.2. _Plot for peridotitic garnets, a scheme originally introduced by Sobolev (1977), with a restricted field for eclogitic garnets, which have <1 % Cr_2O_3. Envelope for P-type inclusions is according to Griffin et al. (1993). Insert in upper right depicts the regions for G9 and G10 "indicator garnets" used in diamond exploration, according to Gurney (1984). The other insert gives the field in Cr_2O_3 – TiO_2 space for chromite as another "indicator mineral" for diamonds, modified after Fipke et al. (1995).

pyrope can be used as a criterion for potential diamondiferousness, as originally suggested and championed by Sobolev (1974).

Ilmenite is relatively rare in mantle xenoliths, but due to its abundance in kimberlites and resistance to weathering, it is an important indicator mineral for kimberlite exploration (Bobriyevich et al., 1959). Ilmenite from ultramafic xenoliths has a wide range of compositions, not only in Mg, Fe, Ti, but also in Al, Cr, and Mn (Spetsius and Serenco, 1990). Higher amounts of a geikielite component ($MgTiO_3$) are typical for ilmenites associated with garnet. There is an apparent correlation of MgO and TiO_2 between different rock types, as well as different kimberlite fields.

Chrome-spinel (chromite) occurs as an accessory phase in most xenoliths of both the spinel and garnet facies, comprising up to 3-7 vol.% in rare cases. In alkremites from the Udachnaya pipe, the abundance of Cr-spinel varies up to 30-40 vol.%, 80 % in extreme cases. Spinels vary from magnesio-ferrites to chromites with Cr_2O_3 contents of >60 wt.%. Chromite is a typical kimberlite indicator mineral, but it should be stressed that some pipes are completely barren in chromite (Cherny et al., 1988). This is particularly true for the newly discovered kimberlites pipes at Botuobinskaya and at the Nurbinskaya pipe of the Nakynsky field,. There are no distinct differences between the composition of spinels in the ultramafic series, although chromites are typical of diamondiferous kimberlites. Low-Cr varieties, generally, with high contents of Al_2O_3, are abundant in ultramafic xenoliths of the non-diamondiferous Obnazhennaya pipe. The study of spinel chemistry has shown that the great majority of compositions in xenoliths from diamondiferous pipes correspond to 60-68 wt% Cr_2O_3. For chromites associated with diamonds, it is typical to have an inverse dependency of Cr_2O_3 content versus Al_2O_3 (e.g., Sobolev, 1974, 1975; Nixon, 1978; Bulanova et al., 1980).

4.6.2. Mineralogy of Eclogite Xenoliths. - Although eclogites contain fewer minerals than peridotites, the abundances of both garnet and clinopyroxene can significantly vary, from 20:80 to 80:20. In addition, eclogites may contain accessory phases such as kyanite, coesite, corundum, sanadine, even diamonds, etc.

Eclogitic garnets have large variations in MgO, FeO, and CaO contents that have been used in the classification of eclogites into Groups A, B, and C (Coleman, 1965). In Group A eclogites, garnets are Mg-rich with high pyropic contents. With progressive decrease in MgO, while FeO and CaO increase, the garnets fall into Group B, and eventually Group C at even lower MgO and higher CaO contents. The garnets from kyanite eclogites of Udachnaya vary widely in Ca# {100x[Ca/(Ca+Mg)]} from 22 to 76%. There is relatively less variation in Ca# of eclogite garnets from the Mir pipe, where kyanite eclogites are rare. There are no clear differences in garnet chemistry of eclogites from different Yakutian pipes, other than the paucity of Ca-rich garnets in Obnazhennaya eclogites (Taylor et al., 2003), in which most garnets have high Mg#.

Eclogitic clinopyroxenes are characterized by high jadeite ($NaAlSi_2O_6$) contents;

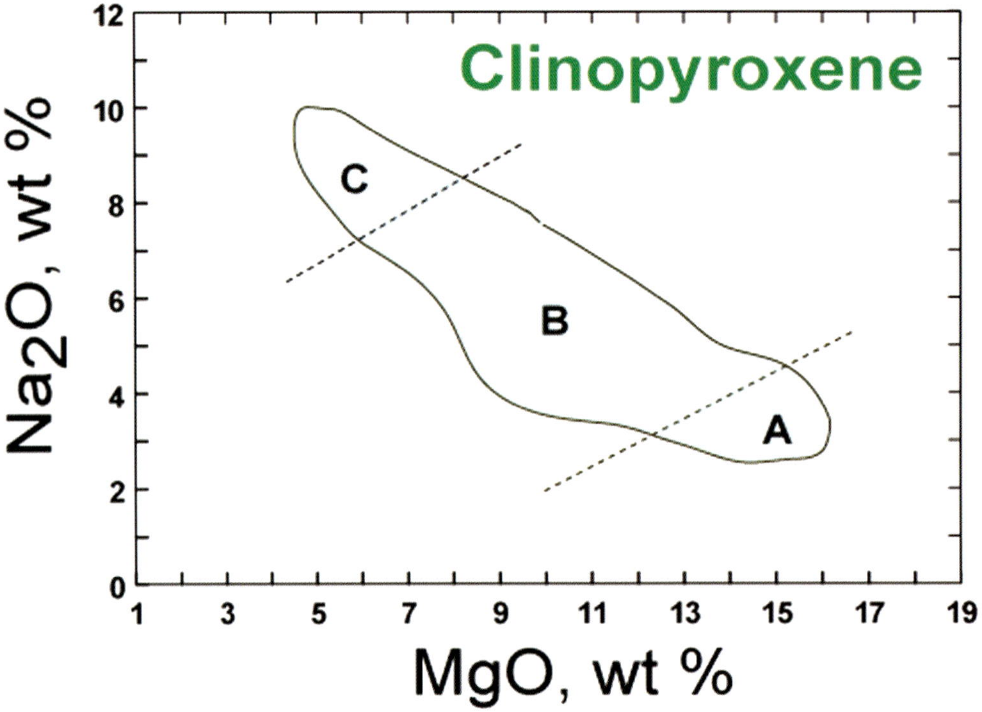

Figure 4.3. A plot of omphacitic clinopyroxenes in eclogite xenoliths from Udachnaya, depicting the three-fold (A, B, C) classification of eclogites, according to Taylor and Neal (1989).

thus, they are correctly referred to as omphacites. There is an inverse correlation between jadeite and MgO contents. As shown in Fig. 4.3, there are no apparent differences between eclogitic garnets from the Mir, Obnazhennaya, and Udachnaya pipes; however, garnet compositions from Udachnaya are somewhat more diverse. The trace-element contents, especially REEs, of clinopyroxenes in garnet websterites and eclogites from the Obnazhennaya pipe are generally higher than those from Udachnaya.

Kyanite is easily recognized by its bluish color and generally unaltered nature. Occasionally, however, it can be replaced by corundum and mullite. Two types of kyanite have been found: 1) tabular crystals with bluish or sometimes greenish tint (the latter due to Cr content), and 2) needle-shaped, colorless crystals, typical in kyanite eclogites, believed to have formed in a late stage. Kyanite is also commonly observed in close association with garnet, where it often occurs as an exsolution product.

Coesite has been found in numerous kyanite eclogites and grospydites (i.e., Group C) from the Udachnaya and Zagadachnaya kimberlites (Sobolev, 1974; Ponomarenko et al., 1977; Spetsius, 1990). The presence and preservation of coesite in eclogites is indicative of pressures in excess of 30 kbars. In fact, Sobolev et al. (2000) in a classic study determined a “coesite-in-diamond” barometer that gave pressures far in excess of this – well up into the diamond-stability field. However besides a pressure indicator, one of the authors of this book believes that the presence of coesite may potentially be used as an indicator mineral for “diamondiferousness”.

Corundum commonly occurs as small needles (0.01-0.2 mm long) replacing kyanite, and more rarely forms separate lamellae up to 2 mm in kyanite eclogites. It has a dark-blue color and is commonly referred to as sapphire. In some eclogites, the corundum is ruby-red and forms elongated crystals up to 1-3 mm. A unique eclogite from Obnazhennaya (0-160; Plate 4a of Sobolev, 1977) contains ruby corundum and large clinopyroxene (Qi et al., 1996). Here, the clinopyroxene has effectively exsolved five distinct minerals: garnet, orthopyroxene, kyanite, corundum, and sanadine. It would appear that the rock originally consisted almost entirely of a high-pressure, non-stoichiometric clinopyroxene, which has subsequently exsolved up to 30% other minerals.

Sanadine has been found in only a few samples of Ky-eclogite from Udachnaya and in an eclogite from Obnazhennaya (0-160). This phase is present only in trace amounts in eclogites worldwide. It has been reported to occur with coesite in a grospydite from Roberts Victor (Smyth and Hatton, 1977), and has also been found as an inclusion in diamond (Prinz et al., 1975).

Rutile, as an accessory mineral, occurs in all varieties of eclogites, most commonly as exsolution lamellae along (111) planes in garnet. The basic trace elements in rutile are Al, Cr, Fe, and Nb. The content of Cr_2O_3 in rutiles varies from <0.03 to 0.72 wt%. Some Al_2O_3 is always present and varies from 0.06 to 1.09 wt.%. The FeO content ranges from 0.27 to 1.42 wt.%. Rutiles in xenoliths from the Mir pipe have higher

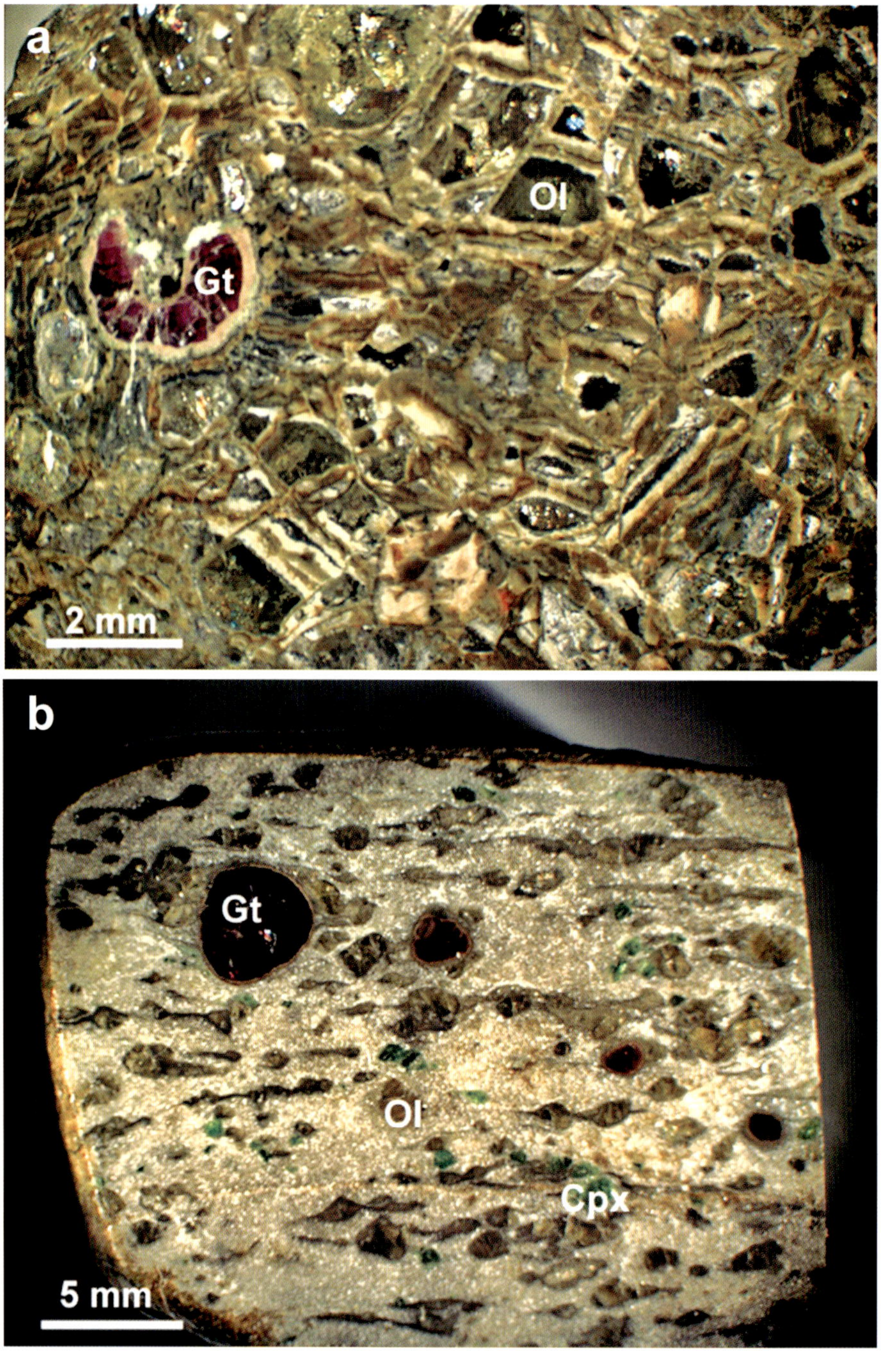

Plate 4. 1. Petrographic Textures in Udachnaya Peridotite Xenoliths. a) Coarse-grained texture of garnet peridotite containing olivine (Ol), orthopyroxene, and garnet (Gt), in a polished section; b) Mosaic porphyroclastic texture of sheared garnet peridotite.

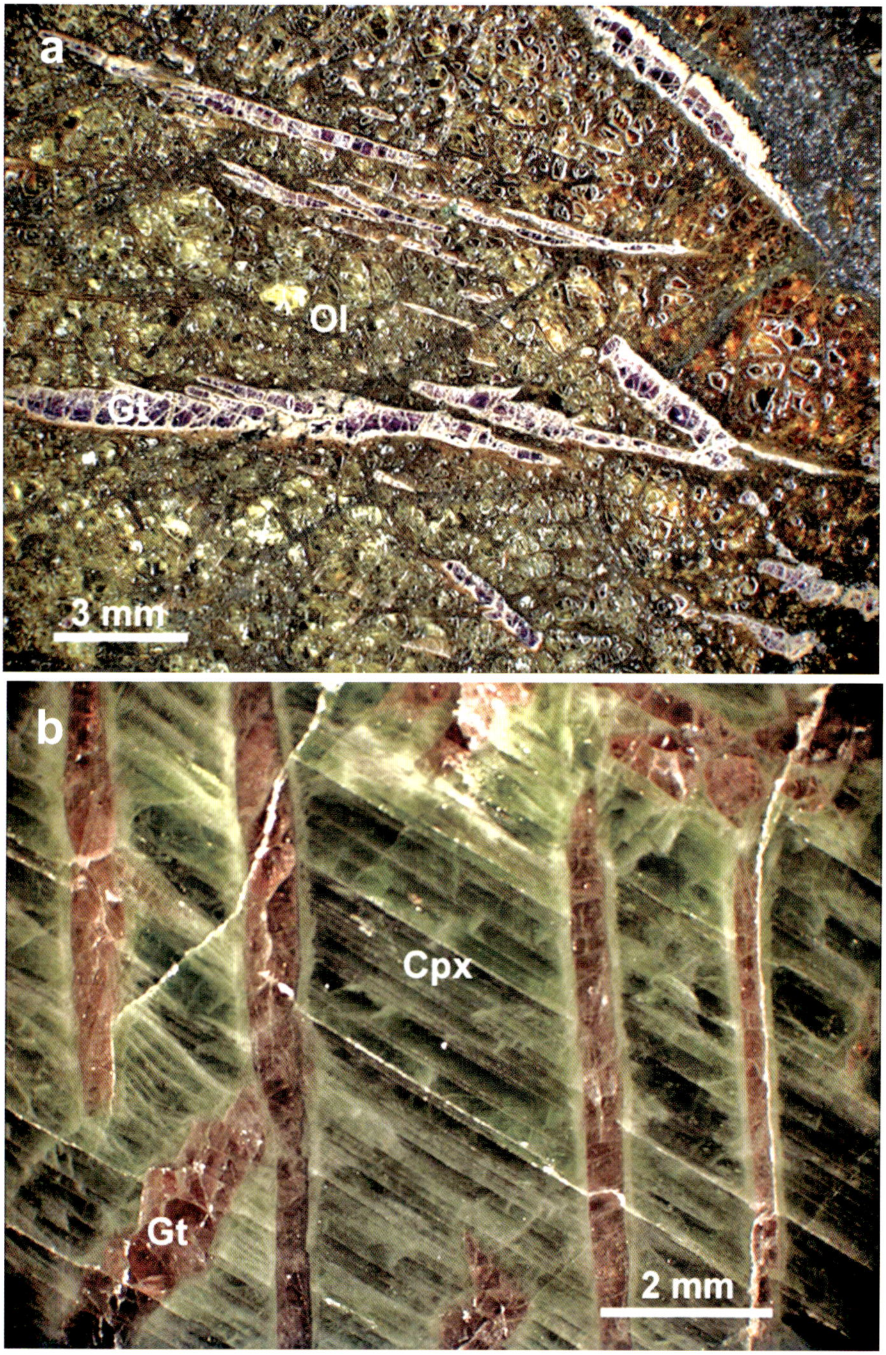

Plate 4. 2. Petrographic Textures in Peridotite Xenoliths. a) Example of garnet exsolution in a lherzolite from Udachnaya; b) Exsolution of garnet in a clinopyroxenite from the Obnazhennaya pipe.

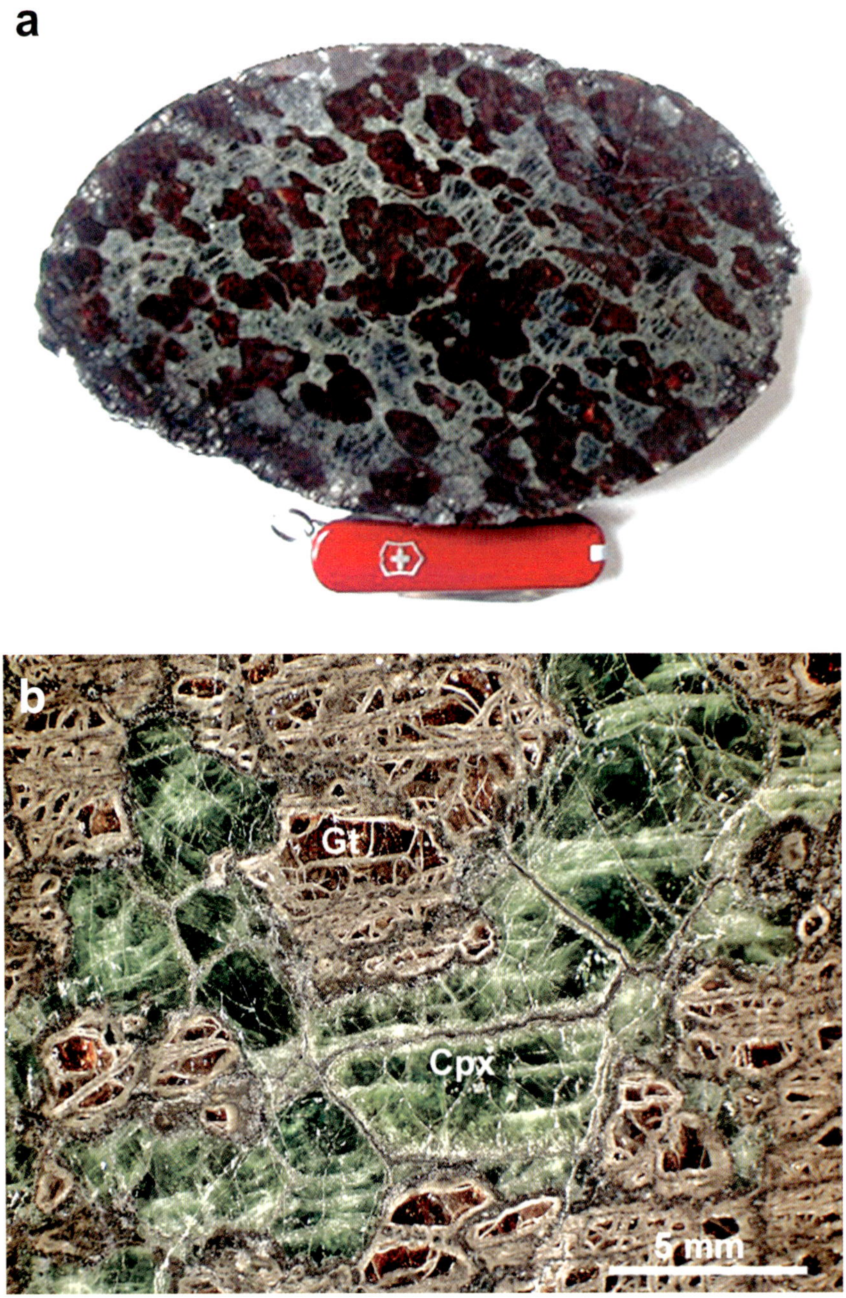

Plate 4. 3. Typical Petrographic Textures in Eclogite Xenoliths. a) Coarse, equigranular texture in bimineral eclogite from the Mir pipe; b) Coarse-grained texture in a Group A eclogite from the Udachmaya pipe. Note the fresh nature of the omphacite (cpx) and garnet, common among xenoliths from Yakutia.

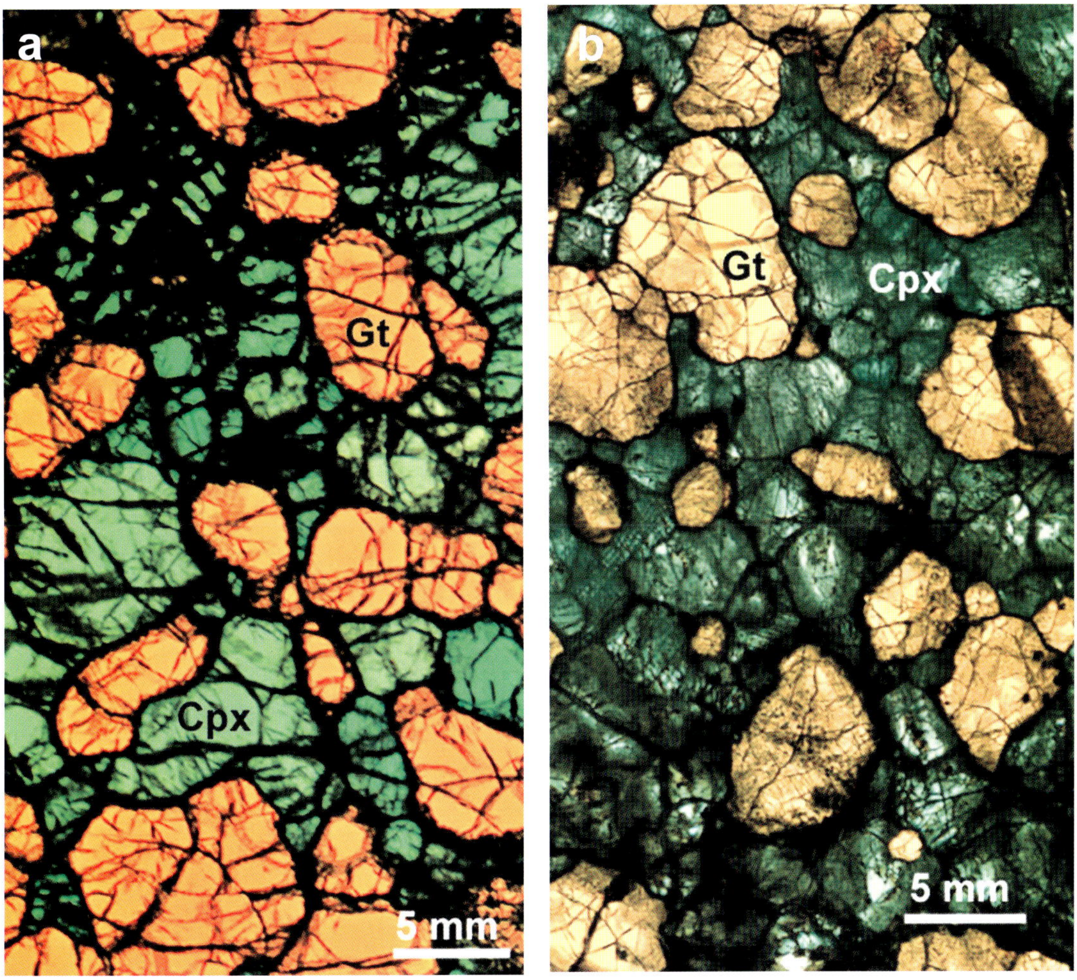

Plate 4. 4. Typical Petrographic Textures in Eclogite Xenoliths. a) Coarse equigranular texture of bimineral eclogites from the Udachnaya pipe; in plane-polarized light.

Plate 4. 5. Unusual Petrographic Textures in Udachnaya Eclogites. a) and b) Inhomogeneous distribution of garnet and clinopyroxene in bimineral eclogites; c) Extensive exsolution of garnet in clinopyroxene, as viewed in plane-polarized light, with blue filters.

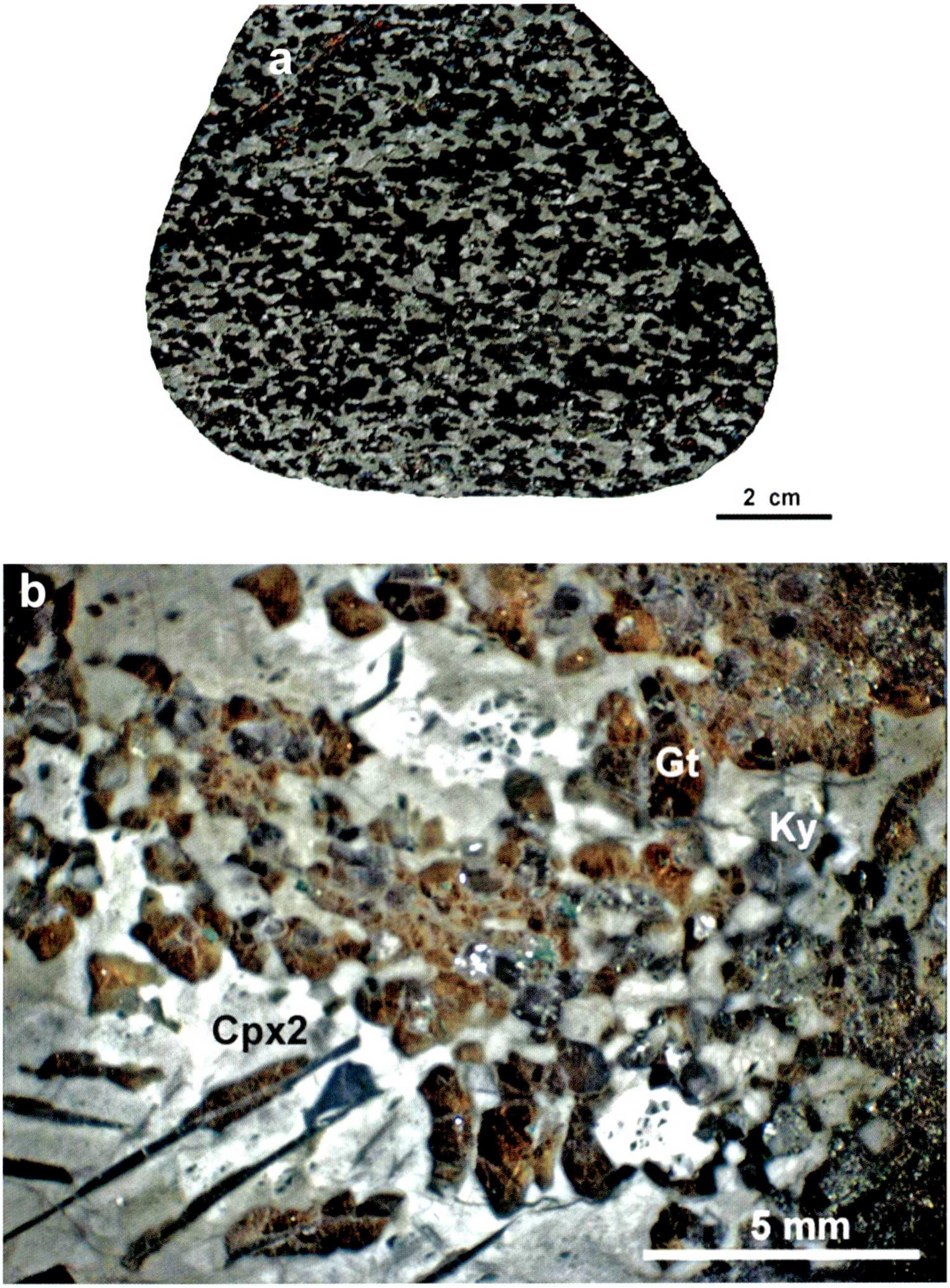

Plate 4. 6. Petrographic Textures in Udachnaya Kyanite Eclogites. a) A kyanite (Ky) eclogite xenolith (wt. = 8.8 kg) with a layered texture, containing coesite (indicative of at least 30 kbar); b) Slightly layered texture, as seen in plane-polarized light, with two varieties of kyanite, prismatic and needle-like; the clinopyroxene (Cpx2) is the result of metasomatism-induced partial melting, as detailed in Spetsius & Taylor (2002).

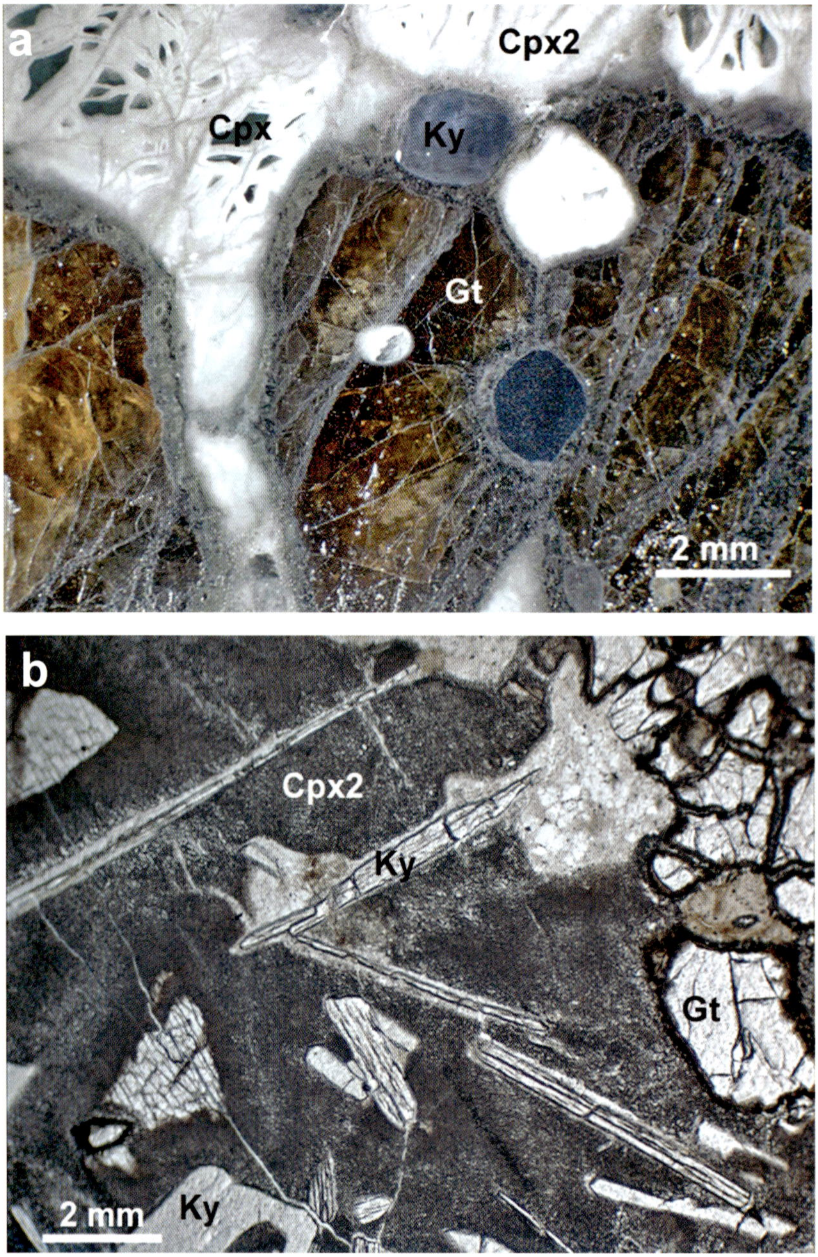

Plate 4. 7. Petrographic Textures in Udachnaya Kyanite Eclogites. a) Coarse-grained, sanidine-bearing kyanite eclogite, with evidence of partial melting of Cpx to form Cpx2, with lower jadeite component; only relicts of primary omphacite are present; b) Kyanite eclogite, in plane polarized light, showing two varieties of kyanite, present as round grains and needles up to 4 mm in length; note the kelyphitic rim around the garnet

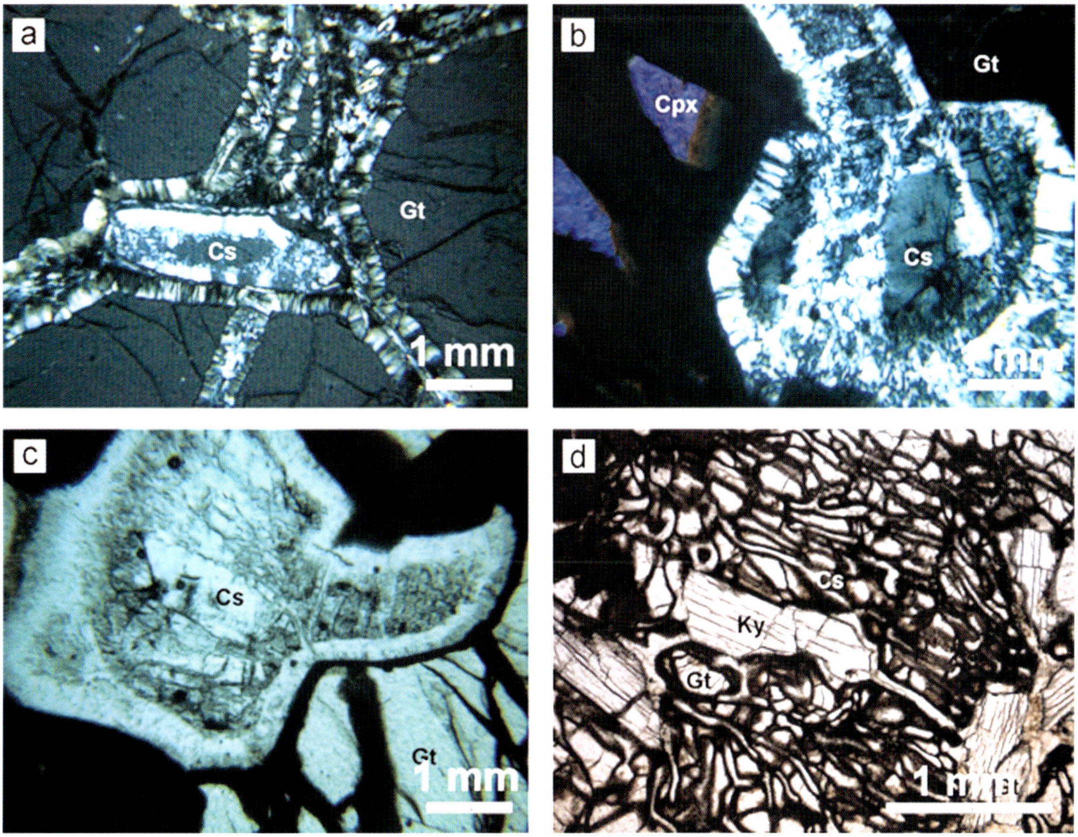

Plate 4. 8. Coesite-Bearing Eclogites from the Udachnaya Pipe. a) View of coesite relict with pronounced texture of surrounding secondary quartz between grains of garnet (Partial XPL, blue filter); b) (Partial XPL) and c) (PPL, blue filter) Relicts of coesite surrounded by rims of quartz, plus additional fine grained quartz; d) Intergrowth of coesite with garnet and kyanite (PPL). Gt = garnet, Cpx = clinopyroxene, Cs = coesite, Ky = kyanite.

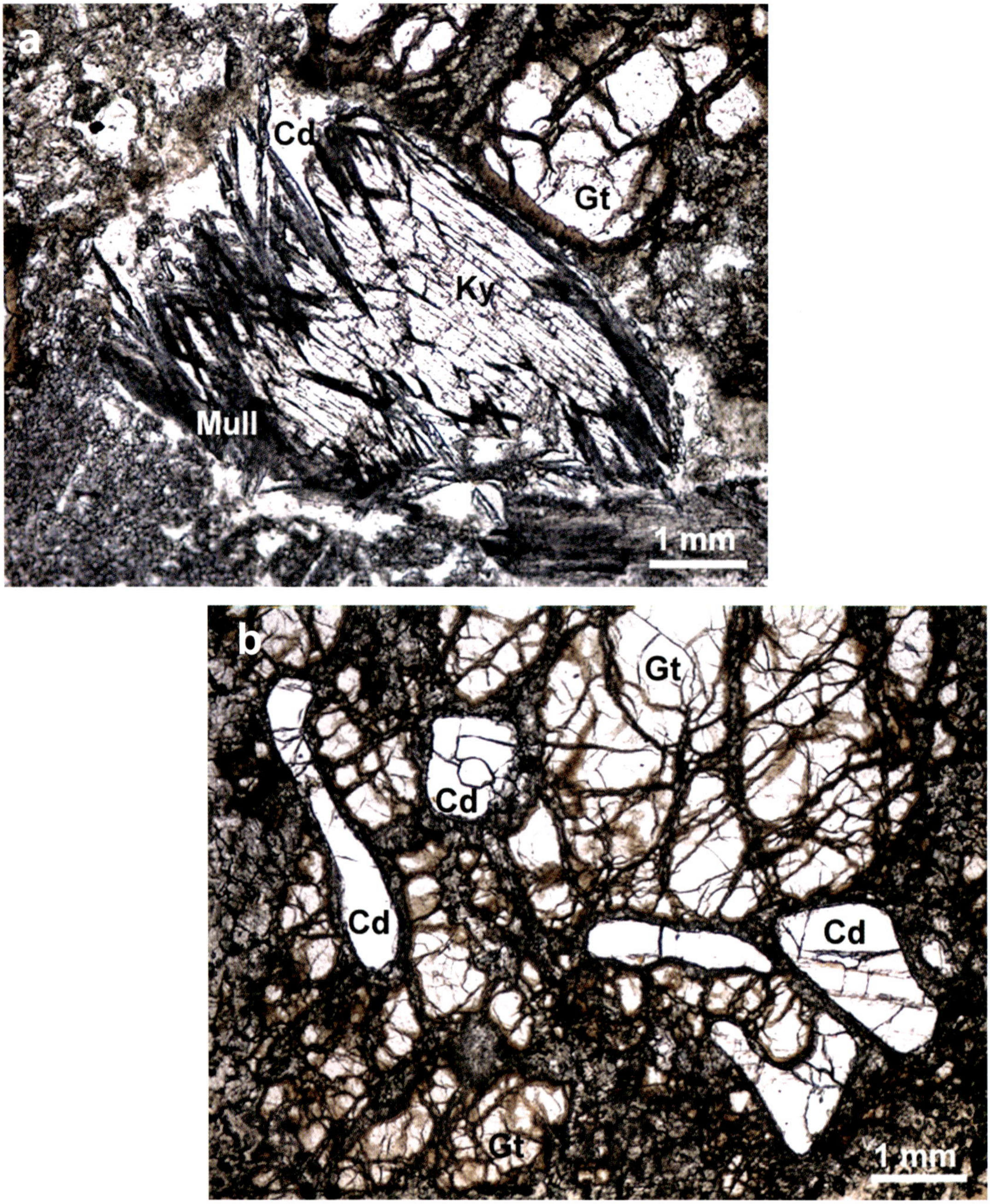

Plate 4. 9. Corundum-Bearing Eclogite Xenoliths from Udachnaya. a) Needles of secondary sapphire corundum in association with mullite and replacing kyanite; also contains small diamonds with cubic morphology (PPL); b) Ruby corundum in a diamondiferous garnetite; note the secondary corundum cross-cutting the garnet (PPL). Gt = garnet, Ky = kyanite, Cd = corundum.

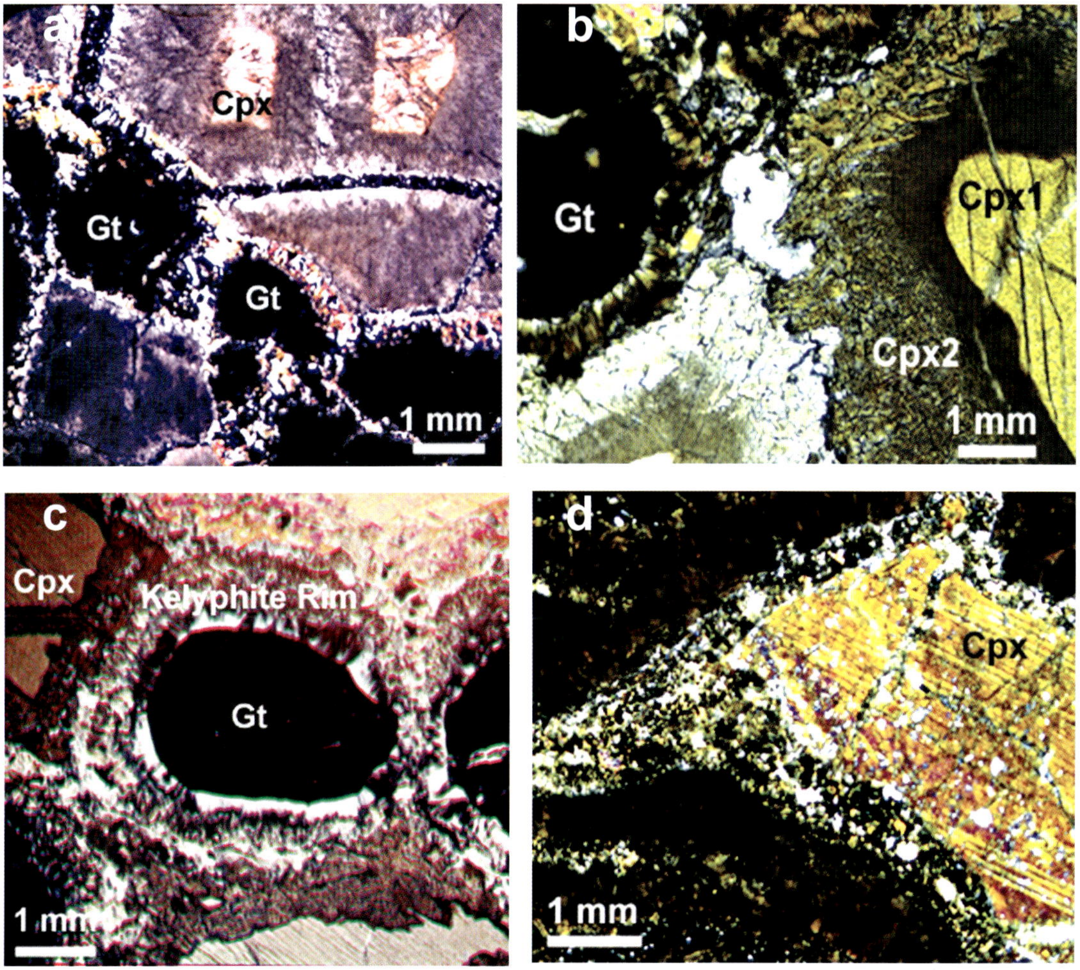

Plate 4. 10. Metasomatism-Induced Partial-Melt Products in Udachnaya Eclogite Xenoliths. All photomicrographs in XPL; a) General view of fine-grained, intergranular, partial-melt products, in a pronounced ophitic texture between grains of garnet and clinopyroxene; relicts of primary omphacite and garnet are present; notice the intersecting partial-melt veinlets in the clinopyroxene; b) Spongy texture of secondary pyroxene (Cpx2) with plagioclase and glass; garnet with a kelyphitic rim is also present; the presence of plagioclase indicates that this process, which may have initiated at higher pressure, was not completed until <20 kb; c) An extensive kelyphitic rim around garnet, consisting of an intergranular mixture of partial-melt products; xenolith is garnet pyroxenite; glass in the kelyphite is evidence for quench late-stage melt due to metasomatism with the enclosing kimberlite, as per Hunter & Taylor (1982); d) Needles of phlogopite intergrown with secondary pyroxene, spinel, and plagioclase in the partial-melt products. Gt = garnet, Cpx or Cpx1 = primary omphacite, Cpx2 = secondary clinopyroxene in spongy texture.

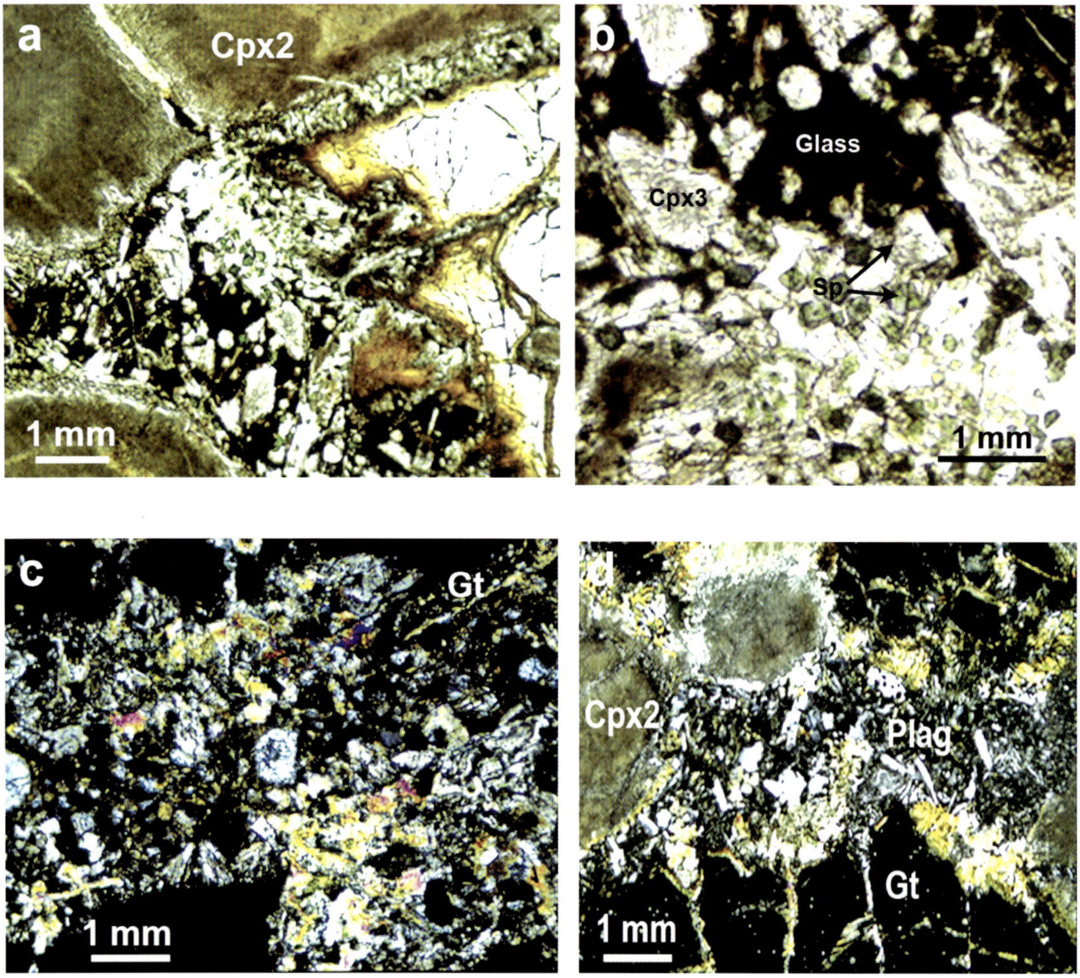

Plate 4. 11. Evidence for Intergranular Partial Melting in Udachnaya Eclogite Xenoliths. a) & b) = PPL; c) & d) = XPL; a) An intergranular vein of partial-melt products between grains of garnet and clinopyroxene; b) Enlarged portion of photo a (rotated 180°). notice the secondary pyroxene (Cpx3) and spinel (Sp) in the glass; c) Crystallized partial-melt products in veinlets, including feldspar and quartz, indicative of low-pressure (<20 kb) reaction; d) A vein of partial-melt products, displaying different degrees of crystallization and metasomatism.

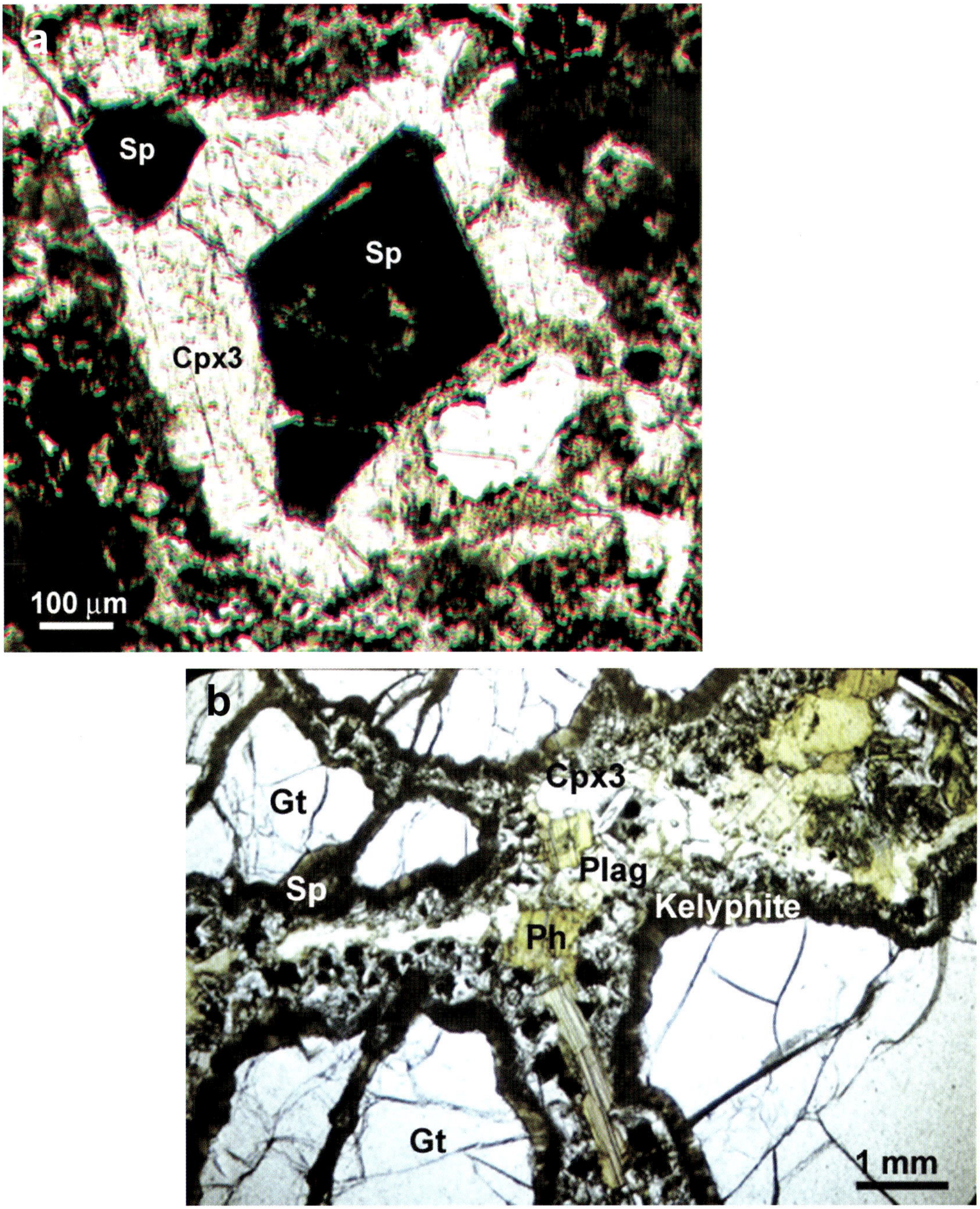

Plate 4. 12. Secondary Spinel and Phlogopite in Partial-Melt Products in Udachnaya Eclogites. a) Zoned spinel in secondary clinopyroxene (PPL); b) Elongated needle of phlogopite encountered with secondary pyroxene, spinel, and plagioclase (PPL); these products are indicative of a low-pressure process (<20 kb).

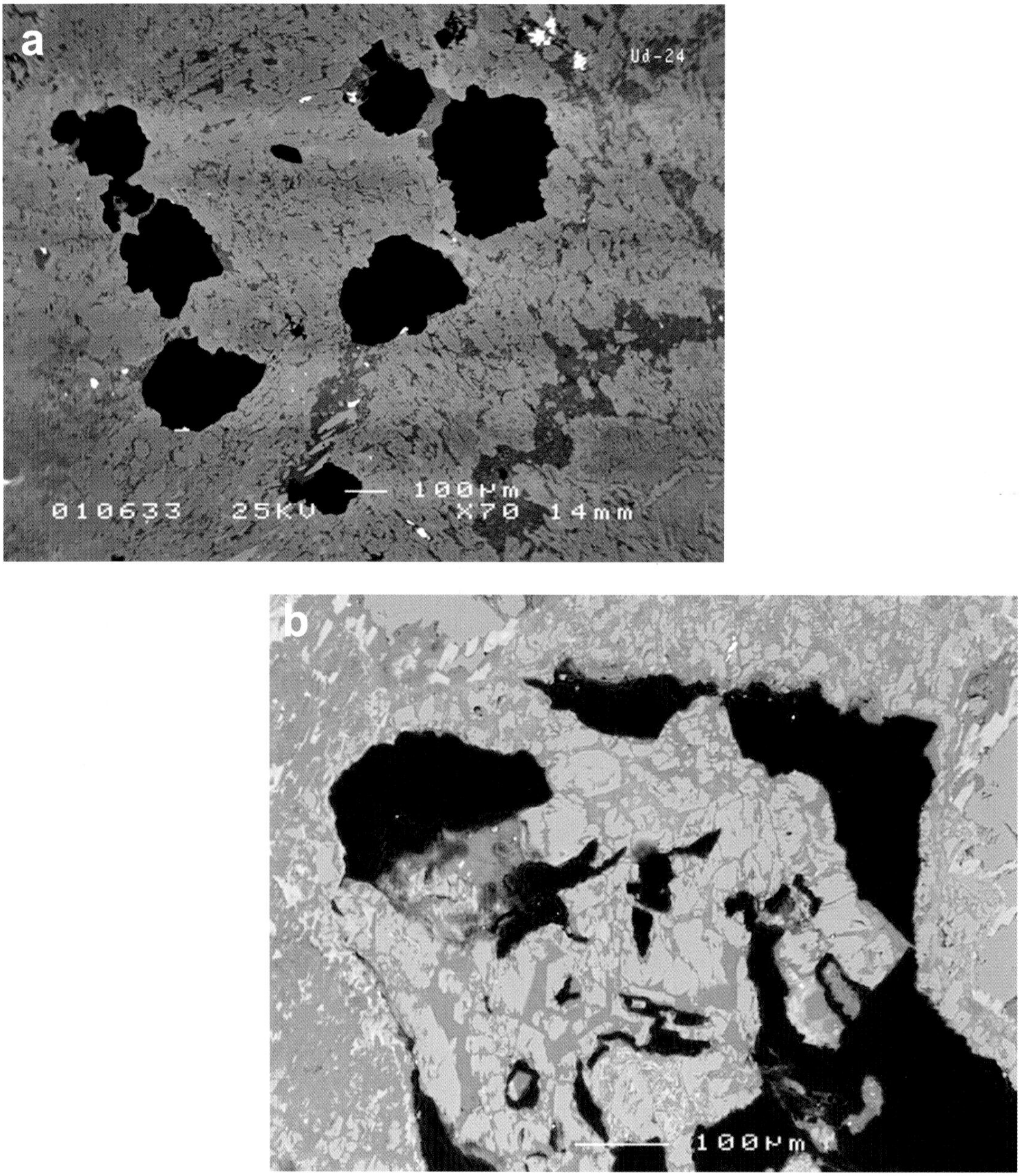

Plate 4. 13. Diamonds Associated with Partial-Melt Products in Udachnaya Eclogites. Back-Scattered Electron (BSE) Images; a) Anhedral, resorbed micro-diamonds (10 black crystals, <1 mm) in secondary clinopyroxene spongy textured partial-melt products; small, bright sulfide grains to upper right; b) Sculptured, resorbed microdiamonds in an atoll texture, with associated with partial-melt products; the obvious corroded diamonds indicate resorption in-situ by the metasomatizing fluids, that probably caused the partial melting of the pyroxene and garnet.

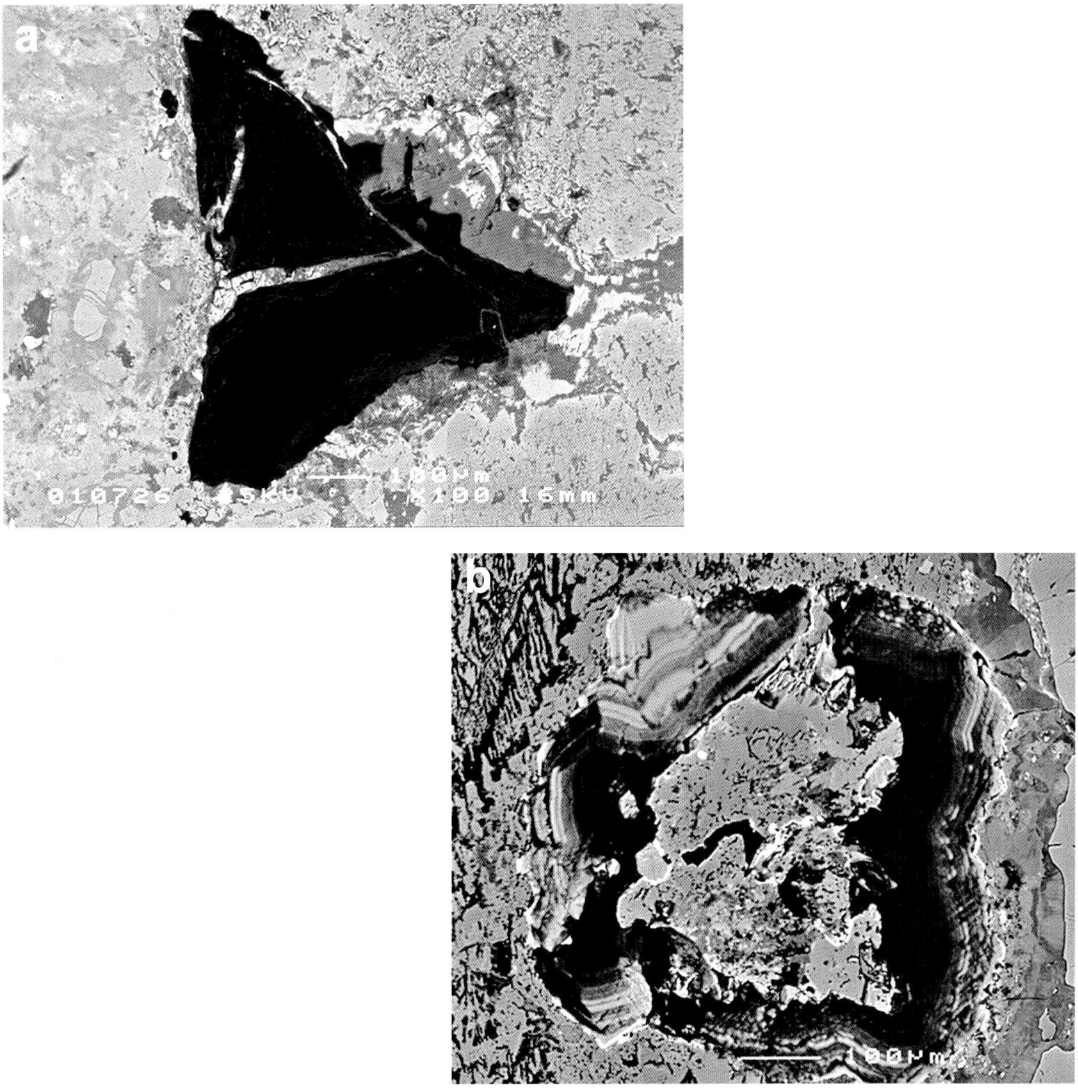

Plate 4. 14. Diamonds Associated with Partial-Melt Products in Udachnaya Eclogites. Back-Scattered Electron (BSE) Images; a) Octahedral face of a diamond cut by secondary clinopyroxene vein containing glass; the diamond itself is situated in products from the partial-melting event; b) Zoned crystal of a hopper diamond showing evidence for resorption by the partial melt; situated between partially melted clinopyroxene, with ‘spongy’ texture, and garnet; The “spongy” texture of the clinopyroxene consists of a central primary Cpx, surrounded by small islands of secondary Cpx, which represent metasomatically altered, jadeite-poor Cpx, however, still maintaining optical continuity with the primary Cpx; the partial-melt products are interstitial, and consist of glass and/or plagioclase and spinel; the unusual feature of the primary and secondary Cpx has been called “spongy texture” (Taylor and Neal, 1989).

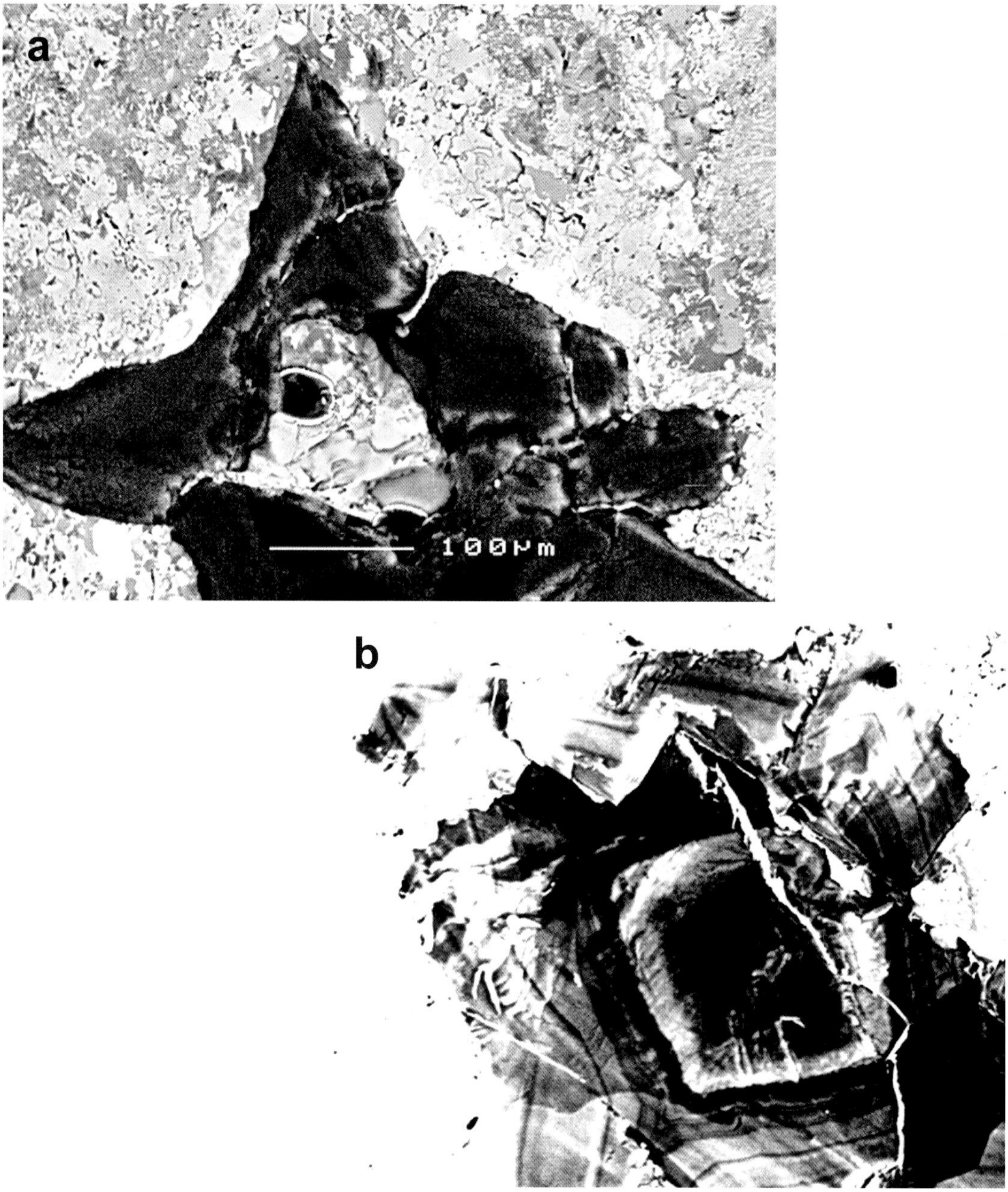

Plate 4. 15. Diamonds with Possible Evidence of Secondary Metasomatic Growth and Resorption in Udachnaya Eclogites. a) Diamond with a complicated growth form, set within partial melt products and with inclusions of secondary Cpx; note the obvious traces of resorption; BSE image b) Octahedral zoned diamond with a complicated growth pattern, set within metasomatic melt products; note the cubic core, with resorbed rims, evidence for several periods of growth, resorption, re-growth, etc; (reflected light).

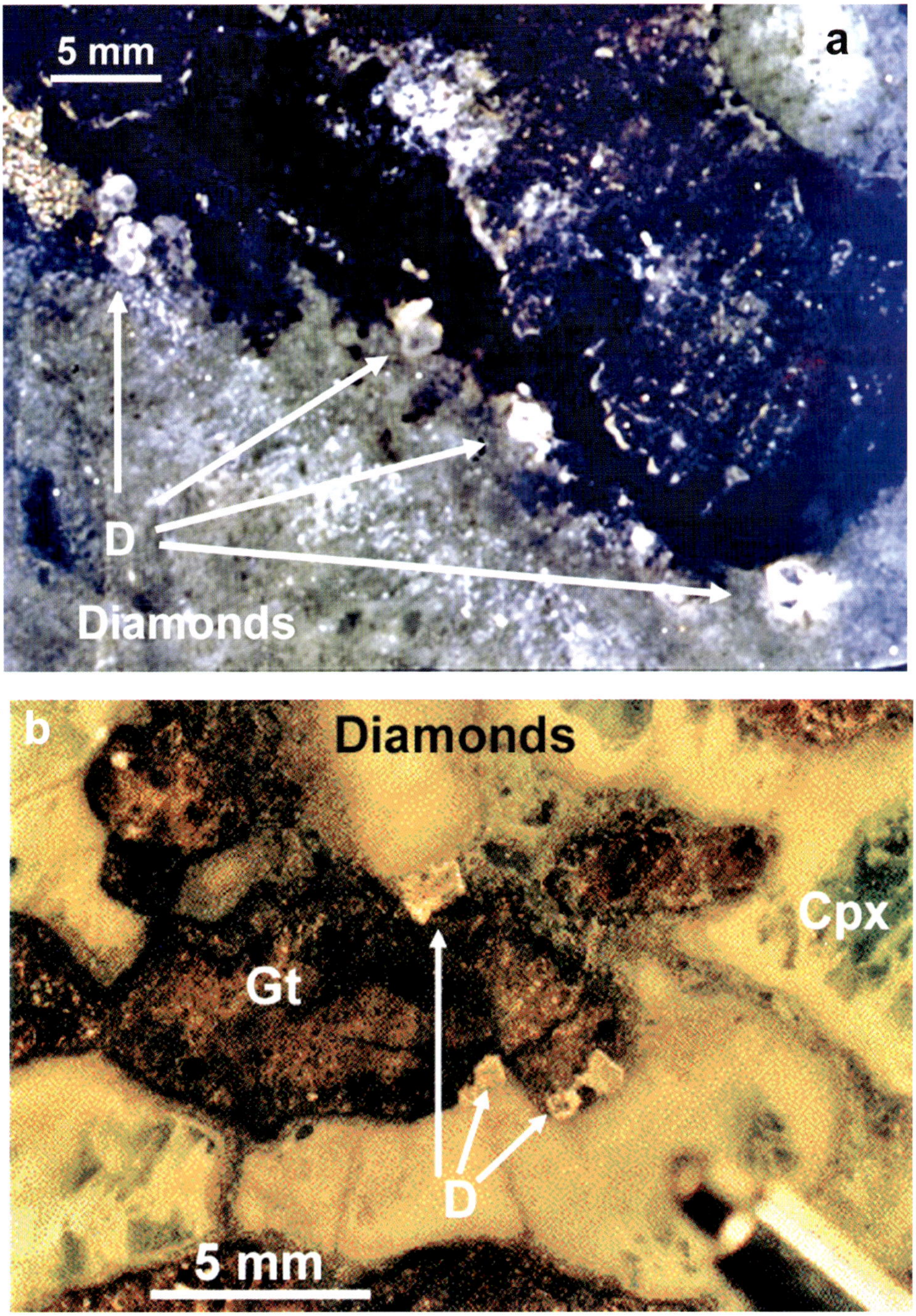

Plate 4. 16. Possible Secondary Origin of Diamonds in Eclogite Xenoliths from the Udachnaya Pipe. a) Six transparent (some partly broken) cubic diamonds present along a linear partial-melt vein between Cpx and garnet; b) Microdiamonds associated with partial-melt products in the kelyphitic alteration surrounding the garnets.

CHAPTER 5
Diamonds *In Situ*

CHAPTER 5

DIAMONDS IN-SITU

5.1. Characteristics of Diamondiferous Xenoliths

The genetic association of diamonds with ultramafic and eclogitic rocks is substantiated by the presence of diamondiferous xenoliths (xeno = foreign; lith = rock) in many diatremes (e.g., Pokhilenko et al., 1982; Viljoen et al., 1992; Jerde et al., 1993; Sobolev et al., 1994). Kimberlites and related mantle magmas commonly entrain fragments of wall rock during their ascent to the Earth's surface. These xenoliths originate from a range of sources, many in the lower and upper crust. However, mantle xenoliths have received the most attention, especially those containing diamonds. Studies of these xenoliths have provided many of the constraints on mantle processes and diamond genesis in the mantle.

As discussed previously, the entire spectrum of xenoliths that are the mantle hosts of diamonds can be divided into peridotite-pyroxenite and eclogite suites. Olivine- and pyroxene-rich peridotites are the most abundant rock type present in the diamondiferous portions of the mantle. This general rock type encompasses harzburgites, clino- and ortho-proxenites, lherzolites, and dunites. It is harzburgites and lherzolites that host the majority of peridotitic diamonds, and the majority of ultramafic xenoliths discovered to date are of harzburgitic affinity.

It should be pointed that the peridotite xenoliths usually contain diamonds only of octahedral morphology (Sobolev et al., 1994). Octahedral forms were found also in garnet pyroxenites and websterites. Rounded dodecahedral diamonds have only been recorded from two lherzolite xenoliths from the Finsch kimberlite, South Africa (Shee et al., 1982), whereas the preponderance of dodecahedrons is the result of resorption of octahedra by the kimberlite.

Eclogites are coarse-grained rocks composed essentially of almandine-grossular-pyrope garnet and omphacitic clinopyroxene. They are mostly the result of subduction of ancient oceanic crust, which has been de-volatilized and possibly partially melted during subduction (e.g., MacGregor and Manton, 1986; Shervais et al., 1988; Snyder et al., 1995; Taylor et al., 2003).

The mineral chemistry of diamondiferous xenoliths indicates that they have been subjected to upper-mantle P-T conditions, consistent with the environment necessary for the formation of diamond (e.g., Boyd and Finnerty, 1980; Dawson, 1980). In addition, the mineral inclusions of many diamonds have similar chemical affinities with the mineralogy of eclogitic and ultramafic xenoliths (e.g., Taylor et al., 1996);

however, some diamondiferous eclogites have diamonds with inclusion chemistry that are totally different from the host minerals.

Diamonds are rare in ultramafic xenoliths, but estimates from geothermobarometry of lherzolitic and harzburgitic xenoliths provide similar equilibration temperatures and pressures to those of mineral inclusions in diamonds (e.g., Boyd and Finnerty, 1980; Jaques et al., 1990). Eclogitic xenoliths predominantly have lower equilibration P-T conditions than those of E-type diamond inclusions. However, modification of eclogites by fluid interaction through metasomatism may have extensively modified the mineralogy of these rocks (Switzer et al., 1969; Ireland et al., 1994; Sobolev et al., 1999; Spetsius and Taylor, 2002). About 95% of known xenoliths with diamonds from Yakutian kimberlites are eclogitic (Spetsius, 1995; Sobolev & Pokhilenko, 1998). Similar estimations are given for diamondiferous xenoliths from South Africa (Kirkley et al., 1992). These data would appear to suggest that the overwhelming majority of diamonds originated in eclogitic environment. In contrast with the distribution of diamonds in diamondiferous xenoliths, however, ~90% of inclusion-bearing diamonds are of peridotitic affinity (e.g., Gurney, 1989; Bulanova et al., 1993; Sobolev et al., 1999).

This apparent paradox between the diamondiferous xenoliths and diamond types is probably due to the method by which the xenoliths are collected. During the production of the kimberlitic materials, the rocks are crushed to about 'golf-ball' size (~50-70 gr) and washed. As they travel along a conveyor belt, they are "sprayed" with X-rays. Any diamond present on the outside of these rocks will fluoresce bright blue, be sensed by a detector, and a shot of compressed air will blow them into a rubber bag. This is the major source of diamondiferous xenoliths. The vast majority of diamondiferous xenoliths so collected are eclogites. Why so few peridotite xenoliths? The major culprit is probably olivine. Because olivine weathers so easily, it renders most peridotites weak and friable and easily crushed. During the production, most peridotites crush and otherwise fall apart, thereby loosing their diamond-bearing identity. It is the bimineralic nature of the eclogites, garnet and clinopyroxene, that does not weather so readily, allowing some eclogites to retain their rock integrity even with initial crushing during production.

The description of diamondiferous samples and diamonds from eclogites of the Yakutian kimberlite province are presented in many works (e.g., Sobolev, 1977; 1994: Spetsius and Bezborodov, 1992). Visual examinations of hundreds of diamonds and >200 diamondiferous eclogites from the Mir, Udachnaya, Sytykanskaya, and other Yakutian kimberlite pipes have revealed the following characteristic features:

1. Diamonds occur in all varieties of mantle eclogites: bimineralic eclogites; kyanite-, corundum- and coesite-bearing xenoliths and garnet websterites.

2. Diamondiferous eclogites vary widely in their abundance of garnet and clinopyroxene and in major- and trace-element chemistry. Coarse-grained and medium-grained equigranular varieties are more abundant than mosaic-porphyritic ones; de-

formed varieties are also common. Ilmenite and graphite occur as accessory minerals. Rutile and sulfides are common; sulfides are virtually present in all the samples, often associated with the diamonds.

3. According to mineral compositions between coexisting garnets and clinopyroxenes that reveal the P-T conditions of their formation, diamondiferous eclogites do not appear to be different from the non-diamondiferous eclogite xenoliths from the same kimberlites. This is reinforced by trace-element contents of eclogitic minerals from the Udachnaya pipe, where it was confirmed that the diamondiferous xenoliths are identical to the general suite of eclogites. Even the geochemical characteristics and trace-element distribution between garnets and clinopyroxene of diamondiferous and barren eclogites are similar (Snyder et al., 1997; Spetsius et al., 1998; Taylor et al., 1998).

4. All the eclogite xenoliths, to greater or lesser extents, have experienced secondary alterations. In particular, a metasomatic-induced partial melting is typical in all kimberlitic eclogites, and is occasionally used as a signature of diamondiferous eclogites, especially from the Daldyn-Alakyt region pipes (Spetsius and Serenco, 1990).

Diamonds in eclogites occur mostly in the form of single crystal, >1 mm. In separate samples, up to 1000 micro-diamonds (i.e., < 1 μm can be present). In high-resolution X-ray tomography of several diamondiferous eclogites from Udachnaya, Taylor et al (2000) and Anand et al. (2003) have reported macro-diamond (>1 mm) abundances of up to 144,000 carats per tonne in individual 50-70 g eclogites. The size of diamonds varies from <1mm to 5mm, rarely over 7-8 mm. All diamond morphologies occur in eclogites. There are different growth patterns that are developed on the diamond crystals; the various forms of dissolution are typically not observed. As a rule, in one eclogite, diamonds have the same morphology, color, and size, although there are minor cases when two forms occur in the same xenolith. In many samples, diamonds are present as regular octahedral forms and show evidence of their formation along selected paths of secondary alteration in the eclogites (Taylor et al., 2000; Anand et al., 2003; Taylor and Anand, 2004), perhaps during the metasomatism that caused some of the alteration. That is, the diamondiferous eclogites show evidence for diamonds occuring along zones of permeability where metasomatic fluids caused their formation (Taylor & Anand, 2004).

Diamond crystals in diamondiferous xenoliths almost always experience 100% preservation, except for deformations seen as shifts along glide-planes or when they are deformed and broken along with the host xenoliths. Most commonly, the distribution of diamonds in xenoliths is irregular and not confined to the surface; in some xenoliths that have been subjected to X-ray tomography, there are some linear distributions of crystals, usually corresponding to the development of intensive zones of partial melting, alteration, and rock deformation (Taylor et al., 2000; Anand et al., 2003). In rare samples, even large and well-formed crystals are surrounded by veins of secondary products. The eclogitic diamonds always show obvious evidence of their multi-stage formation (see the diamond-cathodoluminescent Plates at the end of Chapter 2). Visible inclusions of silicate and oxide minerals are not abundant in the crystals of E-type diamonds, whereas sulfide minerals are most common.

There exists some dependence of diamond morphology on the nature of the host eclogite. In particular, cubic diamonds are developed mainly in high-aluminous varieties of Group C eclogites (Spetsius and Serenco, 1990). Additionally, there is a lateral heterogeneity of diamond populations from different pipes on a geographical basis from the north to the south of the Siberian Platform. For instance, cubic diamonds are present in Udachnaya and other pipes of Daldyn-Alakit region, but virtually absent in pipes of the Malo-Botuobinsky field (Spetsius, 1999).

5.2. Association of Sulfides and Diamonds.

As shown in Chapter 3, the investigations of solid, "syngenetic" inclusions in diamonds indicate the presence of peridotitic or eclogitic paragenesis in diamonds from all kimberlite or lamproite pipes worldwide. The study of diamondiferous xenoliths has also confirmed the host/medium in which natural diamonds were formed — a mantle substrate of mafic or ultramafic composition. It is generally thought that most mineral inclusions in diamonds are "syngenetic" with the diamond formation, a concept seriously doubted by Taylor et al. (2003) and expounded on in Chaper3. The majority of these inclusions in diamonds include silicate, oxide and sulfide minerals, with sulfides comprising the most abundant inclusion in many pipes (Deines and Harris, 1995). Based upon instrumental neutron activation analysis (INAA) (Fesq et al., 1975; Harte et al., 1999) of single, clean and clear diamonds, it was determined that nearly all diamonds contained trace elements common to the typical inclusions, particularly the sulfides. This intimate association of sulfides with diamonds has naturally led to discussion of their possible involvement in the process of diamond formation (Yefimova et al., 1983; Bulanova, 1985). However, such analyses of relatively large pieces of diamonds have been questioned as to the reliability of the data obtained. At the Diamond Finger Printing Workshop in Belgium in October, 2007, a new approach towards analyzing diamonds was presented, using an LA-ICP-MS at Macquarie University. This was judged to have great possibilities for "finger-printing" the locations of diamonds, but still needs considerable study, particularly on representative populations of diamonds from the "outlaw" countries.

Detailed investigations of the mineralogy and chemistry of sulfides in mafic and ultramafic xenoliths, including diamond-bearing ones, from the Udachnaya, Mir, Obnazhennaya, and other pipes led Bulanova et al. (1990) to conclude that sulfides in peridotites and eclogites differ in their chemistry. This was, of course, only a follow-up of the earlier work of Yefimova et al. (1983) on sulfide diamond inclusions. In addition, the abundances of sulfides in eclogites and garnet websterites vary considerably more than in peridotites. In eclogite xenoliths, their abundance is usually 0.25-1.0 vol. % and in garnet websterites, sometimes up to 3.0 vol. %. Ultramafic xenoliths, both with or without garnet, contain not more than 0.5 vol % sulfides. Normally these rocks contain only single dispersed grains. It should be stressed that in some diamondiferous varieties of eclogites, sulfides are more widespread than in samples without diamonds (Bulanova et al., 1990). In both types of rocks, sulfide phase assemblages are represented by Po ± Pn ± Cpy (Table 5.1). In eclogites, pyrrhotite is the most abundant sulfide phase, whereas in peridotites, pentlandite is abundant. The sulfides were initially present in the xenoliths, at the high temperatures (>1000 °C) of the mantle, as Fe-Ni-Cu monosulfide solid solution (Mss). Upon cooling to lower temperatures, aided by the rapid kinetics of diffusion normal for sulfides, the Mss underwent exsolution of both pentlandite and chalcopyrite leaving a residual pyrrhotite phase relatively low in Ni and Cu. It must be emphasized that many sulfide analyses were performed before it became customary to use back-scattered electron (BSE) imaging on phases to discern homogeneity. Therefore, the exact composition of a "Mss-type mineral" is actually that of an intimately intergrown exsolution of two or more phases, because the microprobe beam is sampling the fine-grained intergrowth of phases (Taylor and Liu, 2008 IKC). Consequently, it is not surprising that major-element chemistry of sulfides in peridotite and eclogites xenoliths often overlap in composition.

There does appear to be a real difference in the composition of the sulfides in these xenoliths. Yefimova et al. (1983) reported an arbitrary division of sulfides, as inclusions in diamonds, into P-type with >13 wt% Ni and E-type with <10 wt% Ni. This division is possibly useful for discerning P- versus E-type diamonds; however, only when the compositions are not near the 10-13 wt% boundary. There are too many overlapping compositions near this intermediate composition to be useful, and the probability of compositions being those of fine-grained intergrowths always exists. This uncertainty in intermediate compositions was confirmed by Pearson et al. (1999), using Re-Os chemistry.

Comparison of sulfides in diamonds and xenoliths shows that most sulfides in diamonds are closely related to similar sulfide associations in eclogite xenoliths from diamondiferous pipes (Bulanova and Spetsius, 1986). Based upon numerous analyses of sulfides in diamonds, it appears that the major portion of single diamonds with sulfide inclusions in kimberlites is related to eclogite paragenesis (Table 5.1). Although it is generally thought that the majority of diamonds from most pipes are P-type, it is becoming more apparent that *E-type diamonds are far more abundant than previously assumed.* In fact, eclogitic diamonds may well constitute the real majority of all diamonds, we speculate

It is generally agreed that the combination of sulfides phases (po+pn+cpy) present in xenoliths and diamond inclusions was originally present in the mantle as an immiscible sulfide melt. However, the nature of this sulfide mineralization is further

Table 5.1. Characteristic Associations of Sulfides in Diamonds and Xenoliths. It should be appreciated that many of these compositions probably represent a fine-grained combination of sulfide minerals, not easily resolved by optics. Likewise, several of these minerals may be epigenetic. Data are based upon observations by Z.V. Spetsius.

Characteristics		Sulfides in diamonds	Sulfides in ultramafics	Sulfides in eclogites
Phases		**Po, Mss, Pn, Cpy, Py, Vio, Dj**	**Po, Pn, Dj, Hz, Vio**	**Po, Pn, Cpy, Dj, Mss, Py, Bn**
Prevailing association		**Po + Pn**	**Po + Pn**	**Po + Pn ± Cpy ±Dj**
Prevailing phase		**Po**	**Pn**	**Po**
Presence of Mss		**Abundant**	**Rare**	**Suspicious**
Pyrrhotite	**Ni**	**<0.02 - 7.0 wt%**	**<0.02 – 9.6 wt%**	**0.1 - 10.5 wt%**
	Co	**<0.02 - 0.65**	**<0.02 - 0.92**	**<0.02 – 2.50**
	Cu	**<0.02 - 2.32**	**<0.02 – 0.08**	**<0.02 – 1.42**
Pentlandite	**Ni**	**23.4 – 41.9**	**27.0 – 35.1**	**23.0 – 43.8**
	Co	**0.11 – 1.05**	**<0.02 – 6.07**	**<0.02 – 1.44**
Chalcopyrite	**Ni**	**0.5 – 1.1**	**0.7 – 8.2**	**<0.02 – 0.6**
	Co	**<0.02– 0.07**	**<0.02 – 0.14**	**<0.02 – 0.09**

Note: **Po**- Pyrrhotite, **Mss**- monosulfide solid solution, **Pn**- pentlandite, **Cpy**-chalcopyrite, **Py**- pyrite, **Vio**- violarite, **Dj**- djerfisherite, **Hz**- heazlewoodite, **Bn**- bornite. Content of Ni, Co and Cu in wt.%.

complicated by other processes, such as mantle metasomatism that caused partial melting of clinopyroxene and garnet (Spetsius and Bulanova, 1980, 1987; Bulanova et al., 1990; Spetsius and Taylor, 2002, Misra et al., 2003). There are data that relates not only the spatial but also possible genetic relationships of sulfides and diamonds: 1) the presence of worldwide occurrences of sulfide as one of the most common mineral inclusion in diamonds; 2) sulfide chemistry as found by INAA in clear diamonds (Sellschop et al., 1980); and 3) the close association of sulfides with diamonds in some eclogite xenoliths (Spetsius and Bulanova, 2002).

It has been reasoned by several investigators that the major components of the metasomatic fluid thought to be responsible for the majority of the diamonds in eclogites are C-O-H-N-S (e.g., Haggerty, 1986; Taylor et al., 2000). This reflects the close spatial association of sulfides with diamonds, both inside and out. However, the real nature of this assumed fluid is not known quantitatively at present. It is probably a super-critical watery fluid, containing abundant carbonate (Palyonov et al., 2006), perhaps akin to the low-temperature melting fraction of portions of the mantle. As such, it may well have compositional affinities to kimberlitic fluids, yet not in a genetic sense.

With respect to a sulfide component in the diamond-forming fluid, Miyazaki & Okushi (2002) found two different geometries of S_2 that were in apparent substitutional positions in synthetic diamond. They felt that the formational energies of S-H complexes may be dramatically lowered such that they play a role in the formation of natural diamonds. Chepurov et al. (1985) had earlier stated that sulfides may have some effect in the preservation of diamonds during their transport to the surface by kimberlitic magma. Their experiments on the interaction of the diamond and graphite with sulfide melt at high pressures showed that the preservation of diamond is influenced by the sulfur content of the transporting melt. It is possible that this function of sulfur for diamond preservation is important during initial stages of kimberlite formation; indeed, it may have continued importance afterward, when the kimberlite traps xenoliths and liberates diamonds during movement upward towards the surface (Bulanova et al., 1990). This influence could be one of keeping the oxidation state of the kimberlite sufficiently low so as to prevent destructive oxidation of the diamonds, as substantiated by experimental studies of Palyanov et al. (2005, 2007). In summary, the close spatial association of sulfide minerals with diamonds leads naturally to the conclusion that there may be a genetic connection between these two inherently different types of minerals.

5.3. Multi-Stage Diamond Formation in Eclogites

As mentioned previously, diamonds in eclogites occur mainly as single octahedral crystals, with a heterogeneous distribution within a given xenolith (Taylor et al., 2000). High-resolution X-ray tomography on numerous diamondiferous eclogites has demonstrated that there is close correspondence between the diamond locations and zones of intensive alteration, including partial melting of pyroxene and garnet. These zones are commonly associated with rock deformation (Taylor et al., 2000; Anand et al., 2003; Taylor and Anand, 2004).

During examination of several diamondiferous xenoliths from Udachnaya and Sytykanskaya, the zones of intense alteration were observed to contain abundant micro-granular sulfide segregations, and the diamond was of micro-size and commonly present as cubic crystals (Spetsius, 1980). It has already been mentioned that the high-alumina Group C eclogites have more of a tendency of having cubic diamonds than the Group A and B eclogites.

The complex and multi-staged growth history of diamond has been established for diamonds from both peridotitic and eclogitic paragenesis. Several observations over the years that directly bear on the origin of diamonds are given below.

1) Diamonds are never found in primary garnet or clinopyroxene, either in xenoliths or in kimberlitic megacrystic minerals, thus supporting a secondary origin.

2) Diamonds possess both normal and reverse zoning that is not typical for minerals of mantle xenoliths, especially eclogites. Such zones are highlighted by CL and by elevated $\delta^{13}C$ values from the core to the rim of crystals and vice versa (e.g., Hauri et al., 1999; Kinney et al., 1999; Taylor et al., 2000). Sharp boundaries are present between zones in diamond having different nitrogen contents and aggregations states.

3) There have been several reports of both E-type and P-type inclusions within a single diamond (e.g., Prinz et al., 1975).

4) Large variations in Pb isotopic compositions of sulfides can occur within a single diamond (Rudnick et al., 1993).

5) Overall, there is a large range of isotopic compositions of nitrogen and carbon in diamonds worldwide (e.g., Javoy et al., 1984; Deines et al., 1989, 1993, 2002, 2004, etc.; Cartigny et al., 1999, 2005).

6) There is an apparent difference in "ages" of diamonds of peridotite and eclogite paragenesis (2-3.3 and 1.1-1.5 Ga, respectively); a close similarity exists in the ages of E-type diamonds from Premier and Argyle pipes to their emplacement ages, based on Sm-Nd isotopic determinations (Richardson, 1984, 1985; Richardson et al., 1997, 2001), although the use of hundreds of diamonds to gather sufficient masses of garnet and clinopyroxene is questionable (e.g., Navon, 1999).

7) There is an abundance of sulfides as inclusions in diamonds, and heterogeneities of sulfur isotopes in sulfides from diamonds occur in one given pipe, for example Premier (Chaussidon et al., 1989) and Udachnaya (Ruzicka et al., 1999).

8) Individual sulfide inclusions in diamonds from the Koffiefontein (Pearson et al., 1998) indicate at least two episodes of sulfide crystallization and, perhaps, two ages of diamond formation. The investigation of large diamonds (2-8 carats) of octahedral shape has shown that their central zones could be of different type: cubic, round, cubo-octahedral, or octahedral form (Bulanova, 1995). Even micro-diamonds show zonal and complicated growth patterns, as seen in terms of CL zonation and internal variations in $\delta^{13}C$ of crystals from both kimberlites and lamproites (e.g., Trautman et al., 1997, 1998; Kinny et al., 1999; Bulanova et al., 2005). This observation confirms multi-stage growth of diamonds, with interruptions for resorption, followed by changes in chemistry of the fluids, and perhaps P-T conditions (e.g., Taylor et al., 2000).

The studies of Spetsius (1995), Spetsius and Taylor (2002), and Misra et al. (2003) have emphasized the alteration zones in eclogites with various types of partial melting caused by the late-stage introduction of metasomatic fluids. The close association of small diamonds to these zones leads us to a hypothesis for late-stage diamond formation in some eclogites. Another situation involved small diamonds associated with the kelyphitic rim around garnet and the veins of metasomatic-induced, partial-melt products in clinopyroxene. However, the most important evidence for late-stage diamond growth is the actual confinement of diamonds to alteration zones, perhaps made possible by earlier rock deformation that increased permeability for the invading metasomatic fluids (e.g., Taylor et al., 2000).

During a detailed investigation of diamonds in eclogite xenoliths from the Udachnaya pipe, two distinct types of diamond crystals were observed. These crystals have different morphology, color, and physical features (Table 5.2), which could be identified with two growth generations, separated in time, P-T conditions, and medium of diamond-formation (Spetsius, 1999). Bulanova (1995) examined octahedral diamonds in diamondiferous eclogites and concluded that these diamonds were crystallized, under stable conditions, in the same melt that simultaneously formed the eclogitic garnets and clinopyroxenes. This would appear to be highly unlikely, especially since the eclogites did not originate from a mantle melt, but as the result of subduction of ancient oceanic crust (e.g., McGregor and Manton, 1986; Taylor and Neal, 1989; Jacob et al., 1994). Granted, the crustal protoliths experienced de-volatilization during subduction, and partial melting (leading to the TTG magmas), but little to no subsequent major melting to form a silicate melt from which the diamonds precipitated.

The most accepted scenario for eclogites, and perhaps, peridotites, involves the metasomatic growth of diamonds in the upper mantle (e.g., Haggerty, 1986; Taylor et al., 2000, 2005). The early-formed diamonds exhibit signs of a complicated, sometimes torturous youth, commonly starting with cubic growth, followed closely by resorption, plastic deformation, breakage, renewed growth, sometimes with repeats of these steps. After the initial cubic growth, the renewed growth typically is octahedral, with larger clean zones, as seen by CL. The final stage of diamond growth can be represented by rims of cubic morphology and coated crystals, frequently with fibrous diamond. It is within this late-stage that micro-diamonds are commonly formed. This stage occurred at lower P-T conditions, and may also have resulted in metastable diamond growth. The growth of these late-stage diamonds finds evidence in the presence of microdiamonds with complicated forms and with inclusions of secondary minerals. Some of these diamonds also show signs of resorption by the fluids that formed the partial-melt products of eclogites. Age determinations (Richardson et al., 1984, 1993, 1995, 1998, 2001; Pearson et al., 2003) point to the fact that diamonds can be formed over a long geologic period of the Earth's history, especially the diamonds of eclogite paragenesis.

Summary: There is abundant proof for a multi-staged formational history for natural diamonds. Birefringence, laser-, and CL-investigations show that <<50% of all diamonds grew in a single-stage process. There is evidence in the metasomatic-induced, partial-melt zones of diamondiferous eclogites for a separate late-stage generation of small diamonds, as well as resorption. This appears to be separated in time and space from the earlier main period of diamond formation in these rocks.

Table 5.2. Comparison of two diamonds generations in eclogite xenoliths from the Udachnaya kimberlite pipe.

First generation of diamonds

	# of crystals	Size, mm	Morphology	Color	Photo-luminescence
Ud – 24	12	1-3	oct.	colorless, brown	blue
Ud – 28	2	2-3	comb. oct.	colorless, light-grey	blue
Ud – 45	4	>2	oct.	grey	blue
Ud – 151	4	2-5	oct.	brown	blue
Ud – 161	5	2-7	oct.	smoky grey	light-blue
Ud – 200	6	3-5	oct.	colorless	blue

Second generation of diamonds

	# of crystals	Size, mm	Morphology	Color	Photo-luminescence
Ud – 24	>10	<0.5	oct.	light-pink	light blue
Ud – 28	2	0.2-0.3	oct., aggregates	colorless	none
Ud – 45	4	0.5-1.0	coated	light-yellow	light blue, to none
Ud – 151	2	0.3-0.5	oct.	white	yellow-blue
Ud – 161	6	1-1.5	coated, aggregates	pale rose	none
Ud – 200	8	<1	stepped oct., coated	smoky grey	yellow

5.4. Metasomatic Origin of Diamonds in Eclogites

After the initial formation of eclogite, diamonds grew by secondary metasomatic processes. This is supported by Re-Os and Sm-Nd model ages for eclogite and peridotite xenoliths. The ages of some diamondiferous eclogites from South Africa, Canada, and Yakutia vary between 2.8-3.5 Ga. Estimations of the ages of diamonds, based on their inclusions (see Chapter 3) suggest that diamonds may have formed over a long period of Earth's history, especially diamonds of eclogite affinity. The only reasonable explanation for this is that they continued to grow during different times as a result of secondary processes, not associated with production of silicate melts.

In order to evaluate the relationships of diamond with each other, as well as with the garnets, pyroxenes, and secondary phases in diamondiferous eclogites, our group at the University of Tennessee has performed high-resolution, X-ray tomography on several specimens from Udachnaya (Keller et al., 1999; Taylor et al., 2000, 2003; Anand et al., 2002, 2003; Taylor and Anand, 2004) in collaboration with W. Carlson & R. Ketchum at the University of Texas, Austin. These unique samples were selected from a suite of over 150 diamondiferous eclogites. With this tomographic procedure, which is similar to CAT scans performed in hospitals, up to 360 'cuts' were made through each 50-70 gram eclogite, and 3-dimensional images and movies were made of each entire xenolith. The absolute location of each diamond was noted relative to the garnets, pyroxenes, sulfides, and secondary mineralization. It is readily apparent that many of the diamonds are associated with zones of prior deformation and alteration in the eclogites, immediately leading to the conclusion of a secondary process for the diamond heritage. These maps also permitted detailed plans to be made for the subsequent dissection ("Pull-Apart") of each eclogite. These tomography-based dissections have revealed some tremendously enriched xenoliths (up to 144,000 carats/tonne; Anand et al., 2003).

The diamonds from these dissections were polished, so as to obtain several mineral inclusions on one polished surface, and examined by CL; the nitrogen aggregation state determined for each zone by FTIR; zoning of the diamond (i.e., stratigraphy) was noted, relative to the mineral inclusions; carbon and nitrogen isotopes and nitrogen total were analyzed on the diamond zones; and a comprehensive study of trace- and major-element chemistry of the mineral inclusions were performed and compared to the host-xenolith mineral chemistry. In this detailed, multi-faceted approach, knowledge has been accrued of the petrogenesis of diamond-bearing mantle eclogites from Yakutia.

It was shown that diamonds are found almost exclusively in association with altered clinopyroxene or abutting kelypthitic rims on garnet (Keller et al., 1999; Taylor et al., 2000; Spetsius & Taylor, 2002; Anand et al., 2002, 2003). ***Clinopyroxene and garnet inclusions in diamonds from the same xenolith can be chemically distinct***

and, indeed, can differ markedly from the minerals in the xenolith host. Moreover, inclusions within a single diamond can be extremely variable in both major and trace-element compositions. Overall, it would seem that ***each diamond may indicate a possible different set of P-T-x conditions for its formation***, all of which are not now represented by the host xenolith chemistry. The diamonds appear to not have been in communication with the same mineralogy during their formation, and formed possibly by different waves of metasomatism (Sobolev et al., 1998). It is possible that the host eclogite may only record the most recent residence locale and mineralogy for the diamonds and do not record the original conditions of its formation. Such strong, durable, and hard material as the diamonds may be moved from their original sites of growth while in the asthenosphere.

In another study of mineral inclusions in diamonds by Sobolev et al. (1998), a single diamond was examined that contained 35 garnet and 6 clinopyroxene grains. ***The range in major-, minor-, and trace-element abundances and patterns of a suite of 35 garnet inclusions from a single diamond encompasses nearly the total range in compositions of garnet from all Yakutian eclogite xenoliths examined to date*** (Sobolev et al, 1998). A situation of metasomatic fronts was also advanced to account for this extreme variation in garnet compositions.

5.5. The Largest diamondiferous Eclogites

Recently, two large diamondiferous eclogites were found in the ore stockpiles at the Udachnaya kimberlite pipe. The largest of these weighed 8.8 kg (Taylor et al., 2005). Diamondiferous eclogites provide a unique opportunity for studying the processes that control the formation of diamonds. Three-dimensional, high-resolution computed X-ray tomography (HRCXT) was performed on a 'smaller' portion (UDE2005, 485g), which has successfully revealed ~100 macro-diamonds and their relationships with surrounding minerals of the host eclogite (Taylor et al., 2006, Fall AGU). The size of the diamonds ranges from 0.8 to 5 mm, with the majority being ~1 mm. The majority of diamonds occur as octahedral with step zones on their faces. Most inclusions in diamond are sulfide minerals; compositions of silicate inclusions in these diamonds provide additional information on the formation of eclogite and diamond.

Based on the tomography, a detailed dissection (pull-apart) was conducted on the UDE2005 sample to recover diamonds and to observe their settings within the overall texture of the host eclogite. The nature and physical properties of minerals, in direct contact with diamonds, were especially noted. Locations of diamonds relative to each other and to the outside surface were also recorded. Primary minerals show features consistent with metasomatically induced partial-melting event(s): garnet grains with thin kelyphitic rims (<0.1 mm) and omphacite associated with secondary pyroxenes of lower Na_2O and Al_2O_3 and higher CaO contents in a typical "sponge texture".

Diamonds were never found in direct contact with the primary minerals (garnet and omphacite); their locations are characterized by their association with the metasomatic-alteration veins mostly within the primary omphacites, and with their close proximity to garnets. Clustering of diamonds is also common in this sample. Sulfide minerals (po with pn, and cpy exsolution) are widely distributed through out the sample either in alteration veins or as inclusions in garnet and omphacite. There is no clear association of host sulfides with diamonds. These features taken together support our contention that ***diamonds formed later than the primary silicate of the host eclogite – they are not synchronous with the eclogite formation***. It is becoming increasingly evident that in eclogites, diamonds are NEVER in contact with primary garnet and/or omphacites. Instead, they mostly occur along metasomatically induced alteration zones, sometimes in a linear fashion. Such zones would appear to be the avenues for permeation of the C-O-H-N-S metasomatic fluids, as suggested by Haggerty (1986).

Summary: Diamond growth in eclogite xenoliths does not depend on the conditions of formation of the primary eclogite. Diamond formation probably occurred as an entirely secondary process in the already metamorphosed rocks. It is concluded that diamonds grew in eclogite xenoliths due to the secondary introduction of metasomatic fluids. Some late-stage diamond formation/resorption appears to be closely associated with the metasomatism that caused the partial melting of clinopyroxene and garnet. This later event probably occurred near to and/or at the time of kimberlite intrusion and eruption. The real possibility exists for the metastable growth of fibrous coatings on diamonds during kimberlite eruption.

Plate 5. 1. Complex Growth Histories of Diamonds from Udachnaya. a) A well-preserved crystal of a perfect flat-faced octahedron; b) A partially broken and resorbed crystal of a step-faced octahedron, with partly regenerated surfaces. Both crystals are in kimberlite.

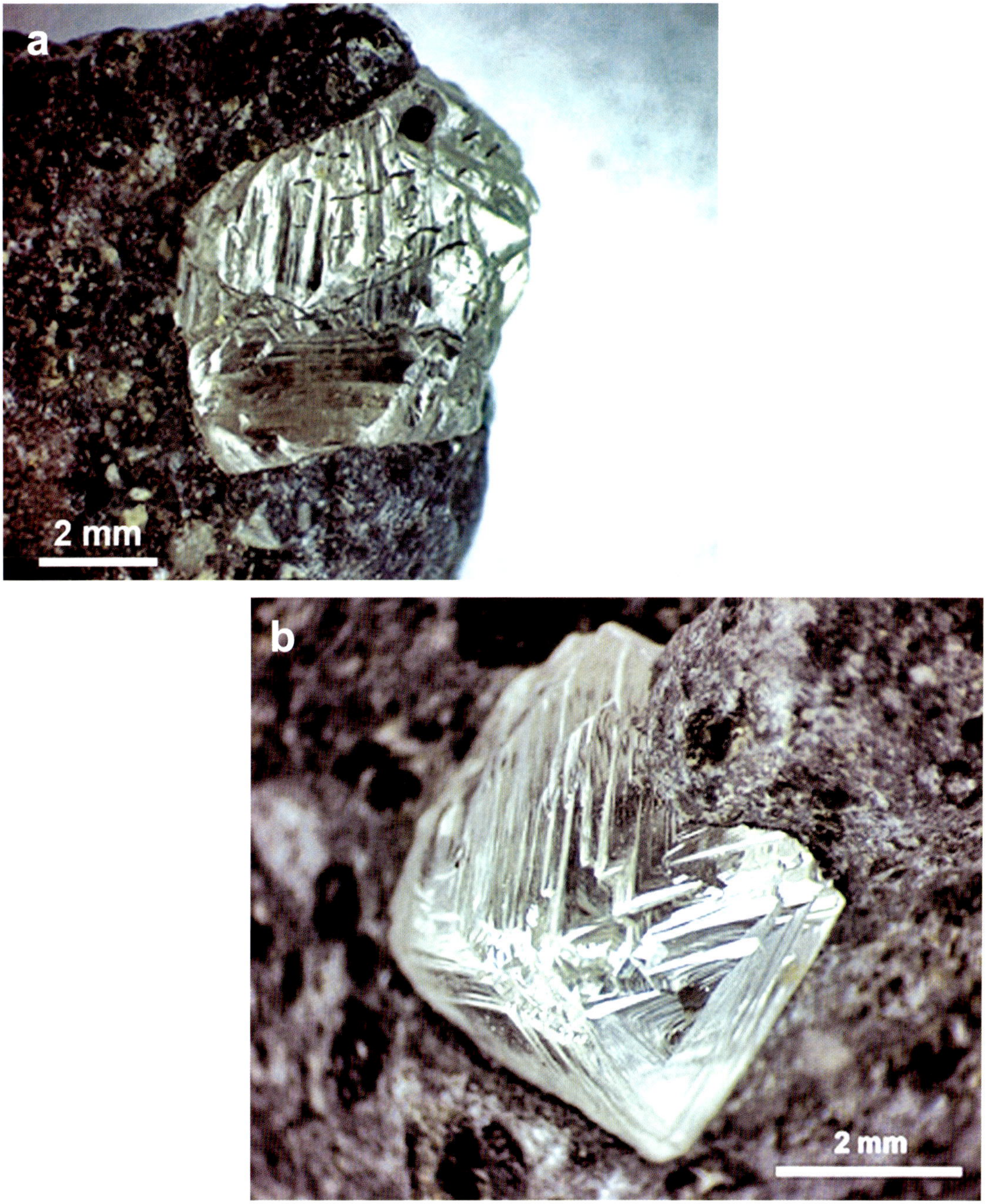

Plate 5. 2. Complex Growth Histories of Diamonds from Udachnaya. Different stages of resorption, corrosion, mechanical distortion, and possible plastic deformation of diamonds. a) A crystal with a combined form displaying strong evidence for resorption and corrosion; late-stage corrosion is evidences by the presence of elongated channels of resorption; b) A step-faced octahedron, slightly broken, in kimberlite matrix.

Plate 5. 3. Complex Growth Histories of Diamonds from Udachnaya.
Both a) and b) show octahedra, with step-growth edges, but with evidence for resorption by the kimberlite, in which they occur.

Plate 5. 4. Complex Growth Histories of Diamonds from Udachnaya.
Both a) and b) show octrahedra, with step-growth edges, but with evidence for resorption by the kimberlite, in which they occur.

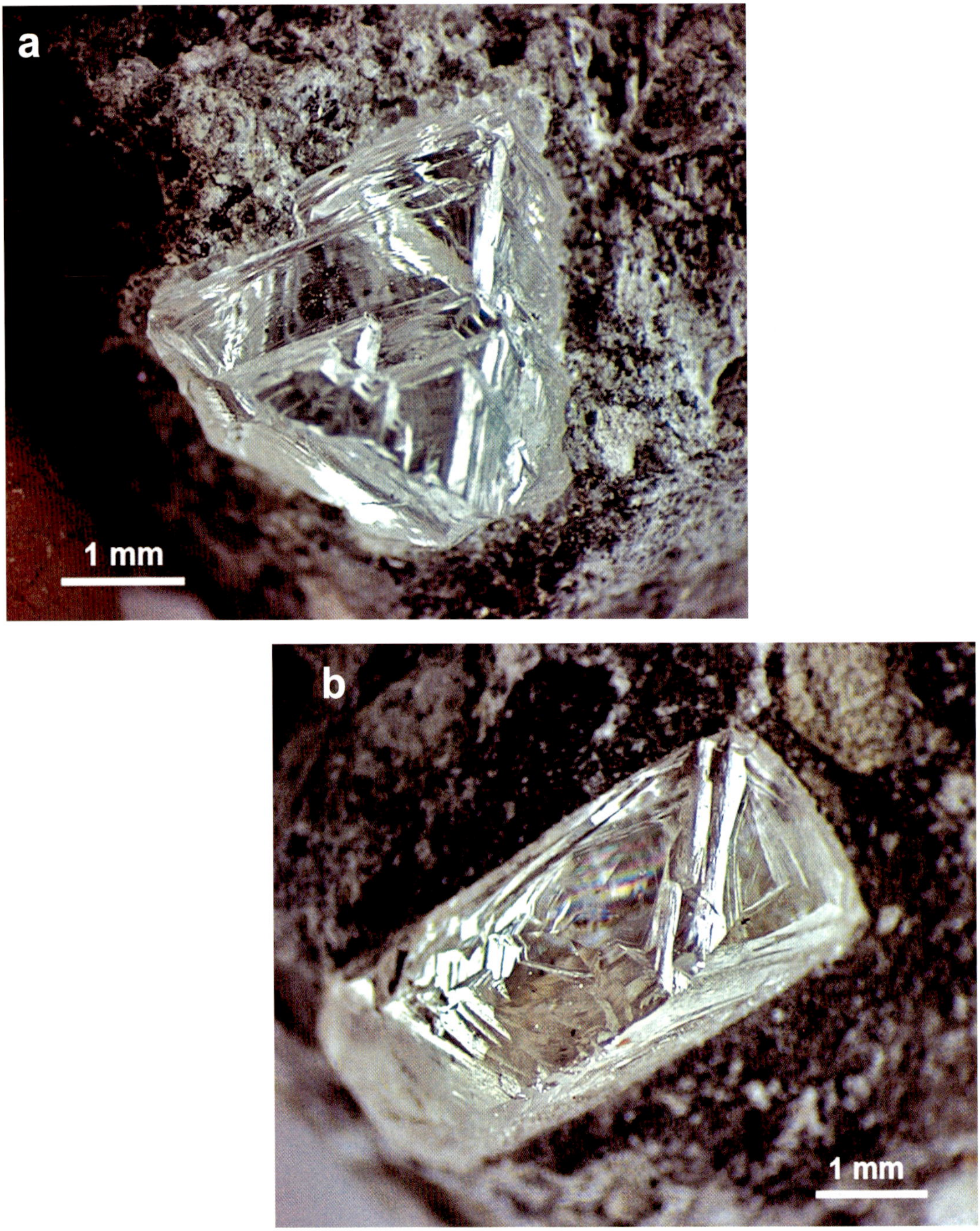

Plate 5. 5. Complex Growth Histories of Diamonds from Udachnaya.
a) A multi-crystalline aggregate of octahedral, all with the same axes orientation, probably as off-growths from a common sub-strate; b) a distorted octahedron of a diamond in kimberlite.

Plate 5. 6. Diamondiferous Eclogites from the Udachnaya Pipe.
a) Truncated, coated octahedron; note the linear alignment of the five diamonds to the left, taken as evidence of the secondary nature of the diamonds (Taylor et al, 2000); b) irregular octahedral step-growth; note the secondary minerals at the contact of the diamond and the underlying garnet.

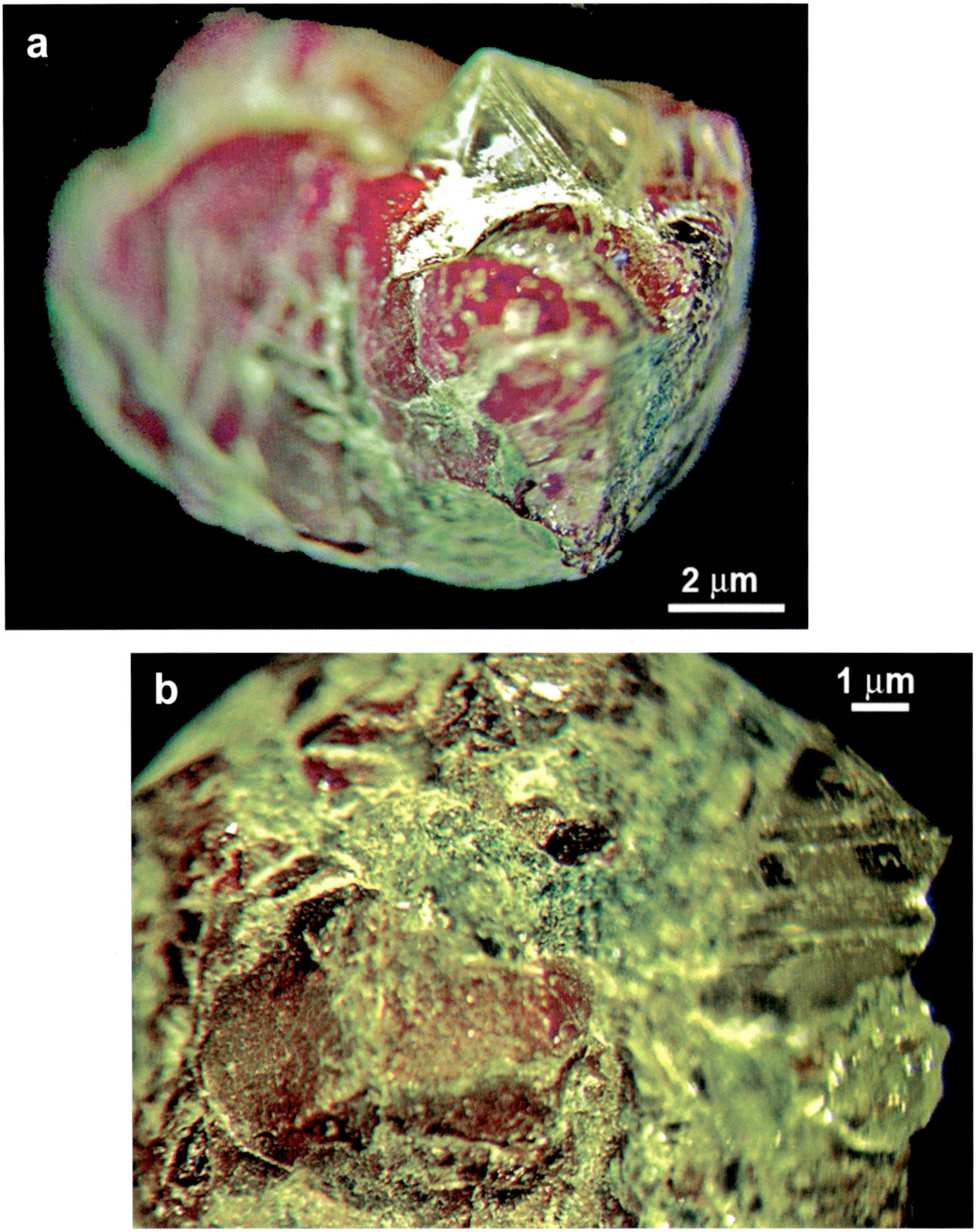

Plate 5. 7. Diamondiferous Eclogites from the Udachnaya Pipe.
a) Yellow octahedron protruding from garnets; b) Diamond apparently wedging a garnet apart in the central zone, a sign of the secondary nature of the diamond.

Plate 5. 8. Diamondiferous Eclogites from the Udachnaya Pipe. a) and b) Octahedra on the surface of eclogites; notice the rounding of the diamond in b), a result of resorption by the kimberlite.

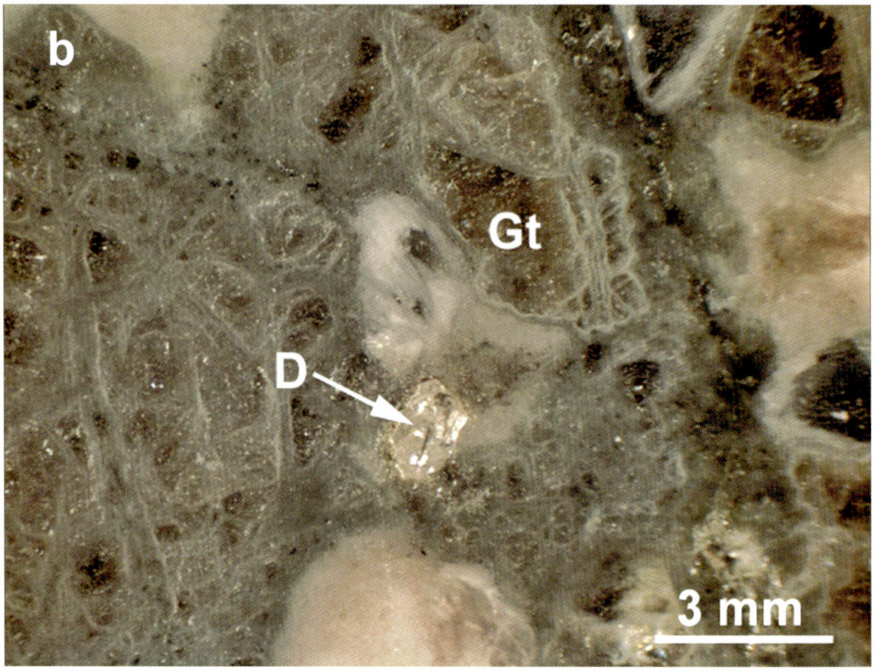

Plate 5. 9. Diamondiferous Eclogites from the Udachnaya Pipe. a) Bimineral eclogite with abraded octahedral diamond, possibly being resorbed; b) Kyanite eclogite with a diamond located in the middle of the metasomatically induced partial-melt products; this is typical textural setting for Yakutian diamonds.

Plate 5. 10. Diamondiferous Eclogites. a) Bimineral eclogite from the Sytakanskaya pipe, with a highly resorbed diamond in a crack along the altered Cpx; d) Garnetite eclogite with two step-faced octahedral diamonds at the surface, from the Botuobinskaya pipe (according to the garnet composition, a garnet websterite).

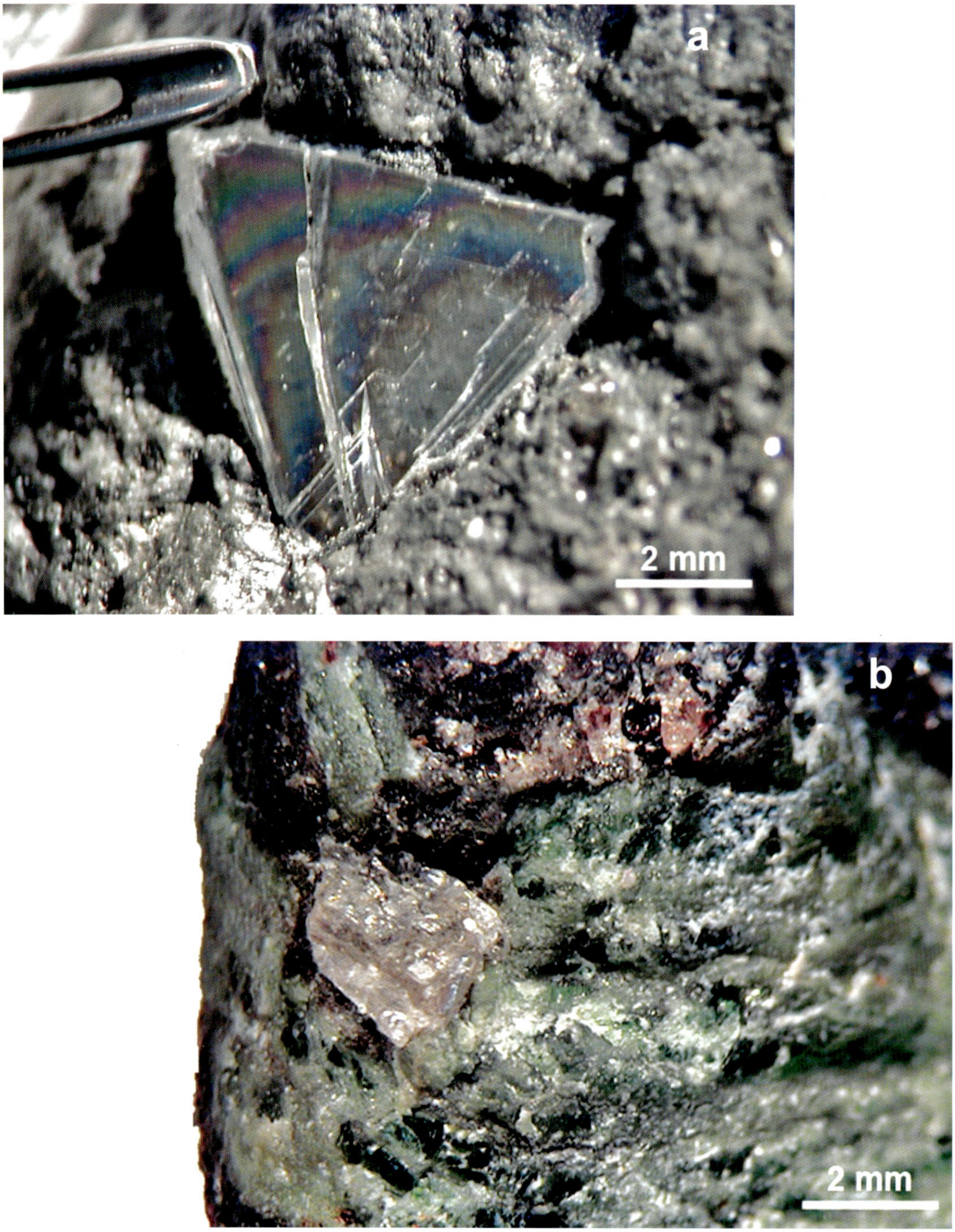

Plate 5. 11. Diamonds in Udachnaya Eclogites.
a) Colorless flat-faced planar octahedron with a mechanically broken surface;
b) Translucent, pink, fractured diamond crystal with relics of octahedral faces.

Plate 5. 12. Diamondiferous eclogites. a) Translucent, step-faced octahedron with minute inclusions of sulfides, from the Sytakanskaya pipe; b) Translucent octahedral spinel twin with step-faced surface from Udachnaya.

Plate 5. 13. Size Distribution of Diamonds in Udachnaya Eclogites. a) Four different size octahedral crystals present on the outer surface of this xenolith; small crystal on the left is perfect, flat-faced, transparent octahedron; other crystals are partly broken; b) Translucent step-faced yellowish octahedral crystal on the left, with a broken, elongated, transparent octahedron on the right.

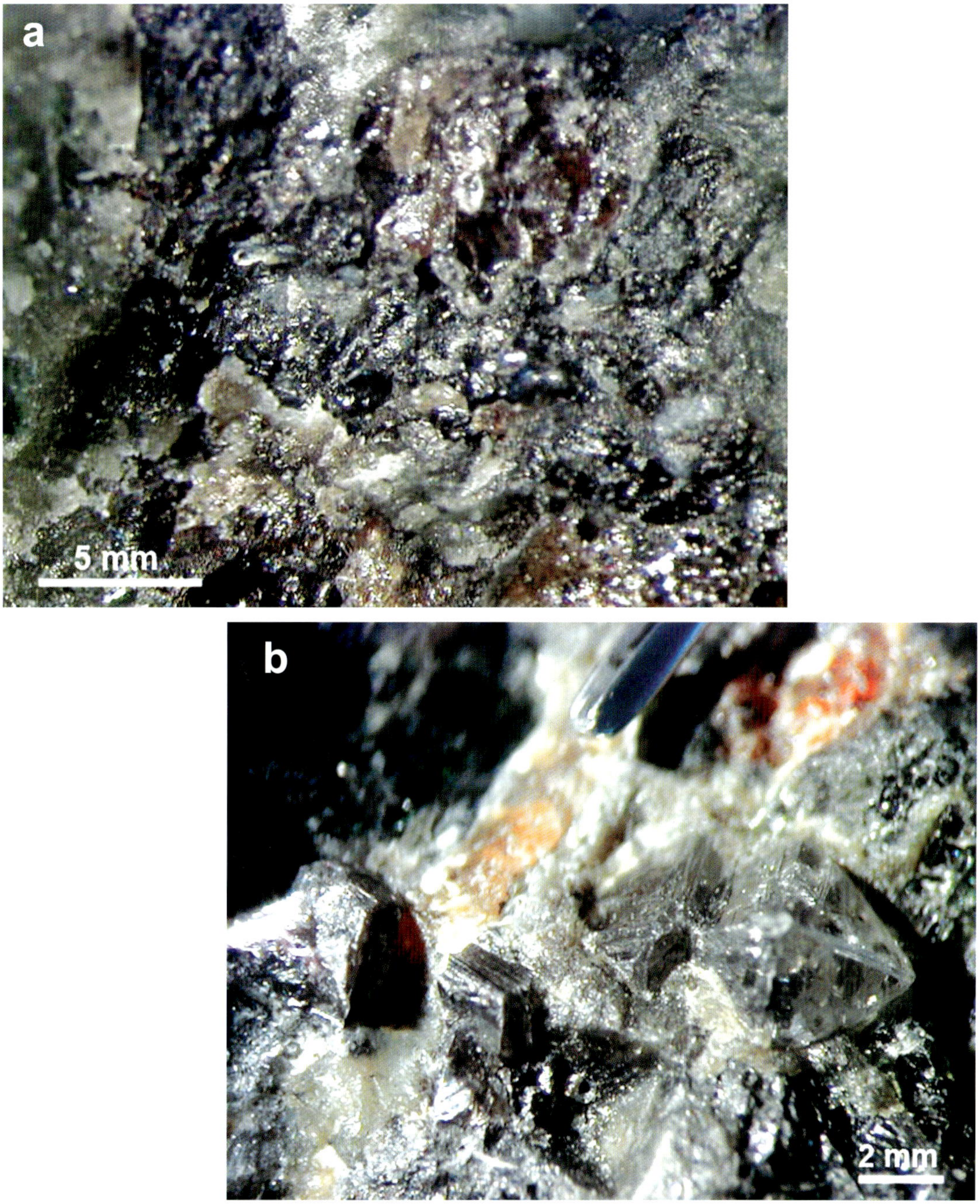

Plate 5. 14. Size Distribution of Diamonds in Udachnaya Eclogites. a) Elongated "vein" consisting of an aggregate of >20 black diamonds, suggestive of the secondary nature of the diamonds; b) Intergrowths of two translucent, step-faced octahedral crystals; some inclusions of sulfides can be seen; evidence of resorption is present on the larger crystal.

Plate 5. 15. Diamondiferous Kyanite-Bearing Eclogites from Udachnaya.
a) Black aggregated diamonds, relics of octahedral faces, mostly round-stepped dodecahedral surfaces from resorption; b) Two intergrown, dark-grey cubic crystals.

Plate 5. 16. Diamondiferous Kyanite-Bearing Eclogites. a) Broken portion of a yellow, transparent octahedron, with regenerated surfaces, from Udachnaya. b) Translucent, yellow cubic crystal from the Sytakanskaya pipe.

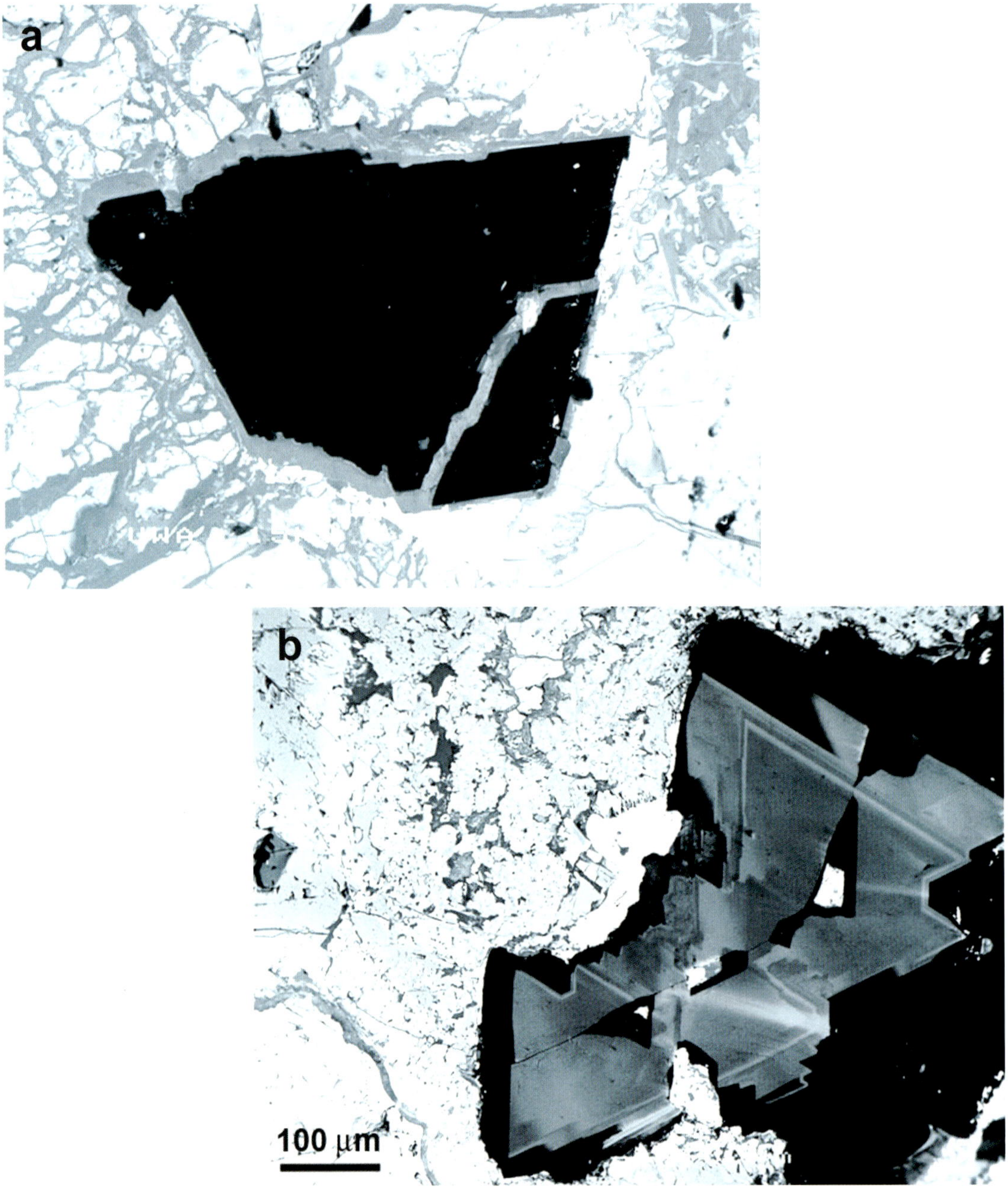

Plate 5. 17. Sulfides Around Diamonds from Udachnaya. Back-Scattered Electron (BSE) Images. a) An octahedral crystal set in the partial-melt products of clinopyroxene; white sulfides can be seen on edges of diamond, but notice the rim of alteration exactly at the edge of the diamond; this is common for virtually all diamonds eclogites, a sign of diamond's secondary nature; b) Octahedral crystal, showing some CL, in partially melted clinopyroxene, with sulfide minerals at the rim.

Plate 5. 18. Two Diamond Generations in Eclogite Xenoliths from Udachnaya. a) Deformed dark-gray diamonds of an early generation represented by crystals > 2mm; diamonds are slightly elongated and show traces of deformation, following the whole-rock deformational fabric; roundness is due to resorption; b) Second generation diamonds represented by a yellowish round crystal <1mm setting within the partial-melt products of secondary clinopyroxene.

Plate 5. 19. Diamondiferous Eclogite from the Udachnaya Pipe. Intergrowth of more than 5 octahedral step-faced crystals, with partly broken surfaces, between garnet grains, with secondary mineral selvages between.

Plate 5. 20. Diamondiferous Eclogites from the Udachnaya Pipe. a) Partly broken twin of step-faced octahedrons in an omphacite matrix. b) Transparent octahedron with broken and resorbed faces, surrounded by a rim of secondary minerals.

Plate 5. 21. Diamonds in Eclogite Xenoliths from the Udachnaya Pipe.
a) Elongated growth of step-faced, planar octahedral crystals. b) String of translucent, irregular-formed, octahedral crystals; such lineations of diamonds along zones of alteration are not uncommon in diamondiferous eclogites (Taylor and Anand, 2004).

Plate 5. 22. Possible Secondary Origin of Diamonds in Eclogite Xenoliths from the Udachnaya Pipe. a) Linear growth of six transparent (some partly broken), octahedral diamonds; note the alignment of the diamonds in a row; b) Irregular growth of octahedral diamonds between grains of the primary eclogite minerals; three crystals in the center have broken surfaces and are elongated in the direction of the fabric of the xenolith.

Plate 5. 23. Possible Secondary Origin of Diamonds in Eclogite Xenoliths from the Udachnaya Pipe. a) Irregular intergrowth of two step-faced, octahedral diamonds surrounded by secondary minerals. b) Diamond of a complicated form, reflecting three stages of growth that are obvious on a broken surface; on the outer portion of the crystal, notice the small linear plates of diamonds that appear to have later overgrown into the surrounding matrix of secondary minerals.

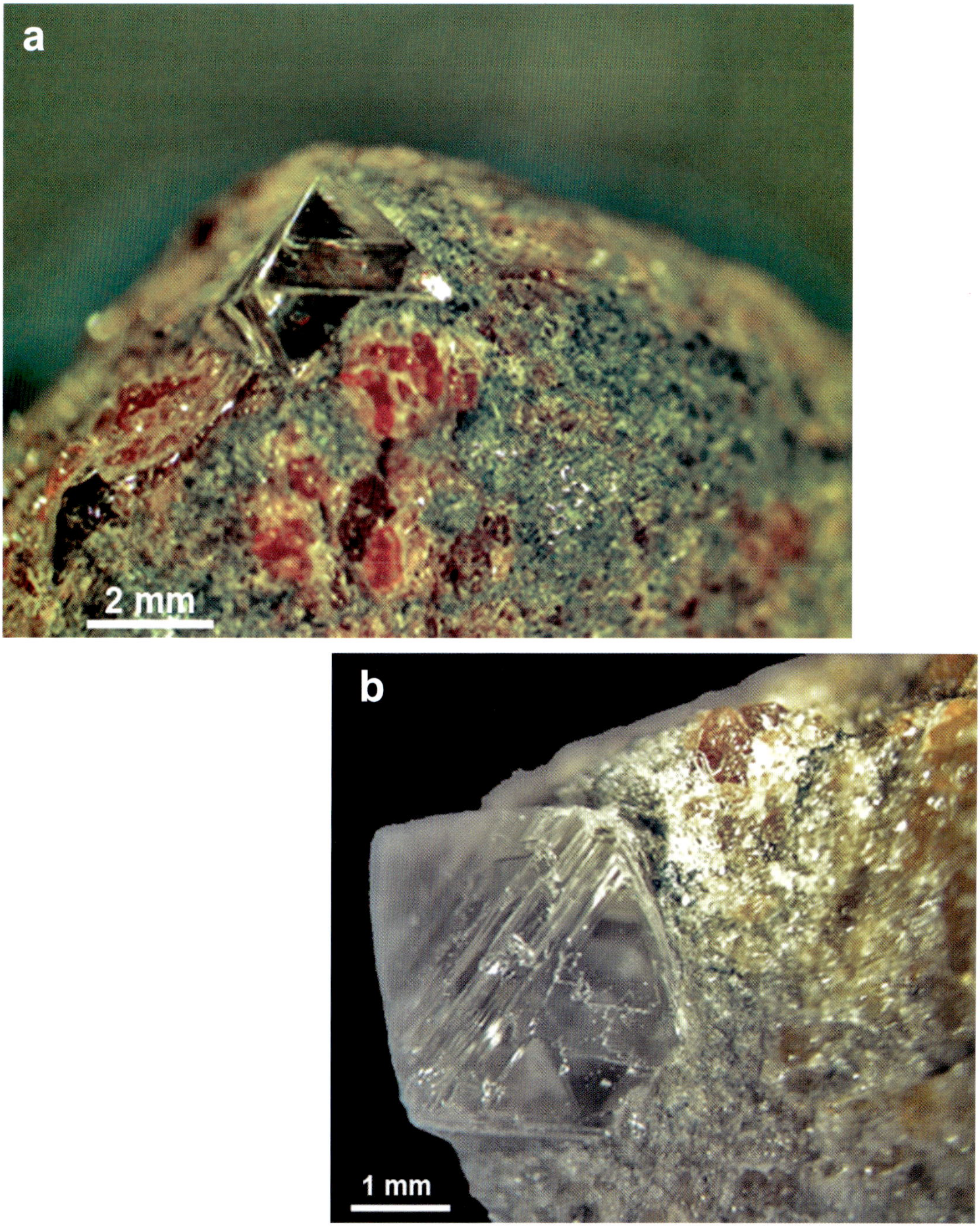

Plate 5. 24. Diamondiferous Xenoliths from the Nurbinskaya Pipe. a) The garnet composition is indicative of a Group A eclogite; the diamond is a colorless, smooth-faced octahedron. b) The garnet composition indicates it comes from a garnet websterite; the diamond is a colorless, polycentric octahedron.

Plate 5. 25. Diamondiferous Xenoliths from the Nurbinskaya Pipe. a) The garnet composition indicates this to a high-alumina, Group C eclogite; the aligned several diamonds are of octahedral form and display rounding from resorption by the kimberlite; b) The garnet composition belongs to the ferroan, Group A eclogites; the diamond is a grey, heavily corroded crystal; obvious channels of corrosion can be seen cutting across the crystal; the rim around the diamond is that of chlorite.

Plate 5. 26. Corrosion and Distortion of Diamonds in Nurbinskaya Xenoliths.
a) Octahedral yellow diamond with parallel striations on edges showing deep corrosion channels that cut across almost half of this crystal, in a garnet websterite; notice the rim around the diamond filled by a chloritic mass possibly replacing primary phlogopite; this same metasomatism may have caused the diamond corrosion. b) A broken octahedral diamond, representing a micro-fault displacement, in alignment with the rock deformation; the surfaces of the diamond are also strongly corroded.

Plate 5. 27. Metasomatic Growth of Diamonds in Nurbinskaya Xenoliths.
a) A literal "vein" of diamonds cutting the entire xenolith of a websterite. b) Higher-power magnification of the same sample; the diamonds in the vein parallel the fabric of the rock deformation, forming a zone of higher permeability for metasomatic fluids.

Plate 5. 28. Metasomatic Growth of Diamonds in Nurbinskaya Xenoliths. The aggregate of pinkish diamonds consists of 6 crystals of combined cubic-dodecahedral forms, in a garnet websterite.

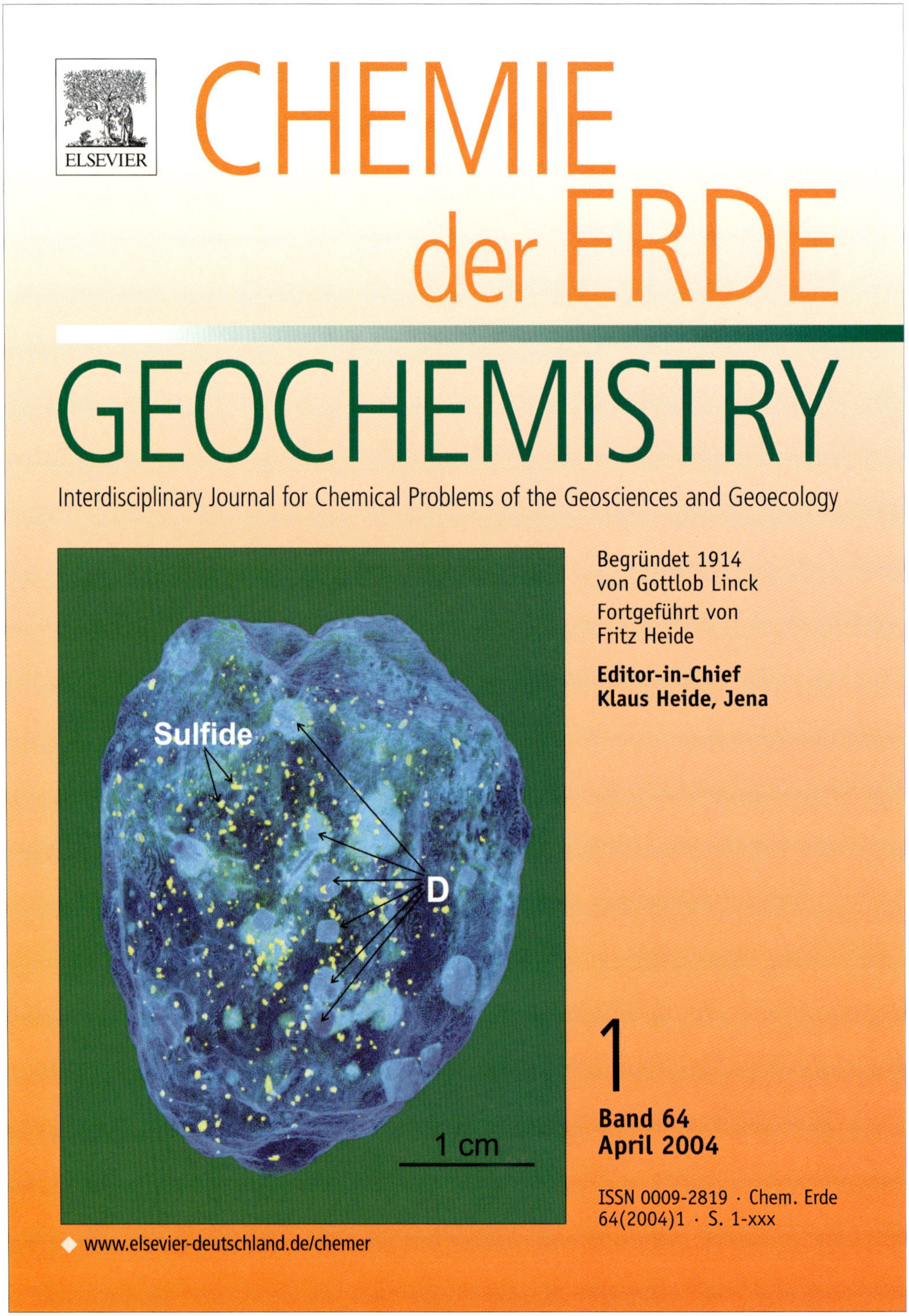

CHEMIE der ERDE

GEOCHEMISTRY

Interdisciplinary Journal for Chemical Problems of the Geosciences and Geoecology

Begründet 1914
von Gottlob Linck
Fortgeführt von
Fritz Heide

Editor-in-Chief
Klaus Heide, Jena

1

Band 64
April 2004

ISSN 0009-2819 · Chem. Erde
64(2004)1 · S. 1-xxx

www.elsevier-deutschland.de/chemer

ELSEVIER

Available online at www.sciencedirect.com
SCIENCE DIRECT

Chemie der Erde 64 (2004) 1–74

CHEMIE der ERDE GEOCHEMISTRY

www.elsevier.de/chemer

INVITED REVIEW

Diamonds: time capsules from the Siberian Mantle

Lawrence A. Taylor*, Mahesh Anand

Department of Earth and Planetary Sciences, Planetary Geosciences Institute, University of Tennessee, Knoxville, TN 37996, USA

Received 15 October 2003; accepted 28 November 2003

Abstract

Diamonds are thought to be "time capsules" from the Earth's mantle. However, by themselves, consisting of nearly pure carbon, diamonds provide little geochemical information about their conditions of formation and the nature of their mantle hosts. This obstacle to studying the origin of diamonds and their hosts can be overcome by using two main approaches that focus on studying: (1) the rocks that contain diamonds, i.e., diamondiferous xenoliths; and (2) mineral inclusions within the diamonds, the time capsule's little treasures, if you will. Diamondiferous xenoliths, their diamonds, and mineral inclusions within the diamonds are the subject of this review, focusing on studies of samples from the Yakutian kimberlites in the Siberian Platform.

Studies of diamondiferous eclogite xenoliths significantly enhance our understanding of the complex petrogenesis of this important group of rocks and their diamonds. Such studies involve various geochemical and petrological investigations of these eclogites, including major and trace-element, radiogenic as well as stable isotopic analyses of whole rocks and minerals. The results from these studies have clearly established that the Group A–C eclogites originate from subduction of ancient oceanic crust. This theory is probably applicable worldwide.

Within the last several years, our research group at Tennessee has undertaken the systematic dissection (pull apart) of diamondiferous eclogites from Siberia, consisting of the following steps: (1) high-resolution computed X-ray tomography of the xenoliths to produce 3D images that relate the minerals of the xenoliths to their diamonds; (2) detailed dissection of the entire xenolith to reveal the diamonds inside, followed by characterization of the setting of the diamonds within their enclosing minerals; and (3) extraction of diamonds from the xenolith for further investigation of the diamonds and their inclusions. In this last step, it is important that the nature and relative positions of the diamond inclusions are carefully noted in order to

*Corresponding author.
E-mail address: lataylor@utk.edu (L.A. Taylor).

doi:10.1016/j.chemer.2003.11.006

maximize the number of inclusions that can be exposed simultaneously on one polished surface. In this modus operandi, cathodoluminescence imaging, plus FTIR/N aggregation and C/N isotopic analyses are performed on polished diamond surfaces to reveal their internal growth zones and the spatial relationship of the mineral inclusions to these zones.

Knowledge gained by such detailed, albeit work-intensive, studies continues to add immensely to the constantly evolving models of the origin of diamonds and their host rocks in the Earth's mantle, as well as to lithospheric stability models in cratonic areas. Multiple lines of evidence indicate the ultimate crustal origin for the majority of mantle eclogites. Similar pieces of evidence, particularly from $\delta^{13}C$ in P-type diamonds and $\delta^{18}O$ in peridotitic garnets lead to the suggestion that at least some of the mantle peridotites, including diamondiferous ones, as well as inclusions in P-type diamonds, may have had a crustal protolith as well.

1. Introduction

Apart from their economic value for jewelry and industrial applications, diamonds are scientifically valuable as samplers of mantle processes at pressures greater than 4 GPa (40 kbars; deeper than 120 km). Whereas the pressure–temperature conditions under which diamonds crystallize are well known, much remains to be learned about *when* the diamond crystallized; *where* the carbon actually came from; *why* the carbon nucleated and crystallized into diamond; *how long* diamonds continued to grow; and the role of tectonics in deciding many of these factors. One fact is certain; diamonds originate in the mantle beneath the cratonic regions of the world (Fig. 1).

Being nearly pure carbon, diamonds themselves contain relatively limited information about the physical and chemical conditions under which they crystallized. The aggregation state of nitrogen impurities in diamond increases with time and temperature, which provides a way to calculate the length of time that a diamond has resided in the mantle at a given temperature (Taylor W.R. et al., 1990; Mendelssohn and Milledge, 1995b), although this temperature is rarely known with any certainty and such estimates are subject to considerable error. All other techniques for determining the growth conditions of diamonds are *inferred* from the minerals that are both inside (inclusions) and outside (host eclogites/peridotites).

To date, a primary source of information on the growth conditions of diamonds has been from investigations of their mantle hosts (xenoliths), as well as the mineral inclusions inside diamonds. Although the host eclogites and peridotites have been the major subject of investigations for many years, it is gradually being realized that they have undergone near-continuous modification by mantle metasomatic processes, such that they may not actually represent the conditions at the time of and during the diamond formation. During the last few decades, and particularly the last one, mineral inclusions have become the subject of considerable research, receiving new impetus from the huge Canadian diamond finds. The inclusions of interest are usually eclogitic or peridotitic minerals, and can represent pristine samples of the environment present during diamond growth (unless the diamond

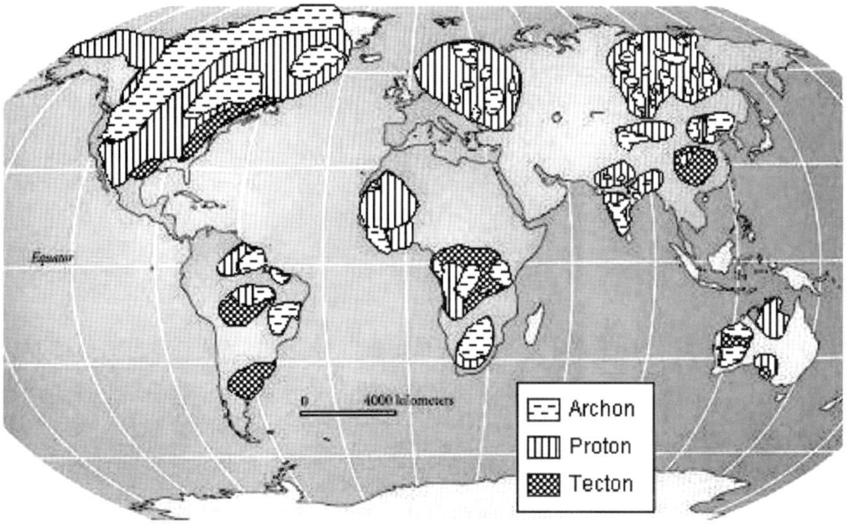

Fig. 1. Cratons of the world, shaded as to ages (Archon >2.5 Ga; Proton = 1.6–2.5 Ga; Tecton = 0.8–1.6 Ga). These are the regions to which the diamondiferous pipes of the world are restricted, adapted after Levinson et al. (1992).

cracked and chemical exchange occurred along the cracks). The subject of this review article includes: the relationships of diamonds to their host rocks; the physical/chemical nature of the diamonds; and the conditions of formation that are revealed by the diamond mineral inclusions. An extensive photographic characterization of diamonds and their mantle hosts is presented in Spetsius and Taylor (2004).

2. Source of diamonds

The hypothesis of diamond crystallization from kimberlitic magma was originally supported by several investigations (e.g., Bobrievich et al., 1959; Harris, 1979). As evidence for this magmatic genesis for diamond, various observations have been utilized, e.g., (a) dissimilar crystal morphologies of diamonds from different kimberlite pipes; (b) commonly found zoned diamonds; (c) "diamond in diamond" inclusions. However, we do not feel that these "lines of evidence" are unique to this kimberlitic magma theory and are equally explainable as resulting from a prior existence in mantle xenoliths. The xenocrystic theory will be detailed in the review presented below.

The hypothesis of the xenocrystic nature of diamonds in kimberlites is not recent, having been proposed Bonney (1899) after the discovery of diamondiferous eclogites as xenoliths in kimberlites in South Africa. Indeed, he reasoned that the kimberlite was merely the carrier of the xenoliths. The numerous discoveries of diamondiferous eclogites and peridotites in kimberlite pipes worldwide further supported this theory (e.g., Sobolev, 1974; Robinson, 1979; Jaques et al., 1986; Gurney, 1989, 1990). This xenocrystic/xenolithic origin is the most widely held theory at present.

The origin of diamond in mantle rocks finds direct evidence in the presence of silicate mineral inclusions in the diamonds from all kimberlite pipes, pioneered by some notable studies (e.g., Sobolev, 1968, 1974; Meyer, 1968; Meyer and Boyd, 1968). These inclusions appear to have originated from two major suites of rocks, ultramafic peridotites (P-type) and mafic eclogites (E-type), and their presence permitted the distinction of P- and E-type diamond genesis, respectively. (It should be noted that P-type is used herein, with U-type for ultramafic; Sobolev, 1977). Other mitigating evidence for the xenocrystic nature of diamonds is the presence of graphite pseudomorphs after diamond, which were discovered in pyroxenites from the delaminated Beni Bousera massif (Pearson et al., 1995). The xenocrystic presence of diamonds is additionally supported by the discovery of diamonds in Australian lamproites. There, numerous diamondiferous mafic xenoliths were found, and the diamond inclusions were also mostly representative of an eclogite paragenesis (e.g., Sobolev, 1984; Sobolev et al., 1989; Jaques et al., 1986, 1989). Perhaps the most convincing evidence, however, in favor of the xenocrystic nature of diamonds comes from the dating of diamonds using their mineral inclusions (Kramers, 1979; Richardson et al., 1984, 1993). The diamonds are reported to be significantly older than the host kimberlite—e.g., 3 Ga versus 90 Ma, respectively. In addition, strong evidence for old diamonds versus young kimberlites has been obtained from the Re–Os isotopic studies on individual sulfide inclusions in diamonds from the Koffiefontein and Udachnaya kimberlites (Pearson et al., 1998b, 1999). Indeed, three sulfide inclusions from the core and intermediate zones of one Udachnayan diamond gave 3.1 to 3.5 ± 0.3 Ga for Re–Os model ages, indicating that the diamond growth possibly did not occur over an extended time period (Pearson et al., 1999). However, the kimberlite eruption occurred only 380 Ma ago.

The host rocks for diamonds in the mantle are the mafic and ultramafic rocks that are common xenoliths in kimberlites, albeit mostly non-diamondiferous. The ultramafic xenoliths are represented predominantly by peridotites, but websterites are common in certain pipes. The mineralogical distinctions between these two ultramafic rock types are shown in Fig. 2. The petrogenesis of these rocks within the mantle will be addressed below. The xenolithic rock type that makes up but a minor abundance (5–20%) in most kimberlites is eclogite. Although not numerous, they can be extremely diamond-rich, as will be detailed below.

2.1. Eclogites

Over the last several years, our research team at Tennessee has comprehensively studied eclogites, both non-diamondiferous and diamondiferous ones, from Yakutian kimberlites in Russia. It is, therefore, natural that the major portion of this review article is directed towards eclogite xenoliths and their diamonds. However, we have made an attempt to incorporate important aspects of peridotite hosts, although it is worth recalling that diamondiferous peridotites are extremely

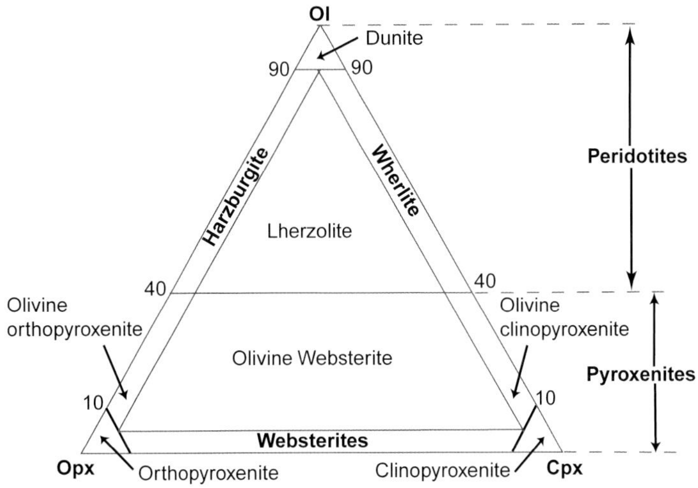

Fig. 2. Streckeisen ternary diagram illustrating the classification of ultramafic rocks (Streckeisen, 1976).

rare, and most are relatively friable and crush easily due to their large amounts of altered olivine.

2.1.1. Mineralogy

Eclogites are basically bi-mineralic, high-pressure rocks (Fig. 3) consisting of garnet and jadeitic clinopyroxene (omphacite). Some eclogites consist of three primary minerals, with the addition of kyanite. In fact, an unusual variety of eclogite with these three minerals, and with largely grossularitic garnet, is called grospydite (gros = grossularite; py = pyroxene; di = diasthene (kyanite), Sobolev, 1977). The classic study of MacGregor and Carter (1970) on the Roberts Victor eclogite xenoliths established Groups I and II as a scheme for the textural distinction of eclogites. This division was later expanded on the basis of the Na_2O contents of the garnets (McCandless and Gurney, 1986), an uncommon trace-element substitution originally discovered by Sobolev and Lavrent'yev (1971) and detailed by Sobolev (1977). However, the textural groups were not in agreement and were effectively dropped. The Group I and II designation, although not very practical in usage, is commonly used today, especially by African geologists.

Coleman et al. (1965) proposed a three-fold classification of eclogites based on the compositions of garnet. However, their groups, A–C, were reflective of different petrogeneses. Taylor and Neal (1989) proposed using the chemical divisions of this garnet classification for mantle eclogites; this is presently used by most other petrologists. The entire eclogite group of kimberlitic xenoliths has been classified into a three-fold division of Groups A–C, based upon the chemistry of their garnets and pyroxenes (Fig. 4), as detailed below. The Group A eclogites, with the lowest Na_2O in clinopyroxene and most pyrope-rich garnets effectively merge into websterites and

Fig. 3. Diamondiferous eclogite xenoliths from Yakutia. (A) Eclogite, UX-1, subjected to tomography and dissection; (B) cut slab, U-119, illustrating the bi-mineralic nature of this Group B eclogite; and (C) polished thin section of U-119 eclogite viewed in transmitted light. Symbols: D = diamond; Cpx = clinopyroxene (omphacite); and Gt = garnet.

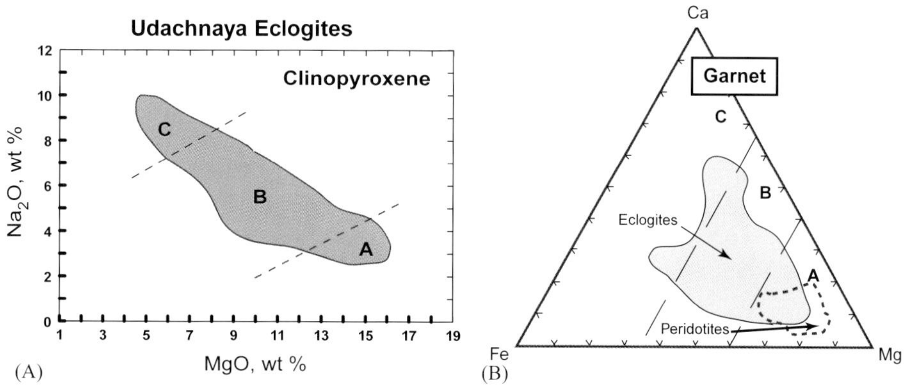

Fig. 4. (A) Plot of omphacitic clinopyroxenes in eclogite xenoliths from Udachnaya, depicting the three-fold (A, B, C) classification of eclogites, according to Taylor and Neal (1989). (B) Plot of Mg (pyrope), Fe (almandine), and Ca (grossularite) of garnets in eclogite xenoliths from Udachnaya, depicting the three-fold (A, B, C) classification of eclogites, according to Taylor and Neal (1989).

pyroxenites (Taylor L.A. et al., 2003b). We have considered garnet websterites as an effective varietal type of eclogites. The boundary between the mafic Group A eclogites and ultramafic peridotites and pyroxenites is based upon the ~1.0% Cr_2O_3 contents of the clinopyroxenes and garnets, with the eclogites being notably Cr deficient.

2.1.1.1. Clinopyroxenes. Clinopyroxenes in mantle eclogites are omphacitic, with Na_2O contents ranging from 1% to over 10%. This is well shown in a plot of MgO versus Na_2O (Fig. 4A), with the group subdivisions of eclogites from Neal and Taylor (1989). In contrast to pyroxene in most peridotites, the eclogitic phases are relatively homogeneous, with many grains displaying no evident chemical zonations. Likewise, the phase compositions are relatively constant throughout a given eclogite. The majority of all mantle eclogites, including diamondiferous ones as well as eclogitic inclusions in diamonds, are of Group B compositions, but some kimberlites are notably enriched in other varieties—e.g., Group A in the Obnazhennaya pipe.

2.1.1.2. Garnets. The garnets in mantle eclogites range from grossular-rich to almandine-rich to pyrope-rich. The ternary plot of garnet compositions (Fig. 4B) also shows the boundaries of the three mantle-eclogite groups: Groups A–C. More grossular-rich garnets, as well as minor amounts of coesite, are typical of Group C eclogites. As with the clinopyroxenes, the garnets are relatively homogeneous, especially compared with those in peridotites.

The general homogeneities of the garnets and clinopyroxenes in eclogites, particularly the diamondiferous ones, result in the assumption that they are in a state of chemical equilibrium, with each other. Inasmuch as pressure and temperature are the most important parameters to constrain the formation of mantle-derived materials, the coexisting garnets and clinopyroxene pairs can be used to make estimates, especially using the Fe/Mg partitioning between garnet and clinopyroxene, based upon the experimental calibration of Ellis and Green (1979). Application of this Fe/Mg exchange geothermometer requires an independent estimate of pressure, which is not directly available from the mineral chemistry of eclogite alone. In most studies of mantle eclogites, either a pressure was assumed (e.g., 5.0 GPa for diamondiferous eclogites) or an equilibrium pressure and temperature were estimated by requiring a simultaneous solution with the ambient geothermal gradient of the craton. Studies of Yakutian eclogites by Sobolev et al. (1994), Beard et al. (1996), and Snyder et al. (1997) have utilized these approaches.

Studies of peridotitic xenoliths (Boyd et al., 1997) and xenocrysts (Griffin et al., 1996) from diamond-bearing Yakutian kimberlites have demonstrated that the geotherm of the Siberia craton is close to 40 mW/m^2. Using this integrated method of temperature estimate coupled with the assumed presence of the eclogite on the geotherm, estimated temperatures range from 950°C to 1280°C, with pressures of about 4.0 to over 6.5 GPa (40–65 Kb), corresponding to depths of >120–195 km in the mantle. These values are well within the stability field of diamond, and consistent with the presence of diamonds in many of the eclogite xenoliths.

2.1.2. Mineral isotope data

A considerable number of radiogenic isotopic studies have been conducted on eclogites from Yakutia (e.g., McCulloch, 1989; Snyder et al., 1993, 1997, 1998a, 1999; Jacob et al., 1994, 1998; Pearson et al., 1995). It seems that the best age data on Yakutian eclogites come from the Re–Os isotopic studies of sulfides (Pearson et al., 1995), which gives an age of 2.9 Ga. The Rb–Sr and Sm–Nd systems have been disturbed, probably as a result of alteration of their crustal protoliths prior to subduction, and do not give consistent results, other than Sm–Nd whole-rock isochron ages of about 389 Ma, the age of the Udachnaya kimberlite emplacement (Snyder et al., 1993, 1997). Sulfide minerals (pyrrhotite, pentlandite, chalcopyrite) are present in most eclogites, some as inclusions in garnet and clinopyroxene. Re and Os isotopic compositions of these phases were determined by Taylor L.A. et al. (2000), and the values of $^{187}Re/^{188}Os$ spread over a large range (from 2.702 to 8.099), yielding an isochron age of 2.6 ± 0.6 Ga, with an initial $^{187}Os/^{188}Os$ of 0.64. Although this age and initial ratio are similar to those for a whole-rock isochron for Udachnaya eclogites (2.85 ± 0.38 Ga and 0.50, respectively; Pearson et al., 1995), this is the first known internal Re–Os isochron for an eclogite. Further discussion of age determinations is given in Section 5.8.

2.1.2.1. Oxygen isotopes.

It was the oxygen-isotope values that gave the first real evidence for the possibility of a crustal origin for mantle eclogites. During the last two decades, oxygen-isotopes data have been accruing (Fig. 5) and are demonstrating that although, on a histogram, the major mode for mantle eclogites is within the mantle region, as determined for peridotites (5.5 ± 0.4‰; Mattey et al., 1994), the outermost regions (tails) of the gaussian distribution display both lower and higher values than the mantle values. It has been the interpretation of Gregory and Taylor

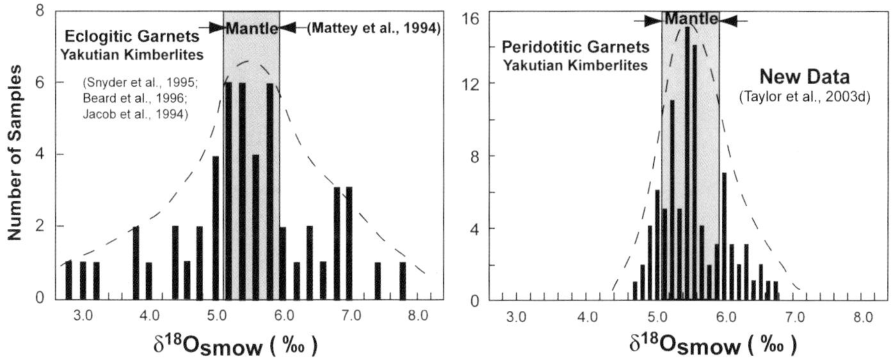

Fig. 5. Comparison of $\delta^{18}O$ of garnets from eclogite and peridotite xenoliths from Yakutian kimberlites. A large number of values from these mantle rocks are outside the mantle region, a signature of their possible crustal heritage. Note the near-Gaussian distribution of the data, with the sharper definition for the peridotites. It is the "tails" for these distributions that are most revealing of crustal protoliths from subduction.

(1981) that the values below and above mantle reflect high- and low-temperature hydrothermal crustal water interactions, respectively. It is now generally accepted that most, if not all, eclogites are of crustal origin, brought to the mantle by ancient subduction (e.g., MacGregor and Manton, 1986; Neal et al., 1990; Taylor, 1993), as detailed in Section 2.1.4.

2.1.3. Secondary partial melting

Spetsius (1995) stated that almost all eclogite xenoliths from Yakutian kimberlites contain evidence of having experienced a partial-melting event, affecting their primary mineralogy. This partial melting commonly manifests itself as "spongy" textured rims on clinopyroxene (Fig. 6), where the primary and secondary pyroxene have the same optical orientation (Taylor and Neal, 1989). The intensity of this secondary alteration varies significantly from grain to grain within an eclogite (Spetsius and Taylor, 2002; Misra et al., 2004). The resulting spongy texture is seen to replace almost all of the large clinopyroxene crystals, leaving only small "islands" of primary clinopyroxene. This secondary spongy texture consists of diopside-rich, Na-depleted clinopyroxene, fine crystals (2–4 μm) of spinel, and interstitial silica glass, rich in K, Na, and Al. This glass is commonly further crystallized to plagioclase, an indication that the reaction, although perhaps initiated at mantle pressures, was still operative at crustal pressures of <10–15 kbars.

This texture was formed by the metasomatic partial melting of the primary omphacitic clinopyroxene by the infiltration of melt/fluid through cracks in the host xenoliths. The "spongy textured" secondary pyroxenes are similar to primary clinopyroxene in terms of Ca–Mg–Fe component but are distinctly less jadeitic. For example, the diopside component ($CaMgSi_2O_6$) in the primary clinopyroxene is ~45 mol%, but in the secondary clinopyroxene it reaches as high as ~76 mol%

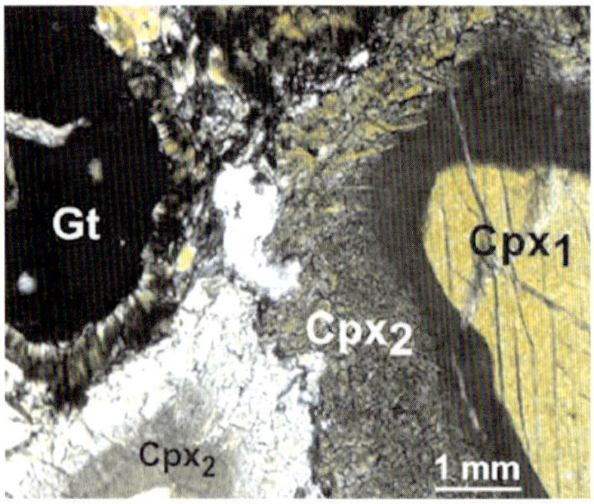

Fig. 6. Photomicrograph of "spongy texture" in an eclogite from Udachnaya. The primary (Cpx1) and secondary (Cpx2) clinopyroxene grains are in optical continuity, but with Cpx2 having lesser Na_2O, as a result of partial melting induced by K-rich metasomatic fluids (Spetsius and Taylor, 2002; Misra et al., 2004).

(Spetsius and Taylor, 2002). The secondary clinopyroxenes are consequently poorer in Na_2O, K_2O, and Al_2O_3. Interstitial glasses between the secondary clinopyroxenes are considerably higher in K_2O (up to 8.7%) and Al_2O_3 (up to 18.9%) compared to the primary clinopyroxene. The overall partial-melting reaction can be written as

$$Cpx_{Primary} + \text{K-rich fluid} = Cpx_{Secondary} + \text{Spinel} + \text{K–Na–Al-rich glass}.$$

Contrasting behavior of Na_2O is evidenced from the partial melting. Contents of Na_2O in secondary clinopyroxene and interstitial glasses are lower than those in the primary clinopyroxene. In general, the Na_2O content in primary clinopyroxene is ~6–7 wt% and decreases to 1–2 wt% in the secondary clinopyroxene. Similar variations were also observed from other samples (Spetsius and Taylor, 2002). In addition, some chemical heterogeneity was also observed in both the secondary clinopyroxenes and the interstitial materials, particularly in Al_2O_3 and K_2O contents. Except for the decreased Na_2O contents, other chemical features of these phases are comparable with those reported by Taylor and Neal (1989), Fung and Haggerty (1995), and Spetsius and Taylor (2002).

Hunter and Taylor (1982) first reported the presence of glass in the kelyphitic rims on mantle garnet and speculated on the K-metasomatic melting of the garnet from kimberlitic fluids. A correlation between the degree of development of kelyphite rims around garnet and the abundance of the intergranular partial melting of clinopyroxene is also apparent. In thin section, there is often a distinct thickening of kelyphite rims on grains in contact with the products of melting and a wide development of kelyphitization with evidence for increased partial melting. It should be emphasized that the formation of kelyphite rims on garnet is evidence for late-stage metasomatism. The following reaction depicts the partial melting of garnet:

$$Gt + \text{K-fluid} \rightarrow Cpx_3 + Sp + \text{K-glass} \pm Opx \pm Phlog.$$

The degree of alteration of garnet (and clinopyroxene) in mantle eclogites is governed by various factors affecting the kinetics of the situation, including availability of fluids and the permeability of the xenoliths. The degree of partial melting of garnet is usually less than that of clinopyroxene, possibly due to the inherent differences in susceptibility to reaction of these two minerals (Spetsius and Taylor, 2002).

The association of some diamonds with secondary minerals is typical of xenoliths from Udachnaya (Spetsius, 1995). This association suggests a metasomatic origin for the diamonds and the secondary minerals, perhaps as a consequence of percolating C-rich fluids in the mantle (Deines and Harris, 1994, 1995; Stachel and Harris, 1997; Spetsius and Griffin, 1998; Taylor L.A. et al., 1998). However, the metasomatic fluids responsible for each product are probably not related.

2.1.4. Origin of mantle eclogites

It is now generally agreed that "mantle" eclogites from kimberlites had their origin in the subduction of oceanic crust beneath the major cratons of the world (e.g., MacGregor and Manton, 1986; Shervais et al., 1987, 1988; Kesson and Ringwood, 1989; Taylor and Neal, 1989, 1993; Jacob et al., 1994; Qi et al., 1994, 1997; Taylor, 1993; Taylor L.A. et al., 2003b; McCandless and Gurney, 1997). One of the first signatures to be identified for the possibility of crustal protoliths was from studies of the carbon isotopes of the diamonds (e.g., Sobolev et al., 1979; Sobolev and Sobolev, 1980; Deines et al., 1984, 1987, 1989, 1991; Deines, 1992; Javoy et al., 1986; Kirkley et al., 1991; Nisbet et al., 1994) (Fig. 7). It is abundantly clear from the $\delta^{13}C$ values that the eclogitic diamonds possess such a wide range (+5 to −35‰) as to have involved processes extraneous to the mantle—i.e., crustal modifications, possibly

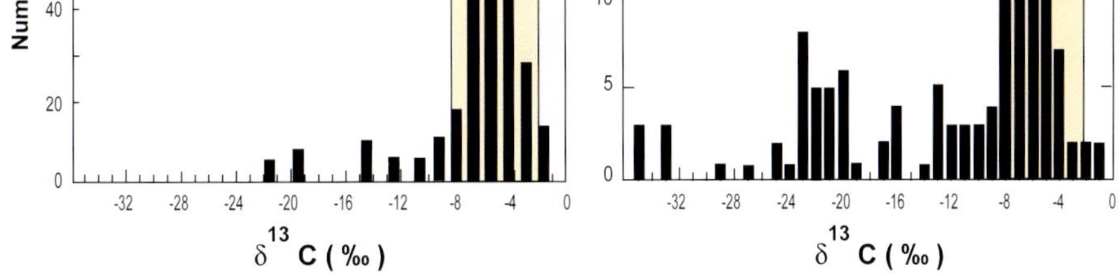

Fig. 7. The $\delta^{13}C$ values of both eclogitic (E-type) and peridotitic (P-type) diamonds. Eclogitic diamonds have a range from +5‰ to −35‰, reflecting a crustal signature. Note also that there have been several P-type diamonds with carbon values outside those of the mantle (e.g., −22‰). Is this also evidence that at least some peridotitic diamonds have a crustal heritage?

through biogenic input. However, suggestions of crustal origins for the carbon went largely unrecognized until a second major signature was brought to bear on the eclogite subject—i.e., oxygen isotopes. It was probably the study of Garlick et al. (1971), as part of the initial study of Roberts Victor by MacGregor and Carter (1970), which first brought the non-mantle $\delta^{18}O$ values from eclogites to the attention of the world, but without an adequate explanation. Subsequently, it was the oxygen-isotope studies by Gregory and Taylor (1981) that laid the groundwork for understanding the $\delta^{18}O$ values of oceanic crust, which were both above and below the mantle range (Fig. 5).

With a decade-plus hiatus, it was not until the classic work of MacGregor and Manton (1986) that real doubt was manifested in the minds of the mantle-eclogite petrologists about the origin of eclogites. Literally, dozens of subsequent studies have confirmed the scenario of subduction of ancient oceanic crust as the tectonic mechanism for bringing the crustal protoliths down into the mantle, into the $P-T$ regime for diamond formation (Jacob et al., 1994). Among these studies by our group are: Shervais et al. (1987, 1988), Neal and Taylor (1989, 1990), Taylor and Neal (1989), Neal et al. (1990), Jerde et al. (1993a, b), Taylor (1993), Taylor L.A. et al. (1996, 1998, 2000, 2003a, b), Sobolev V.N. et al. (1994, 1999b), Pearson et al. (1995), Snyder et al. (1995, 1997, 1998a, 1999), Beard et al. (1996) and Sobolev et al. (1998a, b). These subsequent studies of diamondiferous eclogites, particularly from Yakutia, have revealed several additional crustal signatures, such as +Eu anomalies in garnet and omphacite, both in diamond inclusions and in the host eclogites; ε_{Nd} values that are sometimes as high as +250 in the eclogites; decoupled Rb–Sr and Sm–Nd; and unusually high $^{87}Sr/^{86}Sr$ ratios.

The mantle-eclogite, crustal-protolith hypothesis (e.g., Beard et al., 1996) is envisioned as the subduction of an ophiolite sequence beneath a continental craton, whereby the basaltic and mafic components were metamorphosed, devolatilized and partially melted, and otherwise transformed into eclogites (Fig. 8). This tectonic scheme has found support from the studies of Ireland et al. (1994) and Rudnick et al. (1998) with the partial melting of the subducting crust and the formation of source magmas for the Tonalite–Trondhjemite–Granodiorite (TTG) igneous complex. The partially melted, metamorphosed residual crust is converted into eclogites that are effectively "restites." Being within the diamond-stability field, the eclogites later experienced metasomatic diamond formation (Taylor, 1993).

With this scenario for crustal protoliths for eclogites, it is surprising to us that the closely associated diamondiferous peridotites are considered to be of original mantle origin. Could the ultramafic portion of the subducted crust account for at least some of the mantle peridotites? This is the query at the heart of the study of Taylor L.A. et al. (2003d). They took the approach that history has shown eclogites to have had crustal protoliths, why not peridotites? The evidence may lie in the nature of ophiolite complexes. It is considered probable that a continuum of xenolith compositions exists from eclogites to websterites and harzburgites and peridotites. Evidence for unusual hybrid, mafic–ultramafic xenoliths has been described by Sobolev V.N. et al. (1997). The degree of mantle metasomatism may well be masking some obvious connections between these rock types (Sobolev V.N. et al., 1999a).

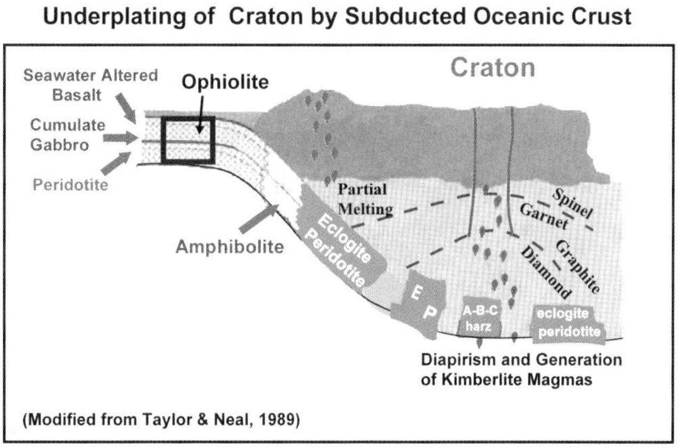

Fig. 8. Cartoon depicting the subduction and under-plating of ancient oceanic crust beneath a craton. This is the tectonic regime (modified after Taylor and Neal, 1989) that gave rise to mantle eclogites and peridotites, including their diamonds, recovered as xenoliths in kimberlites worldwide.

2.2. Origin of mantle peridotites

It has always been assumed that peridotites are of ultimate mantle origin (e.g., Boyd et al., 1997). However, Taylor L.A. et al. (2003d) have addressed the ultimate origin of all mantle peridotites based upon oxygen isotopes and other lines of evidence. They have discussed the nature of the oceanic crust that was subducted beneath the growing cratons of the world to form eclogites, and have placed this within the overall concept of crustal subduction. It is generally accepted that the ancient oceanic crust is represented by an ophiolite complex (Coleman, 1977). The term "ophiolite" refers to an assemblage of mafic to ultramafic rocks in the following sequence (from bottom up): Ultramafic Complex, consisting of variable proportions of harzburgite, lherzolite, and dunite, with metamorphic overprint; Ultramafic–Mafic Gabbroic Complex, containing cumulate peridotites and pyroxenites, leading upward into gabbroic varieties; Mafic Sheeted-Dike Complex; Mafic Volcanic Complex, consisting of pillow basalts with sodic felsic intrusives and extrusives, and an overlying sedimentary section with cherts, shales, and limestones (Coleman, 1977). The Ultramafic Complex contains some intercumulus plagioclase in the peridotites, and crystal fractionation from the melt produces peridotites interlayered with plagioclase-rich gabbros, troctolites, and anorthosites. This is a generalized sequence of an ophiolite complex, realizing that each ophiolite has its own variations on this central theme (Fig. 9). The important point is that ophiolites represent oceanic crust developed at a spreading oceanic ridge and are considered representative of oceanic crust.

In a classic study of the Semail Ophiolite, Oman, Gregory and Taylor (1981) demonstrated that this ancient oceanic lithosphere provides a framework for the time-averaged effects of seawater circulation in "fossil" hydrothermal systems

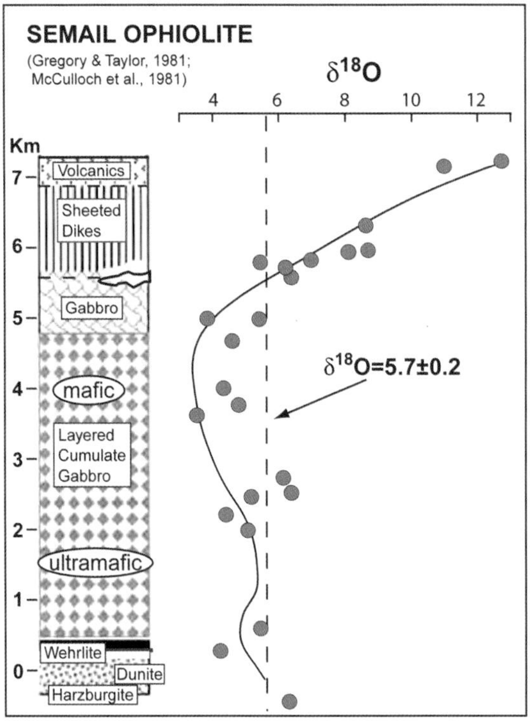

Fig. 9. Cross-section of the Semail Ophiolite Complex, Oman, showing the variations in $\delta^{18}O$ values as a function of position within the complex. The higher than mantle values near the top are due to low-temperature hydrothermal alteration by seawater, whereas the less-than-mantle values are due to high-temperature alteration (Gregory and Taylor, 1981). Note the ultramafic rocks at the lower portions of this ophiolite.

associated with spreading centers. The bottom portion of the Semail complex includes tectonized harzburgitic peridotite; a basal zone of Ol–Cpx cumulates (wehrlites); followed upward by 3–5 km of ultramafic to mafic cumulate gabbros, grading into a km-thick zone of Plg–Hbl–Cpx–Ol–Mt non-cumulate gabbro; with an abrupt transition interval of gabbro, with 10% diabase dikes; to a 1–2 km sheeted-dike complex, overlain by pillow lavas, interbedded with cherty sediments (Fig. 9). The study of McCulloch et al. (1981) further constrained the Rb–Sr and Sm–Nd systematics of this ophiolite complex as these related to the fluid flow of the hydrothermal modifications to the oxygen isotopic values.

Gregory and Taylor (1981) demonstrated that the low-temperature hydrothermal alteration of seawater, in the upper portions of the ophiolite, raised the $\delta^{18}O$ above the average MORB value of 5.7 ± 0.2‰ (later refined to 5.4 ± 0.2‰; Eiler, 2001), as shown in Fig. 5. High-temperature alteration in the lower portions of the complex reduces the $\delta^{18}O$ values. It is envisioned that most mantle eclogites come from the upper two-thirds of the ophiolite complexes (= mostly mafic). It is the lower 1/3rd portion of the oceanic crust, with its ultramafic to mafic compositions that has been often neglected. Below, we address this issue, first from the carbon isotopic compositions of the peridotitic (ultramafic) diamonds; then, from the oxygen values of the mantle peridotites.

2.2.1. Carbon isotopes

The restricted $\delta^{13}C$ values for P-type (peridotitic) diamonds is commonly used as evidence for the mantle origin of peridotites. However, a compilation of $\delta^{13}C$ data (Fig. 7), from Deines (2002, and references therein) for P-type diamonds, mainly from several South African pipes, shows a significant number of values that are more depleted in $\delta^{13}C$ (to −21‰) than the mantle values. The major compilation of carbon isotopes of diamonds by Kirkley et al. (1991) also showed several $\delta^{13}C$ values that are outside the mantle range of −1‰ to −10‰.

As can be readily seen in Fig. 7, the $\delta^{13}C$ values for eclogitic diamonds are predominantly within the mantle region, but almost one-third of the data are clearly lighter in carbon. In contrast, 90% of the $\delta^{13}C$ values for peridotitic diamonds are in the mantle range. However, it has been these few lighter carbon values that have provided the important clues towards understanding the ultimate source of some of the peridotites.

2.2.2. Oxygen isotopes

Since the advent of the laser-fluorination technique for oxygen-isotope analysis, several elegant and extensive studies have been performed on mantle-peridotite minerals (e.g., Mattey et al., 1994). Such studies have greatly broadened our knowledge of the range of $\delta^{18}O$ values in mantle materials. Indeed, some mantle peridotites are enriched in $\delta^{18}O$, with olivine possessing $\delta^{18}O$ values extending from +4.5‰ to +7.2‰ (Mattey et al., 1994). In a remarkably delicate and exacting study of diamond inclusions by the laser-fluorination technique, Lowry et al. (1999) determined that chrome pyrope DIs have a range of $\delta^{18}O$ from +4.6‰ to +5.6‰. The possibility of "outliers" to the normal mantle value of +5.5±0.4‰ being from a non-mantle oxygen source is intriguing, just like the case was for the few $\delta^{13}C$ values that fell "outside the box." With this in mind, Taylor L.A. et al. (2003d) determined the oxygen isotopes in a large population of peridotitic garnets from the archives of the ALROSA Co. in Mirny, Yakutia (Fig. 5).

Taylor L.A. et al. (2003d) carefully selected garnets from over 100 peridotites collected from several Yakutian kimberlites, many diamondiferous. These garnets were subjected to oxygen-isotope analyses by laser fluorination. The majority of the $\delta^{18}O$ garnet values (Fig. 5) plot within the accepted mantle range of 5.5±0.4‰ (Mattey et al., 1994) for peridotites. However, a significant number (~20%) plot outside this window, both above and below. These values are interpreted to represent the effects of both high- and low-temperature hydrothermal alterations that occurred in the oceanic crust prior to subduction of this slab into the mantle. Fig. 10 indicates that the rocks in the lower portion of the ophiolite should have values just below or equal to mantle. However, it must be emphasized that the oxygen isotopes of each ophiolite have a different values, and it is the principle of the presence of values both

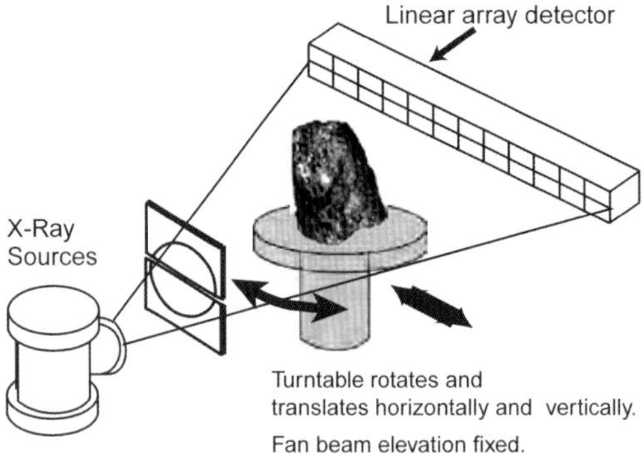

Fig. 10. Schematic representation of the array of the X-ray source, the specimen on a turntable, and the detector, used in the determination of HRCXT imaging.

below and above mantle that is to be valued here. The results of Taylor L.A. et al. (2003d) coupled with the ophiolite oxygen variability are interpreted as indicating that at least some of the diamondiferous peridotites, as well as mineral inclusions in P-type diamonds, owe their ultimate heritage to subducted oceanic crust, possibly represented by the lower portion of an ophiolite complex. The statement that we wish to make with this subduction scenario is that some, if not many, of the mantle peridotites, possibly even those with diamonds, may have ultimately had crustal protoliths, similar to eclogites.

3. Tomography of diamondiferous xenoliths

3.1. Introduction

It is the kimberlitic xenoliths—eclogites, peridotites, harzburgites, websterites—that are the primary source rocks for diamonds in the mantle. Granted, most diamonds are recovered from kimberlite, but this reflects only how the diamonds were transported to the surface. The mantle rocks in which the diamonds actually grew, peridotites and eclogites, are largely broken up by the turbulent movement in the kimberlitic magma, thereby releasing their diamonds. Peridotites are particularly susceptible to this crushing, being composed of easily altered and relatively weak olivine. However, eclogites, although not as numerous as peridotites in most kimberlitic samplings of the mantle, often remain intact, occasionally as diamondiferous xenoliths.[1] These xenoliths contain petrologic and geochemical information on the conditions under which the diamonds grew.

[1] Particularly in Yakutia during the processing of kimberlite, huge amounts of rock are crushed to about 5–6 cm size, washed, and passed along a conveyor belt. The golf-ball sized rocks are exposed to X-rays, whereby any diamonds exposed at their surface will fluoresce a bright blue. Sensed by a detector, a shot of compressed air moves the rock off the belt into a rubber bag. There are diamonds present in this sample,

During the last decade, increasing attention has been given to diamondiferous varieties of these xenoliths, especially to eclogites (Fig. 3A). Our research group at Tennessee has pioneered the use of a non-destructive high-resolution computed X-ray tomography (HRCXT) technique for the initial examination of these diamond-bearing rocks. With HRCXT, it is possible: (a) to create a 3D map of the rock and its constituent minerals; (b) to determine the size, shape, and modal abundances of each mineral; (c) to inspect the relationships of the minerals to each other; (d) to observe unusual features such as mineral lineations or rock fabric; (e) to examine the alignment of diamonds relative to any potential zones of weakness (permeability) in the rock; and (f) to plan for the most effective and efficient dissection of the xenolith, so as to yield a comprehensive data set.

The coarse-grained nature (e.g., 3–10 mm) of the eclogite xenoliths makes it difficult to choose representative samples to study by thin-section or geochemical analysis. Furthermore, determining the 3D aspects of the texture of a xenolith would require slicing, preparing, and examining a prohibitively large number of thin sections. The HRCXT technique, in contrast, provides us a unique opportunity to investigate the spatial and textural relationships between diamonds and other minerals in the host eclogite, by a non-destructive method. Based on these mapping surveys, proper planning is made for detailed dissection of the eclogite. This ensures maximum recovery of diamonds from an eclogite while ensuring minimum loss of material. In addition, this procedure preserves the original fabric and textural features of the eclogites, even after dissecting the sample into several hundreds of pieces.

The diamondiferous eclogites we have studied were first examined by HRCXT to determine a 3D model of the sample. The spatial relationships of the minerals, their textures, and associations were thereby documented. This 3D imaging also permitted precisely locating the diamonds allowing for carefully planned extraction. A series of cuts through the xenolith were made in order to least disturb the diamond-silicate contact, and the diamonds were extracted, leaving behind their molds. Thin sections were made of each diamond mold, preserving the original textural relations. All extracted diamonds were initially examined with a binocular microscope, followed by cutting and polishing several of the diamonds so as to expose their mineral inclusions. The diamonds, their inclusions, and the minerals of the eclogite host were all subjected to detailed chemical analysis (Keller et al., 1998, 1999; Taylor L.A. et al., 1999, 2000; Anand et al., 2004).

The ultimate objective of these 3D studies is to determine if a complete textural characterization of a diamondiferous xenolith could establish any consistent relationships in the xenolith between the diamonds and other minerals, both primary and secondary. This is an independent test of the hypothesis developed from

(*footnote continued*)
and the sample is handled carefully so as to recover these possibly large stones. The diamondiferous eclogites that our group has been studying for several years are from these suites of scientifically invaluable samples.

inclusion studies that diamonds may have grown along metasomatic fronts in the mantle.

3.2. Principles and applications of HRCXT

The industrial X-ray tomographic analyzer used in this study is conceptually similar to a medical CAT scan; however, it is capable of significantly higher X-ray intensity and markedly higher spatial resolution, on the order of microns (Carlson and Denison, 1992; Carlson et al., 1995; Denison and Carlson, 1997; Denison et al., 1997; Rowe et al., 1997; see Fig. 10). Rocks and fossils subjected to HRCXT can reveal the textural features of inside material in these objects, as well as the 3D relationships among their various constituents without destroying the samples.

The HRCXT images used in our studies were obtained at the University of Texas, Austin, as a series of 350–400 2D X-ray slices each, of a diamondiferous eclogite, using a micro-focal X-ray source and an image-intensifier detector system to measure the absorption of X-rays along numerous coplanar paths through the sample. The plane containing the X-ray paths is divided into a matrix of pixels thereby creating a 2D image. Different minerals have different X-ray attenuation values; thus, in the 2D image, they appear as different shades of color (Fig. 11). In this manner, it is easy to differentiate between diamonds, sulfides, silicates, and their alteration products in a xenolith. Several such 2D images are stacked together using volume-visualization software to produce a 3D model of the xenolith. This essentially represents a density map of the sample, from which one can extract the sizes, shapes, textures, and

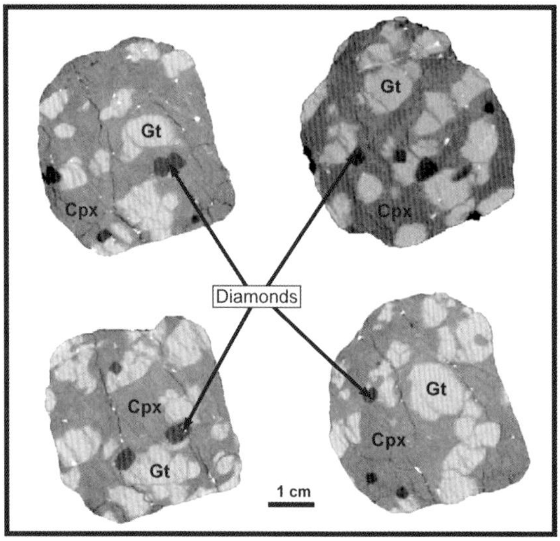

Fig. 11. "Slices" of diamondiferous eclogite UX-1, as determined by tomographic imaging. The gray colors of the minerals correspond to X-ray attenuation, as a function of their densities and average atomic numbers.

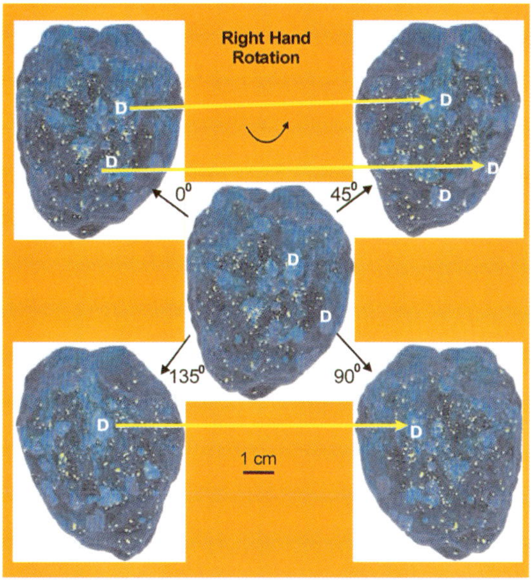

Fig. 12. 3D images of diamondiferous eclogite UX-1, where all the minerals except the diamonds (shades of blue) and sulfides (yellow) have been made invisible. The different images are various positions along a vertical axis. Of particular significance is the near-linear alignment of some diamonds, along zones of secondary alteration (permeability), which may have been the controlling factors for the metasomatic formation of the diamonds. In addition, the sulfide phases are randomly distributed, not associated with the diamonds, as one would expect if they were related in genesis.

locations of individual crystals that have dimensions larger than the spatial resolution of the scans. Such 3D models reveal clearly the spatial relationships between diamonds and their surroundings. This provides clues to the processes that control diamond crystallization. These relationships are determined by rotating and viewing the model at different perspectives to look for any mineral associations, alignments, or fabric. It is also possible to make some of the phases transparent and display only selected mineral phases at a time. For example, a 3D HRCXT model of a diamondiferous eclogite, UX-1, from the Udachnaya kimberlite pipe is shown in Fig. 12 and on the cover of this journal. For clarity, the garnet and clinopyroxene grains are made invisible so that the locations and sizes of diamonds can be easily seen. A total of 69 macro-diamonds ($\geqslant 1$ mm) were located in the interior of this sample, by this technique, which were otherwise invisible. Many of the diamonds in the eclogite seem to be in linear orientation along a vertical axis passing through the central region of the eclogite (Fig. 12). This observation aided our positioning of cuts through this sample parallel to this axis such that a maximum number of diamonds could be exposed on the cut surface.

In Fig. 11, four of the total 360 individual 2D slices, of UX-1, are shown that were originally collected by the X-ray analyzer. The in-plane spatial resolution of this technique is of the order of 10–20 μm, which greatly aided in the imaging of macro-diamonds (⩾1 mm) as well as of some micro-diamonds (<1 mm). These diamonds in the Yakutian eclogites are all associated with alteration zones with prominent subplanar fabric of secondary mineralization. These alteration zones appear prominently on 2D HRCXT images due to their lower densities. It is interesting to note that diamonds were never observed in direct contact or enclosed within fresh garnet or clinopyroxene, confirmed by examination in petrographic thin sections. Furthermore, sulfide minerals are distributed evenly throughout the xenolith and are not preferentially associated with the diamonds.

A complete 3D model (Fig. 12) is more representative of the sample than are thin sections. Modal analyses of five thin sections taken from an Udachnaya eclogite xenolith (U-51) ranged from 25% to 40% garnet. The entire xenolith is actually 25.9% garnet by volume, so most of the thin sections are poor representations of the complete xenolith. This is partly due to the coarse grain size of the xenolith compared to the size of a thin-section slide. Also, the location of a thin section is commonly chosen to include some interesting feature, rather than to accurately represent the entire volume of the rock. Based on these HRCXT images of eclogite xenoliths, it is possible to determine the mineralogical constitutions of the xenolith with much higher precision than by normal point-counting techniques on thin sections. For example, U-51 consists of 25.9% red-orange garnets up to 1 cm in diameter, 0.5% Fe–Ni sulfides up to 3 mm in diameter, and 0.5% macro-diamonds (⩾1 mm) up to 4 mm in diameter (total weight: >3.0+carats). All of these diamonds are dispersed in a variably altered matrix of dark green (where unaltered) clinopyroxene that makes up 73.1 vol% of the sample. Diamonds and sulfides are in unusually high proportions in this sample, which makes them ideal candidates for studying the potential genetic relationships between these two minerals.

In total, 30, 74, and 47 macro-diamonds (⩾1 mm) were found in Udachnaya eclogites U-51, UX-1, and U-33/1, respectively, with calculated grades from 30,000 to over 140,000 carats per tonne, the richest ever reported. The eclogite U-51 is the least altered of all the eclogite samples and contains diamonds with optically visible inclusions of garnets and clinopyroxenes. Thus, diamonds from U-51 were comprehensively studied for their mineral inclusions. Of these 30 diamonds, 18 were wholly within or between clinopyroxene, 12 occurred between garnet and clinopyroxene, and none was entirely within the garnet. Through a combination of HRCXT and hand sample/thin-section observations, it is evident that the diamonds are always surrounded by secondary minerals (Fig. 13), and are nowhere in contact with fresh garnet or clinopyroxene. The sulfide minerals, while similar to the diamonds in total volume, are smaller and more numerous, and occur within and between both garnet and clinopyroxene, not associated with the diamonds (Figs. 11 and 12). It has been reasoned (Bulanova, 1995; Spetsius, 1995) that some diamonds formed from sulfide melts that occurred in the eclogites in the mantle. However, in order for this to be a viable mechanism for diamond formation, evidence should include an absolute association of sulfides and diamonds, and the amount of sulfide should be present in amounts exceeding that of the diamonds by a factor of many. No evidence for either situation has been documented with our tomography (Anand

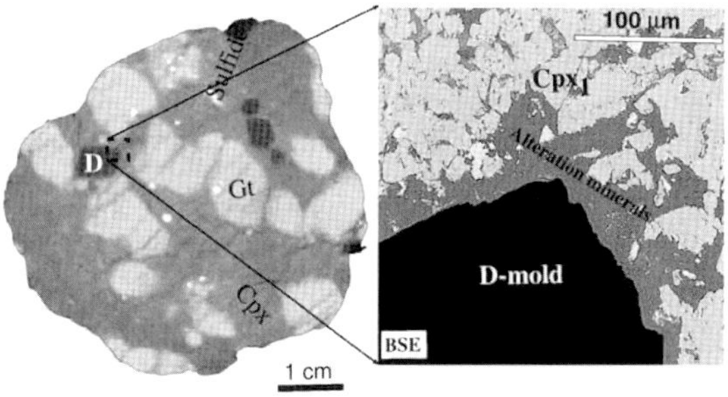

Fig. 13. X-ray tomographic "slice" of diamondiferous eclogite U-33 showing the position of the thin section from which the back-scattered electron imaging (BSE) was taken (Misra et al., 2004). Note the mold where the diamond has been removed prior to polishing. Diamond is never in contact with fresh primary garnet or clinopyroxene.

et al., 2004) or eclogite dissections of many other diamondiferous eclogites from Yakutia (Sobolev et al., 1994).

In hand sample, all of the garnets can be seen to have kelyphytic rims, but are otherwise fresh. Garnets that abut diamonds do not appear to be different from those that abut clinopyroxene, and the same is true for pyroxenes that abut diamonds. Also, clinopyroxene occurs as inclusions in garnet and vice versa. The majority of the clinopyroxene is fresh, but is permeated by a fabric of subplanar cracks. A narrow zone of secondary mineralization occurs along each of these cracks. The lower density of these alteration zones appears as thin dark lines through the clinopyroxene in the HRCXT data (Fig. 11).

The real power that lies in the use of tomography to image, dissect, and aid the unraveling of the mysteries of the host rocks for diamonds in the mantle should be readily apparent. Although far less abundant, diamondiferous peridotites are currently being subjected to a similar modus operandi. Preliminary results have demonstrated that the peridotites are not as rich as eclogites in the abundances of diamonds, but certain similarities in rock fabric, diamond alignment, and secondary mineralization are apparent (Taylor L.A. et al., 2003e).

4. Physical and chemical properties of diamonds

4.1. Crystal morphology

As a single crystal, diamonds occur in several forms such as octahedron—the most common form, cube, cubo-octahedron, and dodecahedron (Harlow, 1998). Experimental studies have suggested that the morphology of natural diamonds may be a function of pressure and temperature, with octahedra supposedly the high-temperature growth form, whereas cubo-octahedra form at lower temperatures (Chepurov et al., 1997). An important process affecting external morphology is dissolution, presumably in the kimberlitic magma during transport to the surface. This produces rounded forms

Fig. 14. Morphology of Yakutian diamonds. (A–B) Octahedra of gem-quality diamonds. Note the presence of mineral inclusions, sulfide and garnet; (C) unusual cube of diamond; (D) 80–100 carat gem-quality diamonds, displaying curved surfaces indicative of resorption leading to the development of dodecahedral morphology; (E–F) diamond octahedra in eclogite xenolith; a stepped octahedral growth mode.

(dodecahedra) and numerous etch features (trigons) on the diamond surface (Robinson et al., 1989). Examples of diamond morphologies are presented in Fig. 14.

The classification schemes adopted for describing diamond morphology are based on external form and surface features (Orlov, 1977; Robinson et al., 1989). The morphological features are the results of primary growth, but can be influenced by subsequent deformation, resorption, and etching by corrosive agents (Mendelssohn and Milledge, 1995a; Orlov, 1977; Robinson et al., 1989).

4.1.1. External form

Natural diamonds exhibit a number of primary forms that can be the result of the conditions at the time of their crystallization (e.g., Chepurov et al., 1997). Classification schemes for diamond morphology have been developed by several investigators (e.g., Orlov, 1977; Sunagawa, 1984a, b). The scheme of Orlov (1977) is based on a combination of the morphology of the diamonds and other properties such as color. For example—Variety I consists of smooth to stepped surfaces of octahedra, with minor modifications by cubic forms, and are largely transparent, with clear to slight yellow colors. This accounts for >90% of natural diamonds; Variety II consists of cubes, with transparent, plane-faced crystals that display no internal structure and are typically yellowish-brown, and variety III, where the cubes and imposed additional forms, such as octahedra, are semi-transparent to opaque crystals with complex internal zonation. Variety IV consists of "coated" diamonds with a variety of cubo-octahedral forms, distinguished by the presence of {1 0 0} and {1 1 1} faces, as a result of a sectoral growth mode. The fibrous coating, thought to be a late-stage growth of diamond from kimberlite melt, is best seen in broken crystals. Several other varieties occur, with increasing complicated morphologies and polycrystalline intergrowths, such as bort. This classification includes transitional forms that combine single crystals and crystal intergrowths, thereby expanding morphological diversities of diamond. In some cases, morphology is assigned to a diamond using its most dominant form shown by the crystal faces.

The development of crystalline or fibrous diamond is dependent on the degree of fluid/volatile supersaturation in the source during their crystallization (Sunagawa, 1984a, 1990), i.e., factors affecting the kinetics of crystal growth. Under low-carbon saturation conditions, single-crystal diamond favors tangential growth and regular atomic arrangement in the crystal lattice. In contrast, fibrous diamond grows during periods of enhanced fluid and carbon supersaturation, a condition that increases the rate of diamond precipitation, resulting in a structurally imperfect growth mode.

Multiple crystal forms develop through twinning or aggregation of diamond during growth, but the conditions that produce these forms are largely unknown. Twins and aggregates are composed of diamonds with either primary or secondary (e.g., dodecahedron) crystal forms, and are classified according to the form of their constituent crystals. Diamonds can display either interpenetrating or contact twinning, but the most common one is a macle twin, which is a contact twin whose twin axis is perpendicular to the octahedral plane (spinel-twin law) with a 60° rotation and often resembles a six-pointed star.

4.1.2. Surface features

Diamonds possess a range of surface textures that reflect source conditions both during and following the growth of a diamond—i.e., paragenetic history (e.g., Orlov, 1977; Robinson et al., 1989). These surface features are largely controlled by growth, deformation, resorption, and late-stage etching and/or corrosion. Surface textures fabricated by deformation imply that the diamond has at some point resided in a solid medium. A commonly observed feature in diamonds is plastic deformation, which is caused by movement along glide planes parallel to octahedral cleavage, giving rise to lamination lines. Usually, the deformation feature is not obvious unless the affected diamond crystals have been significantly resorbed. Textures resulting from resorption are, in general, influenced by the underlying growth structure of affected crystals and are different for octahedral and cubic surfaces (Robinson et al., 1989). However, rounded faces of dodecahedral forms are typical as resorption features on many diamonds.

In contrast to etch features, corrosion produces irregular surface textures. Some late-stage etch and corrosion textures are manifested by truncating features that develop through a series of growth and resorption. Diamonds can also be eventually destroyed by graphitization, dissolution, and oxidation, although only the latter is believed to be important in the mantle environment (e.g., Robinson et al., 1989). The most likely oxidizing agents affecting diamond are volatile species (CO_2, H_2O) carried by mantle fluids/melts and/or by kimberlite magmas.

Investigation of external morphology and surface features alone is of limited value for clarifying the processes of diamond genesis, and thus, may not provide accurate representation of the growth history of diamonds. It is the internal morphology of the diamond that can reveal the details of diamond formation, complicated by several growth and resorption stages, and possibly periods of plastic deformation (Varshavsky, 1968; Bulanova, 1995; Taylor L.A. et al., 2000).

4.2. Internal morphology: cathodoluminescence

The influence of various processes during the development of diamond is clearly visible in polished sections of diamonds as viewed with cathodoluminescence (CL) imaging (Mendelssohn and Milledge, 1995b). Figs. 15 and 16 show several examples of CL patterns in polished diamonds, in which the internal morphology of the diamond reveals details of its formation, complicated by several growth and resorption stages, and periods of plastic deformation (Bulanova, 1995; Taylor L.A. et al., 2000). CL has proven to be a valuable petrographic tool and has been utilized by many workers (e.g., Milledge et al., 1993; Chinn et al., 1995; Taylor L.A. et al., 2000; Bulanova et al., 2002) to investigate the growth history of diamonds. The

Fig. 15. (A–E) CL imaging of Yakutian diamonds, mostly cut and polished parallel to (1 1 0). The various zones reflect degrees of nitrogen aggregation and total nitrogen contents. These zones, akin to the growth features of tree rings, reveal that these diamonds have experienced periods of growth, interrupted by resorption, with continued growth, sometimes changing growth modes, brittle as well as plastic deformation, and a complicated, complex, often tortuous life. The black rounded forms are mineral inclusions. (D) The green CL color of diamond is unusual. (E) An extreme case of complicated zonation; the image at the left shows the position of SIMS analyses for $\delta^{13}C$, $\delta^{15}N$, and N_{total}; note the order of magnitude difference in N within ~600 μm. Each square in the CL mosaics is 300 μm across.

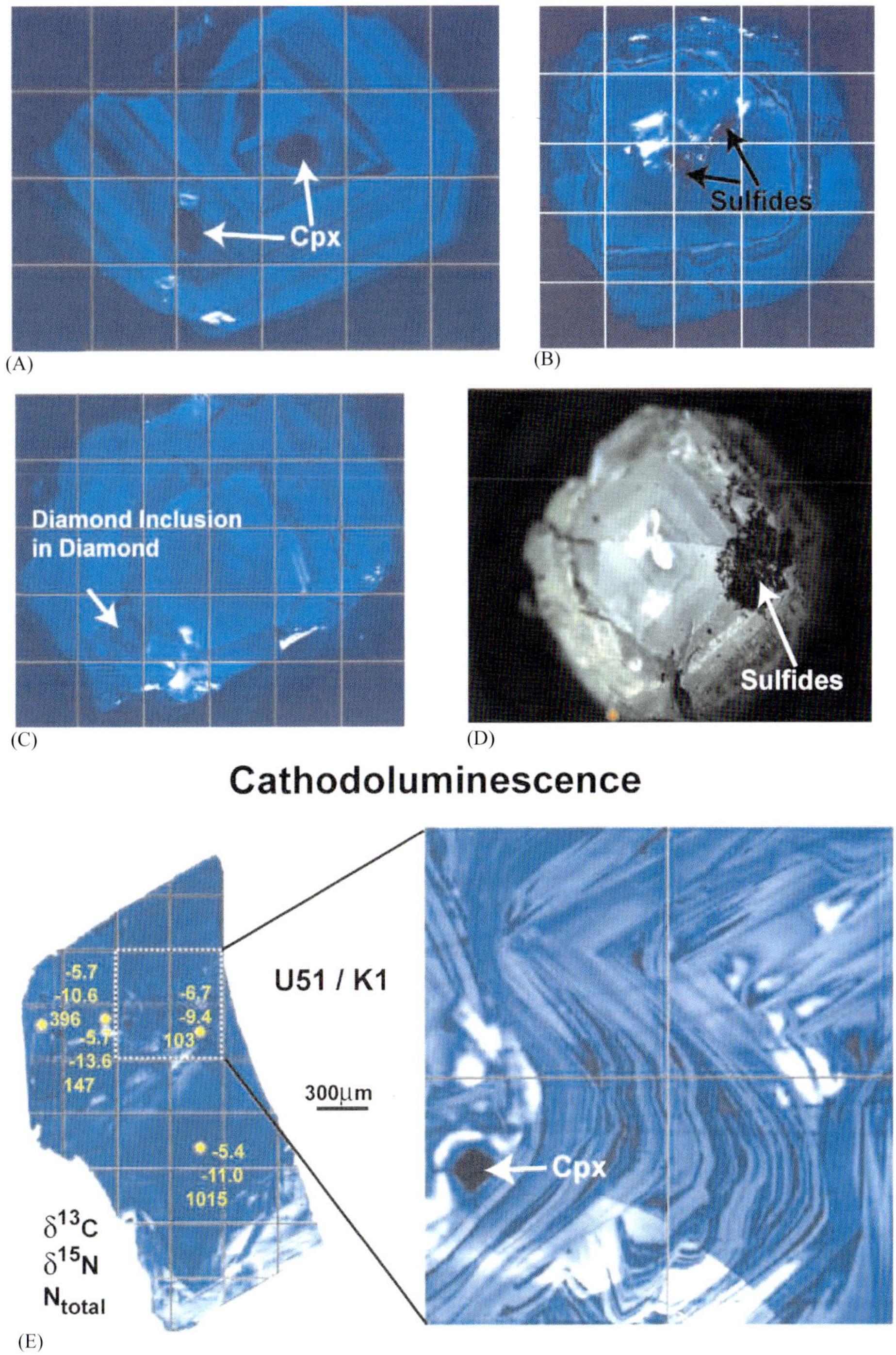
Cpx
(A)
Sulfides
(B)
Diamond Inclusion
in Diamond
(C)
Sulfides
(D)
Cathodoluminescence
-5.7
-10.6
396
-5.7
-13.6
147
-6.7
-9.4
103
-5.4
-11.0
1015
$\delta^{13}C$
$\delta^{15}N$
N_{total}
U51 / K1
300µm
Cpx
(E)

luminescent properties are revealed when a sample emits non-thermal radiation following excitation by some form of energy, particularly by high-energy electrons from a cathode-ray tube, thereby producing luminescent images. Most CL is

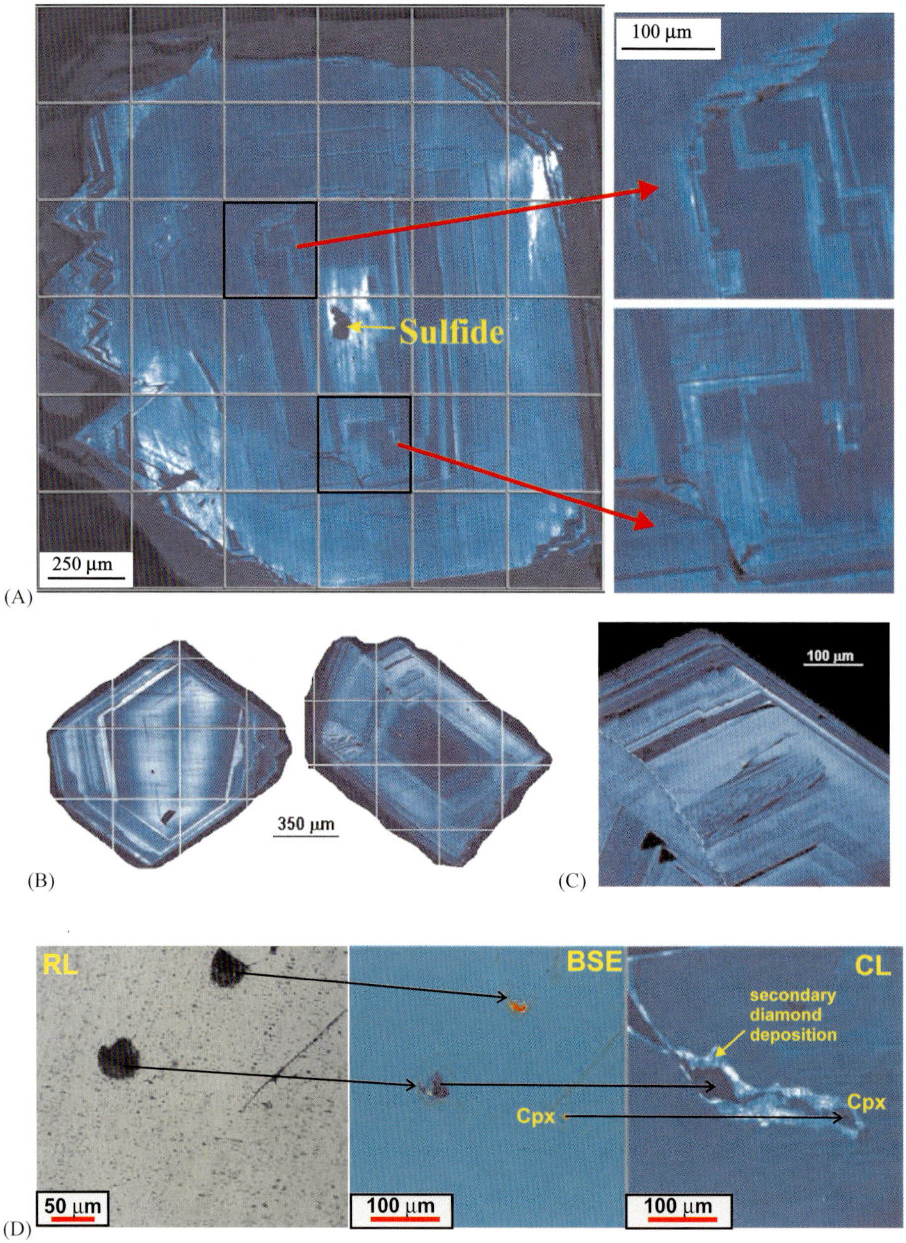

produced "near-surface" at the maximum depth of beam penetration in the sample ($\sim$1 mm). The intensity of the emitted spectrum is dependent on the depth at which the CL is generated, i.e., the initial beam energy, and on the extent to which internal absorption takes place before emission at the surface.

Impurities and radiation damage are both activators of extrinsic CL that contribute towards the variation in luminescent properties of diamonds or any other minerals. As with structural defects, the CL intensity generally increases with the concentration of activator impurities. Natural diamond is rarely pure, and thus, generally displays some degree of CL due to the presence of impurities, particularly nitrogen. The recognition that diamond may have complex zonation based on its nitrogen distribution has initiated the application of CL to investigate diamond internal structure (e.g., Trautman et al., 1998; Taylor L.A. et al., 2000; Bulanova et al., 2003).

The investigation of CL zones in diamonds is somewhat similar to examination of "tree rings" in that they reflect the progressive growth and associated events of the diamond as recorded in successive rings. These CL patterns are commonly referred to as the "stratigraphy" of a diamond. The increasing number of studies of internal morphology has improved the recognition and understanding of diamond growth patterns (e.g., Bulanova and Milledge, 1995 and reference therein). An additional benefit of CL studies of polished diamonds is the recognition of minute, optically invisible, cracks (Fig. 16), that sometimes extend from the diamond inclusions to the surfaces of the diamonds, thereby providing evidence of possible "open-system" behavior of the inclusions (Anand et al., 2004). Electron microprobe (EDS) analyses can rapidly confirm if the inclusions are associated with secondary alteration minerals, resulting from "open-system" behavior. These minerals are generally located in the cracks, which are only discernible under CL images (Fig. 16). This observation highlights the importance of CL imaging in the study of diamonds and diamond inclusions. Caution must be used in significant conclusions based upon the chemical composition of inclusions, without close examination of the diamond hosts. This is especially pertinent for studies where the inclusions are retrieved by burning or crushing of the diamond, thereby destroying any evidence of possible secondary processes possibly rendering many inclusions epigenetic.

Fig. 16. CL imaging of Yakutian diamonds, mostly cut and polished parallel to (110). Insets for (A) and (B) show details reflecting growth modes, resorption, brittle fracture, healing through new growth, new resorption, more growth, etc. (C) an enlargement of part of the diamond in (B), showing the fine detail of the micro-fault and its healed margin. (D) identical portions of a diamond as viewed in reflected light (RL), back-scattered electron (BSE), and CL imaging. Optically, there is little evidence of a crack in the diamond; even the BSE imaging does not show anything anomalous; however, the CL image shows a large crack, connected to the surface of the diamond, that has been completely filled in with secondary diamond. It is such cracks, "hidden" from optical sight, that can provide "open-system" behavior to modify pristine inclusions, perhaps significantly. This amply illustrates the power of *in situ* inclusion examination, a point emphasized by Taylor L.A. et al. (2000) and Anand et al. (2004).

Different stages of diamond growth are usually accompanied by marked changes in zonation that can be readily identified by CL imaging of sectioned and polished crystals (e.g., Bulanova, 1995). The variation in signal intensity as a function of different nitrogen defects and substitutional forms is presently not well understood; therefore, the application of CL to diamond internal morphology is largely qualitative. Luminescence centers may also form in response to radiation damage and deformation. Slip planes are visible under CL in diamonds that have undergone plastic deformation, whereas evidence of brittle deformation and subsequent annealing is usually observed as brecciated regions in affected crystals (Bulanova, 1995), as shown in Fig. 16.

The physico-chemical properties of diamonds support a model for periodic and continuous growth of octahedral forms in the mantle environment, although late-stage participation of mantle melts and fluids, including kimberlitic, cannot be precluded. Inhomogeneities of nitrogen and carbon isotopic characteristics within and between growth zones of octahedral diamonds also testify to time-dependent changes in the source region (e.g., Taylor L.A. et al., 2000; Bulanova et al., 2003). Notably, these variations are represented by abrupt steps in compositional gradients that are most easily modeled on the premise that there are intermittent changes in source composition and/or growth conditions.

4.3. Nitrogen aggregation

The presence of impurities within the atomic structure of diamond is reflected in many of its chemical and physical properties (Davies, 1984; Trautman et al., 1998). Despite its relative purity, natural diamond contains a variety of contaminants which can be grouped into: (1) impurities that occupy interstitial sites or substitute for carbon in the diamond lattice (trace elements), and (2) impurities such as fluids, melts, or mineral inclusions enclosed within the diamond body (Meyer, 1985). Both types of impurities provide information on growth processes and compositional variations in the diamond source regions.

Nitrogen is the major substitutional impurity found in diamond (Sellschop et al., 1980). Two classes of substitution, Types I and II, are distinguished by the presence or absence of nitrogen in the crystal lattice (e.g., Harris, 1987). For nitrogen-bearing specimens, both the nitrogen content and the type of nitrogen defect have proven useful in defining discrete populations with similar genetic affinities (Taylor W.R. et al., 1990, 1996). For this reason, diamond is typically described in terms of the "defect type" created by substitutional nitrogen.

Nitrogen-bearing (Type I) diamonds typically accommodate between 10 and 5000 atomic ppm of nitrogen, and contain defects that form part of a progressive aggregation sequence (Davies, 1976; Harris, 1987). Nitrogen is considered to have been incorporated into the diamond structure during growth when individual carbon atoms are substituted by single nitrogen atoms. The use of fourier-transform infrared spectroscopy (FTIR) enables one to detect the variations in nitrogen aggregation (Fig. 17). Type Ib diamonds are dominated by singly substituted (C center) defects that are present in synthetic, cubo-octahedral diamond. In nature, C center defects

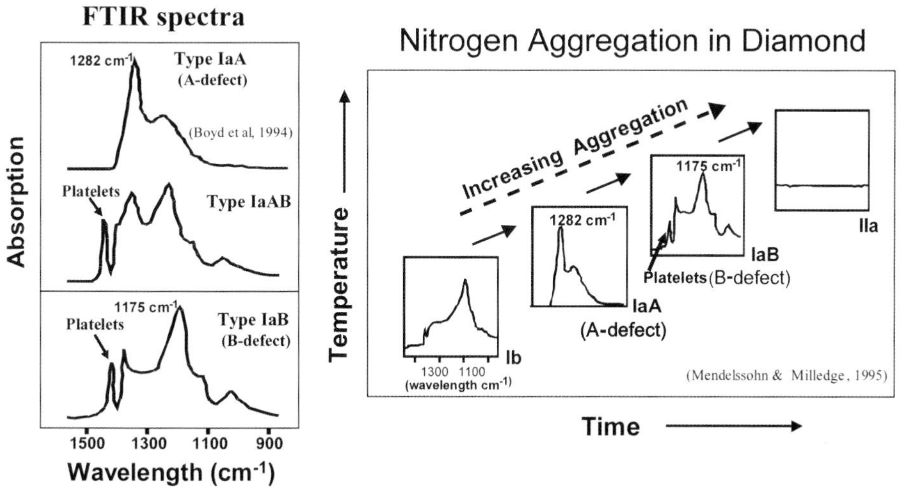

Fig. 17. FTIR spectra of diamond illustrating the different types of nitrogen substitution for carbon in the diamond structure. The development of the type and degree of nitrogen aggregation as a function of time and temperature is illustrated, as modified after Mendelssohn and Milledge (1995a, b).

are common in diamonds derived from ultrahigh-pressure metamorphic rocks. These defects are mostly restricted to fibrous overgrowths (coats) and crystals with cubic habit (e.g., Taylor W.R. et al., 1996). The relative scarcity of natural type Ib diamond in deposits derived from kimberlites implies a low-activation energy required to convert singly substituted C defects into aggregated nitrogen defects. Nitrogen aggregation is thought to be a kinetic process, which is amplified under rising temperatures. Thus, the preservation of type Ib point defects in any diamond requires a cool source region and/or a short residence interval prior to its emplacement at the surface (Taylor and Milledge, 1995; Taylor, 1998).

As depicted in Fig. 17, with extended residence time in the mantle and/or at higher mantle temperatures than those of type Ib, the singly substituted nitrogen atoms progressively aggregate and develop type Ia paired defects. Nitrogen-pair defects (A-defects), i.e., two nitrogen atoms on adjacent lattice sites, define type IaA diamonds. Further aggregation results in four nitrogen atoms arranged in a tetrahedral coordination with a carbon-site vacancy in the center (B-defects), producing type IaB diamonds. At high degrees of aggregation, platelets of solid nitrogen (?) exsolve from the diamond along {1 1 1} planes. The presence of nitrogen in diamond gives rise to characteristic absorption spectra in the infrared (IR) over the 1400–1000 cm^{-1} wavelength region. Types C, A and B centers generate absorption peaks at wavelengths specific to each defect, so that the degree of aggregation can be determined (e.g., Mendelssohn and Milledge, 1995b).

Taylor W.R. et al. (1990, 1995) and Taylor and Milledge (1995) have used the degree of nitrogen aggregation in diamond in attempts to constrain the residence time of diamond in the mantle (assuming a given temperature) or backing out a temperature, assuming a given residence time. However, it is difficult to estimate the temperature at which the diamond has been during the major portion of its existence, especially since coexisting (touching) garnet and clinopyroxene are rare, and there is growing proof that multiple inclusions (non-touching) in diamonds may not be related in $P-T$ conditions, e.g., Taylor L.A. et al. (1998, 2000). Likewise, the time of residence is a direct function of the age of the diamond, another uncertainty. Therefore, because of these large uncertainties in time–temperature estimates, this approach is not well constrained.

4.4. Stable isotopes in diamonds

The chemical compositions of diamond that are most commonly determined are the isotopes of carbon, whereas the isotopes and total abundances of nitrogen have become of increased importance for evaluating the source region for the carbon in the diamonds.

4.4.1. Carbon isotopes

Stable carbon isotopes represent an additional source of data within and between diamonds that can provide important information on the compositional budget and other conditions prevalent during the formation of the diamonds (e.g., Javoy et al., 1986; Deines, 1992, 2002, and references therein; Boyd et al., 1994a; Boyd and Pillinger, 1994). The isotopic compositions of carbon reflect the $^{13}C/^{12}C$ ratio, which is expressed relative to PDB ($\delta^{13}C = 0‰$). The global distribution of $\delta^{13}C$ values for diamond ranges from +5‰ to −34.4‰, with the major mode between −5‰ and −6‰ (e.g., Deines, 2002), as shown in Fig. 7.

The carbon isotopic ratios of diamond broadly reflect the nature of the possible carbon source from which the diamonds were derived. Carbon isotopic compositions have been shown to correlate strongly with mineral paragenesis (e.g., Sobolev et al., 1979; Deines et al., 1993, 1997, 2001). Peridotitic diamonds display a restricted range of $\delta^{13}C$ values, between −1‰ and −10‰, whereas eclogitic diamonds span the known range of carbon isotopic compositions in diamonds (+5‰ to −35‰).

Several hypotheses have been put forth to explain the range and distribution of carbon isotopic compositions in diamonds, taking into account the restricted range of values displayed by diamonds from individual intrusions. Homogenization of a primordial carbon reservoir in the mantle may account for the relatively restricted range of $\delta^{13}C$ values observed for peridotitic diamonds; however, this model finds difficulty in explaining isotopically light crystals. Several authors have invoked Rayleigh fractionation as an alternative explanation (e.g., Deines, 1980; Galimov, 1991), but fractionation of a homogeneous carbon reservoir alone is unlikely to account for the known distribution of isotopic compositions.

Cartigny et al. (1998a, b, 1999) suggested that high-temperature fractionation of mantle fluids might account for the complete $\delta^{13}C$ range for diamonds, given that these fluids evolved separately in peridotitic and eclogitic environments. The loss of CO_2 from eclogitic sources is thought to leave a $\delta^{13}C$-depleted residue that may crystallize diamonds with a broad $\delta^{13}C$ range. Several studies (e.g., Deines et al., 2001; Galimov, 1984, 1985) have concluded that eclogitic diamonds inherit at least some of the carbon isotopic characteristics from crustal carbonaceous materials that display strong enrichment in ^{12}C, possibly from biogenic sources. Crustal carbon present in subducted ancient oceanic crust combined with a primitive carbon source in the mantle are considered capable of generating the observed range of $\delta^{13}C$ values for diamonds (Sobolev and Sobolev, 1980). But much research remains to be done before definitive answers are obtained. Indeed, carbon isotopic compositions of diamonds have been explained by invoking a variety of processes—e.g., intramantle isotopic fractionation during fluid transport and diamond growth (Deines, 1980; Galimov, 1991); primordial mantle heterogeneities (e.g., Deines et al., 1987); introduction of heterogeneous subducted carbon (e.g., Sobolev et al., 1979; Kirkley et al., 1991; Navon, 1999), etc. However, it has become increasingly apparent that carbon isotopic studies alone are not able to resolve the source of such carbon isotopic variability, and the relative importance of these competing processes. Thus, current approaches have used studies of carbon isotopic composition, integrated with N abundance, N aggregation, and N isotopic composition, along with age determinations on the same diamonds.

4.4.2. Nitrogen isotopes

The application of nitrogen isotopic analyses in diamonds is well established (e.g., Javoy et al., 1986; Boyd et al., 1992; Deines et al., 1993, 1997; Cartigny et al., 1998a, b). The data set for nitrogen is less extensive than that for carbon, chiefly due to the lower abundance of nitrogen (<10 ppm to 1000 ppm) that requires low analytical blanks. The nitrogen isotopic composition of diamonds shows more isotopic heterogeneity and does not display any clear paragenetic distinction (P- versus E-type), as observed with the distribution of carbon isotopes.

Except for the unusually large nitrogen isotopic range ($\delta^{15}N$) seen in P-type diamonds from Pipe no. 50 in China (Cartigny et al., 1997) and the "low $\delta^{13}C$ group" of Boyd et al. (1994a, b), P- and E-type diamonds from many localities have overlapping $\delta^{15}N$ values, ranging from –12‰ to +4‰ (Cartigny et al., 1998a). This range is identical to that of the present-day depleted mantle, as measured in ocean ridge basalts (Javoy et al., 1986; Marty and Zimmerman, 1999). A close examination of the nitrogen isotopic values of diamond allows a crude paragenetic distinction to be made. For instance, E-type diamonds are light in carbon isotopic composition, but contain isotopically heavy nitrogen, extending towards $\delta^{15}N$ values characteristic of sediments and S-type granites (+5‰ to +10‰).

4.4.3. Carbon–nitrogen isotopic systematics

The combined systematics of $\delta^{13}C$–$\delta^{15}N$ and N_{total} abundances in P- and E-type diamonds provide the best constraints on the abundance and ultimate sources of carbon and nitrogen in the mantle (e.g., Javoy et al., 1984; Milledge et al., 1989; Fitzsimons et al., 1999; Hauri et al., 1999), an example of which is shown in Figs. 15 and 18. Cartigny et al. (2001) estimated the carbon/nitrogen ratio of a diamond growth medium to be around 200–500 (mantle carbon = 400 ppm; mantle

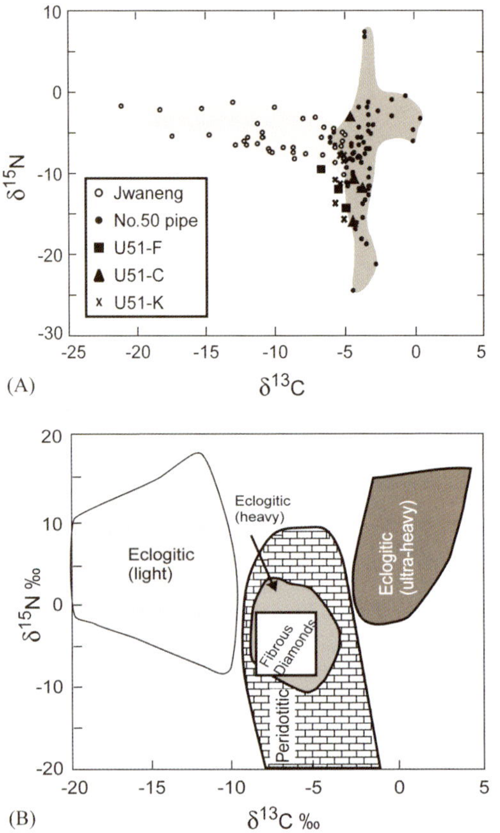

Fig. 18. (A) Plot of $\delta^{13}C$ versus $\delta^{15}N$ for a large selection of diamonds from Jwaneng, Botswana (Cartigny et al., 1998a), No. 50 pipe, China (Cartigny et al., 1997), and three diamonds from the Udachnaya eclogite, U-51 (Taylor L.A. et al., 2000). (B) Schematic plot is of $\delta^{13}C$ versus $\delta^{15}N$ for diamonds worldwide, which puts the E- and P-type diamonds within perspective, as modified after Bulanova et al. (2002).

nitrogen = 1–2 ppm) or as low as 10, depending on the assumptions used for the partitioning behavior of nitrogen. P-type diamonds, based on their mantle-like carbon isotopic composition, must have predominantly mantle nitrogen. Collectively, the entire P-type diamond population, however, shows a large range in $\delta^{15}N$ from −25‰ to +6‰, which has been ascribed to primordial mantle heterogeneity inherited from the large isotopic range seen in meteorites (Cartigny et al., 1997). To explain the diamonds with non-mantle-like carbon and nitrogen isotopes characterized by low $\delta^{13}C$ and high $\delta^{15}N$, Cartigny et al. (1998a, b, 2001) argued for a fractionation process in the diamond-forming fluid, because they reasoned that mixing between mantle and sedimentary end-members could not reproduce key features of their data. Navon (1999) objectively argued against simple recycling models and proposed a model in which the carbon–nitrogen isotope systematics of E-type diamonds can be explained as a mixture of mantle and subducted components.

Bulanova et al. (2002) performed a comprehensive study of the carbon and nitrogen isotopic systematics of a natural diamond with sectoral growth, from the Mir kimberlite in Yakutia. The sample had a graphite "seed" inclusion enriched in K, Ca, Ti, Rb, and Sr, which was interpreted to mean that diamonds may have grown from a carbonate melt/fluid interacting with upper-mantle rocks. A combination of FTIR and secondary-ion mass spectrometry (SIMS) measurements was used to obtain aggregation and abundance state of nitrogen in this Mir diamond. Carbon and nitrogen isotopic determinations along specific traverses were also made using SIMS. This study revealed that there were distinctive isotopic trends that existed in the diamond. These are expressed by an increase in $\delta^{13}C$ and $\delta^{15}N$ values and a decrease of nitrogen concentrations and nitrogen aggregation from the core region towards the intermediate regions of the diamond. This progressive trend in carbon and nitrogen isotopic relationship is believed to reflect diamond growth from an evolving fluid medium. Furthermore, measurements made in this study did not reveal any significant or consistent fractionation of carbon and nitrogen isotopes between the cubic and octahedral growth zones of the diamond. Fig. 18 shows the $\delta^{15}N$–$\delta^{13}C$ systematics of diamonds from worldwide locations. As seen from this figure, the peridotitic diamonds show a narrow range in $\delta^{13}C$ values as compared to a larger range for $\delta^{15}N$ values. This is consistent with the derivation of P-type diamonds from mainly mantle sources compared to contribution from a variety of source regions for eclogitic diamonds. The complexity of eclogitic diamond formation is highlighted by the $\delta^{13}C$–$\delta^{15}N$ systematics that shows just the opposite relationship to that of P-type diamonds with a very large range in $\delta^{13}C$ at relatively restricted $\delta^{15}N$ compositions.

Recently, Cartigny et al. (2001) have compiled published data from over 1200 analyses of macro-diamonds that have been investigated for both C and N isotopes and N abundance. This compilation revealed that over 90% of natural diamonds show a narrow range in $\delta^{15}N$ values, whereas the nitrogen abundance varied over three orders of magnitude. Cartigny et al. (2001) proposed that individual suites contain diamond populations showing trends of varying nitrogen abundance at nearly constant $\delta^{15}N$ that resulted from kinetic processes related to the very slow diamond growth.

The resolution of the source of the carbon and nitrogen in diamonds lies in detailed *in situ*, high-spatial resolution, carbon–nitrogen isotopic studies on typically complex-zoned diamonds using a combination of CL, FTIR, and SIMS. Such studies may demonstrate large covariations in coupled C–N isotopic compositions that correlate with diamond growth zones, and generally do not corroborate thermodynamic fractionation during diamond growth models (e.g., Bulanova et al., 2002).

4.5. Fibrous diamonds

Many diamonds contain a secondary fibrous diamond overgrowth in cubic form in the outer few hundred microns of their rims. These overgrowths are commonly known as coatings and such diamonds as coated diamonds (Kamiya and Lang, 1965). The cores of coated diamonds are transparent single crystals of octahedral shape. Cubic diamonds form when the cores are small or absent. Optical examination of the coated and many cubic diamonds have confirmed the presence of numerous submicron, subrounded inclusions in the coatings. Further, IR spectroscopic studies have indicated the presence of water, carbonate and phosphate inclusions in cubic and coated diamonds (Navon et al., 1988). Instrumental neutron activation analysis of a coated diamond indicated LREE-enriched patterns, which led Bibby (1982) to suggest that the inclusions contain melt. However, this study was unable to determine the major-element composition of the included matter. The findings of coated and cubic diamonds in eclogitic xenoliths (Hills and Haggerty, 1989; Spetsius, 1995; Izraeli et al., 2001; Klein-BenDavid et al., 2003) and of eclogitic inclusions within coated diamonds (Talnikova, 1995) confirm the close association of these fluids with the eclogitic environment. Two occurrences of chromite inclusions were also reported suggesting that there could be fibrous diamonds of peridotitic affinity (Talnikova, 1995).

Navon et al. (1988) reported that micro-inclusions in cubic and coated diamonds from Zaire and Botswana are rich in water, carbonate, silica, K_2O, CaO and FeO, and their bulk composition roughly resembles those of potassic magmas, such as kimberlites and lamproites. However, these inclusions were significantly depleted in MgO compared to potassic magmas and also had a high K_2O/Al_2O_3 ratio. This led Navon et al. (1988) to suggest that the bulk compositions of micro-inclusions represents a volatile-rich fluid or melt trapped by diamonds during growth. Izraeli et al. (2001) in a further study of fibrous coatings determined that these inclusions are even more complicated, many with non-kimberlitic signatures.

In addition to typical IR spectra of micro-inclusions bearing zones of coated and cubic diamonds, absorption bands for water, carbonate (calcite), phosphate (apatite), Nitrogen platelets were also observed (Navon et al., 1988). These observations indicate the extremely volatile-rich nature of the fluid/melt from which the micro-inclusions crystallized. In addition, greater enrichment in K_2O and LREE, and depletion in MgO compared to kimberlitic or lamproitic melt suggests that mantle fluids/melts rich in volatiles might have been responsible for the formation of such micro-inclusions. However, over 40% of volatile contents are required to account for the measured bulk composition of these inclusions. The nature of the fluid is still debated, and in most cases there has been no direct observation of the existence of such fluids in the mantle. Eggler (1987) has suggested that highly alkalic, dense hydrous fluids, similar in composition to that of micro-inclusions, may exist at high pressures and that the miscibility gap between hydrous fluids and melts may be narrowed or closed in alkalic systems. Such fluids/melts may represent effective metasomatizing agents.

The similarity in composition of micro-inclusions in the coats of coated diamonds and in cubic diamonds is consistent with a common growth environment for the coats and for cubic diamonds, as suggested by the clustering of carbon isotope ratios, from the two diamond types (Galimov, 1984; Boyd et al., 1987). The isotope composition of the cores of the coated diamonds shows no correlation with that of the coats, implying different growth conditions for the cores (Galimov, 1984; Boyd et al., 1987). Based on their studies, Navon et al. (1988) stated that there is a clear compositional distinction between the micro-inclusions and their host kimberlites, strongly suggesting a xenocrystic relationship between the two. They suggested that the cubic diamonds and fibrous overgrowths on coated diamonds formed from the fluids represented by the micro-inclusions that they trapped. Broad resemblance of micro-inclusions and kimberlites may reflect a common origin at depth. It may entirely be possible that the metasomatizing agents in the mantle have kimberlite-like composition, but it is not possible to make a clear-cut similarity or distinction between the two at this moment. Whether the micro-inclusions in cubic and coated diamonds are derived from the host kimberlite or not, based on the fibrous nature of the coats and cubic diamonds, and the nitrogen aggregation states, it is well established that the ages of these cubic diamonds and coats on coated diamonds are close to kimberlite eruption since a longer mantle residence will result in crystalline growth as well as aggregation of nitrogen atoms into complex molecules.

5. Mineral inclusions in diamonds

5.1. Introduction

Historically, mineral inclusions in diamonds (Fig. 19) were originally divided into two groups: *epigenetic* and *syngenetic* (Table 1). Epigenetic phases (literally, of later origin) are secondary minerals, typically associated with crustal processes that have occurred due to "open-system" behavior of the encapsulated inclusion. These minerals are atypical to the primary minerals in mantle xenoliths. All other diamond inclusions were considered as syngenetic, that were formed simultaneously with the diamond. Later, Meyer (1987) suggested the addition of a third class of diamond inclusions, *protogenetic*, for inclusions that had formed before the encapsulation by the host diamond, with the criteria that are elaborated upon below. However, the distinction between the syngenetic and protogenetic diamond inclusions is difficult, but it is important to understand the differences between these types of inclusions. The inclusions are thought to directly reflect the $P-T-X$ conditions at the time of the diamond formation, but this may not be the case for the protogenetic inclusions. In addition, until recently, all mineral inclusions in diamonds were thought to come from lithospheric mantle (<200 km). However, mineral inclusions from the asthenosphere and transition zones have been identified during the last two decades that have shed new light on our understanding of diamond genesis (see below).

5.2. Paragenesis of mineral inclusions

The minerals found as inclusions in diamonds (Fig. 19) are typically those from the host xenoliths in which they formed. There are two main paragenetic classes: (1) Peridotitic (P-type), also known as P/U-type (from Ultramafic; Sobolev, 1977). This

Fig. 19. Mineral inclusions in diamonds. (A) Garnet; (B) garnet; (C) clinopyroxene; (D) euhedral pyrope inclusions, displaying cubo-octahedral morphology imposed by the diamond. Typical, non-inclusion garnet, if euhedral, usually occurs as dodecahedrons.

group is represented by olivine, chromite, orthopyroxene, garnet, and diopsidic clinopyroxene. (2) Eclogitic (E-type). This group is mainly represented by pyrope-almandine garnet and omphacitic clinopyroxene. There are several other high-pressure phases that are known to be of E-type paragenesis (see below). The paragenetic association of sulfide inclusions is determined based on their Ni contents (e.g., Yefimova et al., 1983; Bulanova et al., 1996). Sulfides of P-type have higher Ni contents compared to the E-type sulfides. In addition, there is a third, yet rare, group of inclusions of websteritic (W-type) affinity, whose designation is not as simple or certain as the P- and E-types. In fact, the U-type designation of Sobolev (1977) is probably the best alternative to attempting to differentiate the W- and P-type mineral inclusions. The carbon and nitrogen isotopic compositions of P/U- and E-type diamonds are described above in Section 4.4.

5.2.1. Mineral inclusions from the transition zone and the lower mantle

In the last two decades, several studies have identified diamond mineral inclusions that are derived from depths below the base of the cratonic lithosphere. Stachel

Table 1. Classification of mineral inclusions in diamonds (modified from Meyer, 1987)

Pre 1950	Post 1950[a]			
	Protogenetic and/or syngenetic		Epigenetic	Lower mantle phases[b]
	Ultramafic (U or P)	Eclogitic (E)		
Quartz	Forsterite	Omphacite	Serpentine	TAPP[c]
Diamond	Enstatite	Pyrope-almandine	Calcite	$CaSiO_3$
Pyrite	Diopside	Kyanite	Graphite	$MgSiO_3$
Hematite	Cr-pyrope	Sanidine	Hematite	Ferropericlase
Ilmenite	Cr-spinel	Coesite (quartz)	Kaolinite	Majorite Gt
Garnet	Mg-ilmenite	Rutile	Acmite	Stishovite
Graphite	Sulfides	Ruby	Richterite	
Diopside	Zircon	Ilmenite	Perovskite	
Olivine	Diamond	Chromite	Mn-ilmenite	
Calcite	Native iron	Sulfides	Spinel	
Chromite	Silica carbide[f]	Diamond	Xenotime	
Zircon			Goethite	
Biotite				
Phlogopite[d]				
Biotite[d]				
Muscovite[d]				
Amphibole[d]				
Magnetite[d]				
Apatite[d]				
Gas (CO_2)/liquid[e]				
Topaz[e]				
Plant remains[e]				
Gold[e]				

[a] Minerals identified positively by X-ray diffraction and electron microprobe.
[b] From Stachel (2001).
[c] Tetragonal almandine pyrope phase.
[d] Uncertain paragenesis.
[e] Probably misidentified.
[f] Leung et al. (1996).

(2001) presented an excellent review of the current understanding of the lower-mantle associations of diamond inclusions; herein, we only provide a brief description of lower-mantle phases in diamonds. Moore and Gurney (1985, 1989) found majoritic garnet (a high-pressure phase that becomes stable at >200 km) inclusions in diamonds from the Monastery mine, and provided the first evidence that diamonds may contain material from the asthenosphere and transition zone. Earlier, Scott Smith et al. (1984) had raised the possibility of some diamonds being derived from the lower mantle. The evidence for such lower-mantle diamonds was subsequently established through the discovery of inclusions of ferropericlase coexisting with $MgSiO_3$, $CaSiO_3$, SiO_2, and a tetragonal almandine-pyrope phase from the Brazilian mine in Sao Luiz (Harte and Harris, 1994; Harte et al., 1999). Comprehensive reviews are available for the two main sources of lower-mantle diamonds at Sao Luiz (Harte et al., 1999) and Kankan, Guinea (Stachel et al., 2000).

Based on the $P-T$ regime of the Earth's mantle, a number of phase transitions and chemical reactions occur in the asthenosphere (>200 km) and transition zone (410–660 km). Along an adiabatic $P-T$ gradient, α-olivine converts into the spinel-like β-phase (wadsleyite) over a narrow depth interval, at about 410 km. Deeper, it further converts into the spinel-structured γ-phase (ringwoodite) at about 520 km (Ringwood, 1956; Ringwood and Major, 1966; Agee, 1998 and references therein). With depth, solid solution between ortho- and clinopyroxene increases before Ca-poor pyroxene converts from orthorhombic to monoclinic symmetry below 300 km (Pacolo and Gasparik, 1990; Woodland, 1998). Pyroxene also becomes soluble in the garnet structure with increasing depth (Ringwood, 1967). This corresponds to simultaneous accommodation of tetra- and divalent cations on the octahedral garnet sites, resulting in silica excess (Ringwood, 1967; Akaogi and Akimoto, 1977; Irifune, 1987). A second type of pressure-dependent substitution occurs with depth, which enables accommodation of the Na content of omphacitic clinopyroxene in majoritic garnet. The solution of pyroxene in garnet begins at 250 km and is completed at about 450 km for a pyrolite bulk composition (Irifune, 1987). At a depth exceeding 550–600 km, majorite garnet starts to exsolve CaSi perovskite (Irifune and Ringwood, 1987; Wood, 2000), which reverses the increasing accommodation of Si on the octahedral sites in garnets. In basaltic bulk compositions, the dissolution of clinopyroxene in garnet produces a free SiO_2 phase (stishovite) at a depth exceeding 300 km (Irifune and Ringwood, 1987); for harzburgitic bulk composition, stishovite results from the breakdown of pyroxene into ringwoodite (γ-phase) and majorite garnet (Liu, 1976; Ringwood, 1991).

Due to these phase transformations and reactions, the asthenosphere and transition zone may be expected to form source regions that are mineralogically sufficiently different from lithospheric reservoirs, to be recognized in the mineral inclusions of deep-seated diamonds. However, single-crystal, X-ray diffraction studies on inclusions in such deep-seated diamonds (Harte et al., 1999; Joswig et al., 1999; Stachel et al., 2000; Kaminsky et al., 2001) showed that high-pressure polymorphs are eliminated by isochemical phase transitions. Inferred primary stishovite occurs as coesite; former ringwoodite and wadsleyite have reversed to α-olivine; and $MgSiO_3$ compositions are invariably orthorhombic. These observations would appear to indicate that ascent velocities in the mantle are slow compared to the speed of isochemical phase transitions. The recognition of former high-pressure polymorphs based on chemical characteristics has also not been successful so far (see review by Stachel, 2001). The most important observations from these studies, however, are that at least some of the diamond inclusions are of lower-mantle origin, and inclusions in diamonds from the asthenosphere and transition zone are almost exclusively of eclogitic paragenesis. This may have important bearings on the origin of eclogites and E-type inclusions in diamonds.

5.3. Abundances of P-type versus E-type diamonds

Eclogites are one of the major host rocks for diamonds, and E-type diamonds are said to account for an estimated 20–25% of total worldwide diamond production. However, it must be noted that most diamonds that we have investigated from eclogite xenoliths are relatively inclusion free. Thus, there could be a sampling bias towards peridotitic-diamond heritage, and this issue is worth exploring further. Indeed, if the diamonds originating in peridotites have a greater tendency for mineral inclusions compared to eclogitic diamonds, sampling of the same number of diamonds from diamondiferous eclogites and peridotites may well indicate that the P-type diamonds were more numerous. Inasmuch as only about 10% of all diamonds contain identifiable mineral inclusions, it is entirely possible that there is an inherent fallacy in the gross assumption that the statistics from this 10% population with inclusions can be extrapolated to a generalization that includes the remaining 90%.

5.4. Syngenesis

In diamond literature, there is a general axiom that mineral inclusions in diamonds are syngenetic with their host diamonds. Virtually all papers on diamond inclusions start with the major assumption that the diamonds and their mineral inclusions are syngenetic—i.e., formed simultaneously under the same $P-T-X$ conditions and by the same genesis—the requirement of formation of the inclusion and its host diamond at exactly the same instant and by the same genetic process. However, supporting evidence for this assumption is rarely presented. We have recently presented evidence (Taylor L.A. et al., 2003a, c) that questions and negates syngenesis for some types of diamond inclusions, considered by most to be syngenetic. Below, we will address the criteria for establishing the syngenesis of diamond inclusions, especially since the validity of these criteria for the synchronous, syngenetic cocrystallization nature of diamond inclusions is at the very foundation of diamond inclusion studies. Several reviews of the criteria for "syngenesis" of diamonds and their mineral inclusions have been written over the years, e.g., Sobolev (1974, 1977), Harris (1979), and Pearson and Shirey (1999). We will summarize these discussions, especially as it applies to the modern usage of this term.

X-ray diffraction studies of diamonds relative to their mineral inclusions, dating back to the 1950s, demonstrated that, in many cases, the principal surfaces of the inclusions are closely related to {1 0 0} and/or {1 1 1} planes of the diamonds. Mitchell and Giardini (1953) examined olivine inclusions in diamonds and found that the olivines had a specific crystallographic orientation. That is, the {1 0 1} and {0 1 0} planes of the olivine are aligned approximately parallel to the {1 1 1} of the diamond. This apparent epitactic relationship was confirmed and expanded upon by Harris (1968), who confirmed that atomic orientations between inclusions and their host diamonds, although not common phenomena, were observed in many instances. Indeed, this similarity between crystallographic orientations of the inclusion and its host demonstrates an epitactic relationship (Harris, 1979). Pearson and Shirey (1999) summarily stated that "the epitaxial relationship between inclusions and host diamond, as revealed by X-ray studies, is considered as powerful evidence in favor of the syngenetic origin of many inclusions in diamonds."

Other criteria for syngenesis of diamond inclusions come from studies of the morphologies of the mineral inclusions. Sobolev (1968), Sobolev N.V. et al. (1969, 1970, 1972), and Sobolev V.S. et al. (1972), based on the octahedral forms of silicate inclusions, felt that diamond had imposed its growth morphology upon these inclusions. Harris (1979) concluded that the inclusion morphology developed as a result of the greater "form energy" of diamond, thereby imposing its morphology upon inclusion, during mutual growth. The morphology of the inclusion was regarded as indicative of an approach to textural equilibrium during the simultaneous growth of both host and inclusion, with the inclusion nucleating on the octahedral or cubic plane of the diamond substrate (e.g., Harris and Gurney, 1979). For example, cubo-octahedral morphology is a common feature of olivine and enstatite inclusions, even though they have orthorhombic crystal structures. In fact, olivine with its own orthorhombic morphology is rare as an inclusion (Sobolev, 1977). Likewise, the common dodecahedral morphology of "normal" garnet is seldom observed for garnet inclusions; instead, the cubo-octahedral morphology that has been imposed by the diamond is normally present. This is the rarest of all forms ever observed in non-inclusion garnet (Goldschmidt, 1897).

In practice, the most common criterion for syngenesis is the imposition of the morphology of the host diamond on the inclusion (e.g., Meyer, 1985, 1987). Pearson and Shirey (1999) stated that the synchronous growth hypothesis seems the simplest and best explanation for the production of oriented, faceted inclusions in diamond. In summary of their review on the various criteria for syngenetic diamond inclusions, they felt that a substantial weight of evidence is in favor of the syngenetic growth of most diamond inclusions. We do not completely share this opinion, however.

This leaves open the evaluation of criteria for the recognition of protogenetic inclusions; however, little has been published on this subject. Meyer (1987) felt that inclusions that existed prior to their encapsulation in the diamonds typically should have either irregular morphologies or euhedral morphologies that are entirely consistent with the crystal structure of the mineral. Thus, if euhedral, a protogenetic garnet inclusion should have a dodecahedral form, whereas a syngenetic garnet inclusion should have an octahedral morphology. The above discussion on inclusions has emphasized that the most commonly used criteria for the identification of syngenetic mineral inclusions (e.g., Sobolev V.S. et al., 1972; Sobolev, 1977; Bakumenko et al., 1984; Meyer, 1987; Pearson and Shirey, 1999) is the imposition of the morphology of the host diamond on the inclusion, with the added epitactic relationship of the two.

Taylor L.A. et al. (2003a) have presented an example of harzburgitic garnets in diamonds where the cubo-octahedral morphology of the garnets would seem to indicate a syngenetic association for the diamond and the garnet (Fig. 19D), based upon the criteria of most diamond experts (e.g., Harris, 1968; Sobolev, 1977; Harris, 1979; Meyer, 1987; Pearson and Shirey, 1999). However, based on the sinusoidal REE pattern of the harzburgitic garnets (Fig. 20A), Taylor L.A. et al. (2003a), in

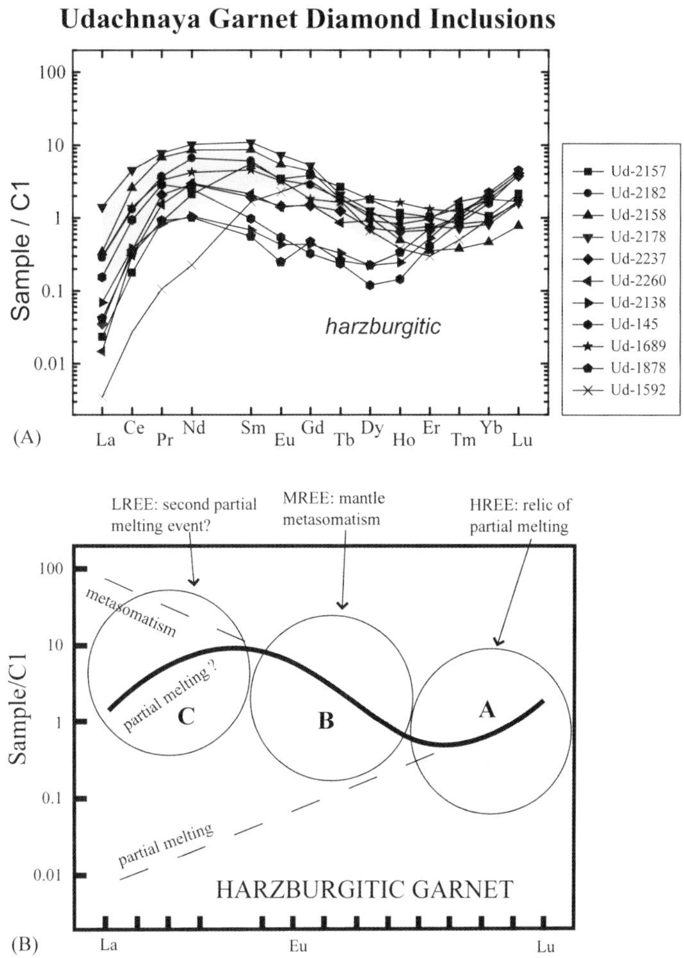

Fig. 20. Sinusoidal rare-earth-element patterns of harzburgitic garnet diamond inclusions. (A) Garnet inclusions from Udachnaya; the shaded area represents the data for all Udachnaya harzburgitic garnets. (B) Schematic diagram illustrating a scenario by Taylor L.A. et al. (2003a), wherein the harzburgite first underwent a partial-melting event (A), followed by metasomatism (B), and finally LREE depletion caused by another process (C), thus illustrating the complex history for harzburgitic garnets. Such patterns are indicative of a complex history of the garnet, prior to its incorporation into the diamond.

agreement with most other petrologists, concluded that the typical REE patterns for harzburgitic garnets worldwide are indicative of a complex genetic history for the garnets *prior* to diamond encapsulation. The REE patterns relate to a scenario in which an ancient melting event initially resulted in the LREE-depletion signature of the garnet, followed by mantle metasomatism, culminating in a possible late-stage, LREE-depletion episode (Fig. 20A–C).

Taylor L.A. et al. (2003a) concluded that the REE evidence casts severe doubt on the morphologic criterion. They stated a new paradigm wherein "most, if not all, harzburgitic garnet inclusions in diamonds, worldwide, are not syngenetic with their diamond hosts." In fact, it is possible that many diamond inclusions do not have the same genesis as that which formed the diamonds. Specifically, because garnet and diamond have such different compositions, it is difficult to envision a fluid that would be capable of producing both minerals. However, the natural follow-up query is, if the inclusions are protogenetic, does this affect the synchroneity of the diamonds and their inclusions?

The ages of diamonds are "inferred" from the age determinations of their inclusions. But, if the inclusions pre-date the diamond encapsulation, the age of the inclusion may give an erroneous diamond age. Fortunately, the blocking (closure) temperatures of most radiogenic systems are well below that at the time of the diamond growth. Thus, it is the encapsulation of the mineral that effectively isolates it from its radiogenic reservoir and permits its "radioactive clock" to start. However, there are probably exceptions to this scenario, most likely for P-type inclusions. In fact, these warnings find support from the recent study of Spetsius et al. (2002), in which it was demonstrated that sulfide mineral inclusions in kimberlitic zircons do not relate to each other—i.e., they are not synchronous. Therefore, the general axiom in the diamond community that inclusions are syngenetic with their host diamonds—i.e., synchronous and same genesis—needs careful reconsideration.

5.5. Chemical compositions of mineral inclusions in diamonds

Mineral inclusions in diamonds exhibit an extensive range of compositions in both P- and E-type diamonds. The sulfide phases, consisting of monosulfide solid solution, pyrrhotite, pentlandite, and chalcopyrite are probably the most common and have also been studied in reasonable detail (e.g., Yefimova et al., 1983; Bulanova et al., 1990, 1996). Garnet and clinopyroxene constitute the majority of silicate inclusions in diamonds (e.g., Meyer and Boyd, 1972; Sobolev, 1977; Sobolev N.V. et al., 1997; Moore and Gurney, 1989; Harris, 1992; Griffin et al., 1993; Taylor L.A. et al., 1996; Kopylova et al., 1997; Stachel and Harris, 1997; Sobolev N.V. et al., 1999b; Viljoen et al., 1999). In addition to these, olivine, chromite, and orthopyroxene constitute a substantial proportion of P-type inclusions and have characteristic geochemical features.

The majority of inclusion studies have been performed on minerals extracted by burning or crushing the host diamond. However, the stratigraphic relationships between the inclusions and the host diamonds are not preserved and the possible presence of minute cracks in the diamonds mostly goes unrecorded. Although this is a less work-intensive method, it is difficult to make any correlation between the inclusion composition and other chemical characteristics of different growth zones of the diamond, such as carbon and nitrogen isotopes, or to further examine the possibility of an "open-system" behavior due to cracks.

Recently, some workers have adopted a more laborious process of making polished plates of diamonds to expose the mineral inclusion(s) on the surface, and to further perform geochemical analyses *in situ* on the diamond host and the inclusions (e.g., Taylor L.A. et al., 1998, 2000; Bulanova et al., 1999; Promprated et al., 2004). In an attempt to correlate the geochemical variation among multiple diamond inclusions with that of the xenolith host minerals, Taylor L.A. et al. (2000) performed the first comprehensive "pull apart" of a diamondiferous eclogite xenolith from Yakutia. A feature of this diamond-bearing eclogite xenolith was that the eight of the total 30 diamonds contained optically visible mineral inclusions (20–100 μm). In total, 28 inclusions were recovered from these eight diamonds, among which three were garnets and 25 were clinopyroxene and were subjected to geochemical analyses.

5.5.1. Major-element compositions

Sobolev (1974, 1977) proposed that P-type garnet inclusions be subdivided into two main types based on their CaO and Cr_2O_3 contents (Fig. 21). These two divisions correspond to two main mantle rock types, lherzolites and harzburgites. Most garnet diamond inclusions are harzburgitic, but this division is not sharp but transitional. Therefore, instead of a line, there is a narrow zone separating the two areas. In essence, this zone in Fig. 21 represents the Ca saturation in garnet

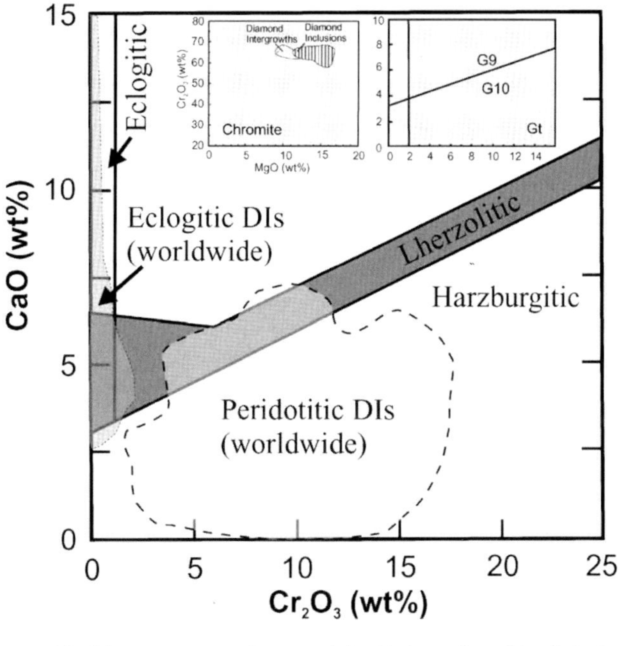

Fig. 21. Plot for peridotitic garnets, a scheme originally introduced by Sobolev (1977), with a restricted field for eclogitic garnets, which have $<1\%$ Cr_2O_3. Envelope for P-type inclusions is according to Griffin et al. (1993). The inset in the upper right depicts the regions for G9 and G10 "indicator garnets" used in diamond exploration, according to Gurney (1984). The other inset gives the field in Cr_2O_3–MgO space for chromite as another "indicator mineral" for diamonds, modified after Fipke et al. (1995).

inclusions. Those data points lying below the narrow band are subcalcic garnets of harzburgitic paragenesis, whereas those that plot above the band are lherzolitic. Usually, the harzburgitic garnets are chrome-rich. Sometimes, when the distinction is not sharp, the Mg♯ of associated olivine inclusions is used to assess the nature of the ultramafic paragenesis. Harzburgitic olivines usually have an Mg♯ of about 94, whereas lherzolitic olivines have an Mg♯ slightly less, around 92. Similarly, enstatites have Mg♯ of 96 and 94, respectively (Harris, 1992). The presence of rare chrome-diopside with ~2 wt% Cr_2O_3 is specific to the lherzolitic diamonds. Chromites, associated with both paragenetic types usually have >60 wt% of Cr_2O_3 (Harris, 1992), and it is practically not possible to classify them separately, as to harzburgitic versus lherzolitic.

E-type inclusions comprising omphacitic clinopyroxene and pyrope-almandine garnets are classified into different groups based on their MgO–Na_2O and CaO–MgO–FeO contents, respectively. As explained in Section 2.1, the three-fold classification (Group A, B, C; Taylor and Neal, 1989) of eclogitic clinopyroxenes depends mainly upon their MgO and Na_2O contents (Fig. 22A). This classification is consistent with the garnet chemical classification scheme of eclogitic garnets by Coleman et al. (1965), albeit without any petrogenetic significance. Clinopyroxene inclusions from Group A eclogitic diamonds have highest MgO and lowest Na_2O, whereas those from Group C eclogitic diamonds have the highest Na_2O and lowest MgO contents. The Group B clinopyroxene inclusions have intermediate compositions. Similarly, different groups of garnet diamond inclusions are characterized by increasing grossular and decreasing pyrope contents from Group A to Group C, respectively (Fig. 22B). In general, the eclogitic garnet inclusions are more Ca- and Fe-rich compared to their peridotitic counterparts. Because of this, the eclogitic garnet inclusions show broad compositional variations in terms of Ca–Mg–Fe contents as seen in the Siberian data (Fig. 22B). Eclogitic clinopyroxenes, when plotted on a similar diagram to that of the garnets, show the Ca- and Mg-rich nature of peridotitic pyroxenes from relatively Fe-rich compositions of E-types. Based on very high Ca contents in pyroxene inclusions from south-eastern Australia (Sobolev et al., 1989), Meyer (1987) proposed the possible existence of a third distinct diamond paragenesis. Additional important minor oxides in the eclogitic inclusion paragenesis are Na_2O (>0.10 wt%) in garnets and K_2O in the clinopyroxenes (up to 1.4 wt%) (Sobolev and Lavrent'yev, 1971; Sobolev, 1977).

The inclusion compositions in diamonds from the U-51 eclogite xenolith (Taylor L.A. et al., 2000) showed that clinopyroxenes are significantly different in chemical composition from those in the host eclogite (e.g., Fig. 23). In spite of the fact that all of these inclusions are in diamonds from only a single eclogite xenolith, the major-element compositions of the inclusions vary substantially. In fact, the multiple inclusions from a single diamond vary significantly as well. Most clinopyroxene inclusions from individual U-51 diamonds show generally constant major-element compositions, except in a few cases. In one instance, two clinopyroxene inclusions from a single diamond had contrasting compositions for almost all elements. Considering all clinopyroxene inclusions from eclogite U-51 as a group, Fig. 23 shows that TiO_2 contents are 0.35–1.06 wt%, with most values higher than the

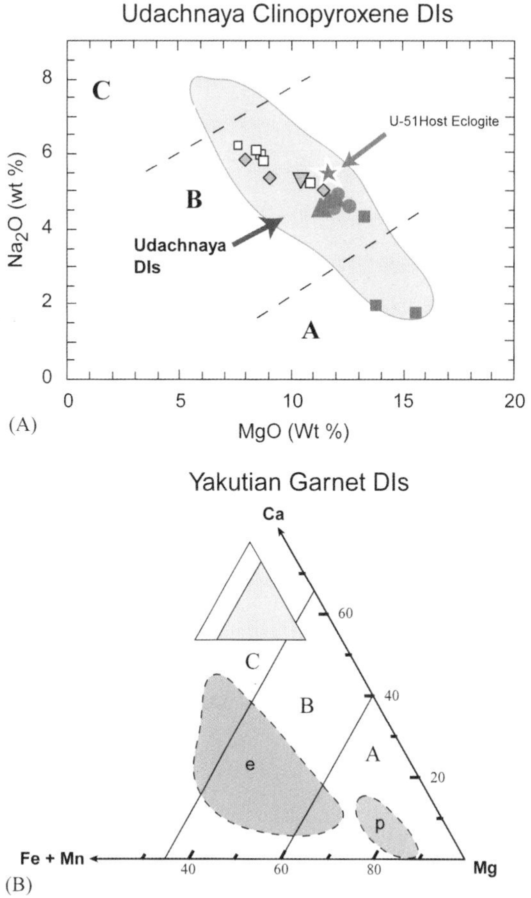

Fig. 22. (A) Clinopyroxene inclusions in diamonds from Udachnaya eclogite, U-51 (Taylor L.A. et al., 2000). Each symbol is for multiple inclusions in a single diamond. Note the spread in compositions of the clinopyroxenes, most of which are even different from those in the host eclogite xenolith. The shaded envelope is for all clinopyroxene inclusions from Udachnaya eclogitic diamonds. The A–C boundaries are for eclogite classification (Taylor and Neal, 1989). (B) Garnet plot with envelopes for diamond inclusions for all of Yakutia. Data from Sobolev et al. (1998a, b). The (A–C) regions are for the chemical classification for eclogites, as modified after Coleman et al. (1965), subsequently adopted as the classification for mantle eclogites by Taylor and Neal (1989).

0.47 wt% TiO_2 of clinopyroxene in the host eclogite, as well as in other diamondiferous eclogites from Udachnaya. The K_2O contents in these clinopyroxene inclusions vary significantly from 0.17 to 0.73 wt%, with an exception of 0.05 wt% in one sample (D-L-DI-30). In contrast, the highest K_2O content in clinopyroxene from host eclogite xenoliths from the Udachnaya kimberlite pipe is only about 0.3 wt%

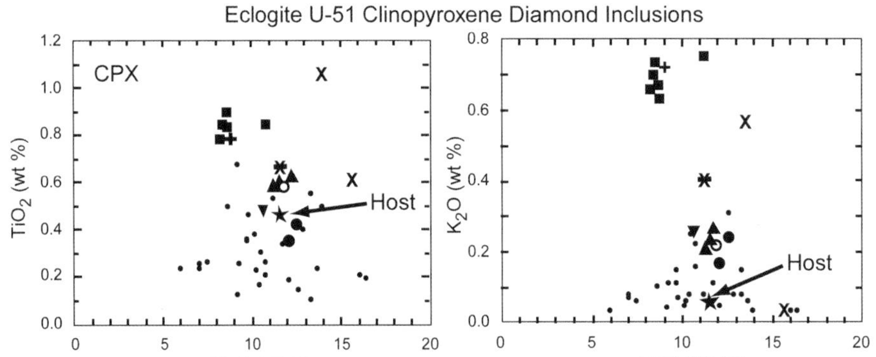

(A) (B)

Fig. 23. (A,B) Clinopyroxene inclusions in diamonds from Udachnaya eclogite, U-51 (Taylor L.A. et al., 2000). Each large symbol is for multiple inclusions in a single diamond. The small dots are for clinopyroxenes from other Udachnaya eclogites (Sobolev et al., 1994). Note the variations in multiple inclusions from the same diamond.

(Sobolev et al., 1991). The considerably higher K_2O in clinopyroxene inclusions compared to those in the host U-51 eclogite are an important feature of this xenolith. Similar findings were also reported by Taylor L.A. et al. (1996, 1998); and Taylor (1998). These differences in K_2O contents of the clinopyroxenes are indicative of re-equilibration of the host eclogite clinopyroxenes to lower pressures (Harlow, 1997).

Most sulfide inclusions in diamonds indicate a subsolidus re-equilibration from an original single phase of monosulfide solid solution (Harris, 1992), which separated chalcopyrite solid solution and nickel-bearing pyrrhotite at about 970°C, before further division into pyrrhotite and pentlandite at temperatures below 610°C (Craig and Scott, 1974). Yefimova et al. (1983) showed that peridotitic sulfide inclusions have much higher Ni compared to eclogitic inclusions in Russian diamonds, averaging about 23 wt%, in the former, and this feature can be used to distinguish between the two paragenetic associations of sulfide inclusions (e.g., Bulanova, 1995). However, such studies are lacking from diamonds on a worldwide basis.

5.5.2. Trace-element compositions

Trace elements, particularly REEs, in minerals from mantle xenoliths are sensitive to conditions during their formation, and can effectively provide signatures of these processes. This is also true for mineral inclusions in diamonds.

Griffin et al. (1992) measured Ti, Zn, Ga, Zr and Y in subcalcic garnet inclusions in southern African diamonds and found that these elements decrease with decreasing Ca contents relative to Cr. This feature was interpreted as being indicative of increasing depletion by removal of silicate melt. However, the same garnets showed enrichment in Sr and LREE. Coupled with the contrasting depletion in Zr, Ti, and Y, this is inconsistent with a single-stage depletion event. Griffin et al. (1992) suggested that the extreme depletion of Ca in most subcalcic garnets and their coexisting olivines and the Sr enrichment was a result of metasomatism by a carbonatitic fluid following depletion. Similarly, in a study of mineral inclusions from Akwatian diamonds, Stachel and Harris (1997) found extremely high Ni contents in harzburgitic and lherzolitic olivines, relatively low Mg/Fe ratios for harzburgitic and lherzolitic olivines and orthopyroxenes, and a higher mean Cr content in chromites. The inconsistency between the low Mg/Fe ratios and the highly refractory compatible trace-element signature was explained by refertilization of a previously depleted source, similar to the metasomatic re-enrichment of deformed, Fe–Ti-rich hot peridotites (Stachel and Harris, 1997). In a comprehensive geochemical study of mineral inclusions in Kankan diamonds from Guinea, Stachel et al. (2000) showed that lherzolitic garnets have extreme Cr_2O_3 contents (up to 17 wt%). They proposed that this anomalous feature could be explained if the garnets experienced a previous melt-extraction event in the spinel stability field, followed by metasomatic re-enrichment. Other trace-element signatures of these garnets indicated that there were at least two such metasomatic events. Stachel et al. (2002) postulated that the first event was associated with LREE and Sr enrichment, whereas the second one introduced mainly MREE-HREE and other HFSE (Ti, Y, Zr, Hf). The Ti- and HFSE-poor nature of stage 1 indicates a C–H–O-rich fluid or carbonatitic melt; the HFSE-rich nature of stage 2 favors a silicate melt as the enriching media.

Shimizu et al. (1997) demonstrated chemical heterogeneities in terms of REE and trace elements (Ti, V, Sr, Y, Zr) of peridotitic garnet inclusions from several Siberian diamonds. Based on diffusion kinetics of Sr in garnet, they suggested that the diamond formation and metasomatism of the continental lithosphere beneath Siberia were events preceded immediately before the kimberlite emplacement. However, nitrogen aggregation studies on Siberian diamonds provide evidence for substantial periods of mantle residence for most Siberian diamonds and are in agreement with the radiometric ages of diamonds (Richardson and Harris, 1997). Thus, strong Sr zonations in garnet inclusions observed by Shimizu et al. (1997) could only be explained if the diamond had cracks such that inclusions behaved as open system and interacted with the kimberlitic fluid on their way to the surface.

Trace-element studies of eclogitic inclusions have revealed involvement of much more complex processes in their formation than in the case of peridotitic inclusions (e.g., Taylor L.A. et al., 1996, 1999, 2000; Sobolev et al., 1998a, b). Also, all super-deep inclusions are solely of eclogitic paragenesis (Stachel, 2001). Taylor L.A. et al. (1996) did the first ion-probe trace-element analyses of clinopyroxene and garnet inclusions, both included within single Siberian diamonds and from the eclogite host xenoliths. They reported non-systematic variations and diversity of relationships between diamond inclusions and eclogite hosts in terms of their chemical compositions. Sobolev et al. (1998a) reported extreme chemical diversity among 35 garnet and five clinopyroxene inclusions from one single diamond from Mir kimberlite. These observations reflect the complex growth history of eclogitic diamonds, at least in Siberia. Studies on African eclogitic diamond inclusions have shown similar results. Stachel et al. (2000) reported that the eclogitic inclusions in Kankan diamonds were derived from large depth intervals ranging from the lithosphere through the asthenosphere into the transition zone. They also found negative Eu anomalies in garnet and clinopyroxene diamond inclusions from both the lithospheric and transition zone, interpreted to be evidence for involvement of subducted oceanic crust in the origin of these diamonds. Based upon Ca and REE contents of minerals, the inclusions were divided into two groups, each corresponding to unique lithospheric source regions. In a recent trace-element study of eclogitic and websteritic inclusions in diamonds from the Venetia mine in South Africa, Aulbach et al. (2002) found significant variations in REEs, Sr, and Zr between the two groups. Both positive and negative Eu anomalies were also found in garnets and clinopyroxenes from both parageneses, which were interpreted to indicate feldspar fractionation and accumulation in a magmatic precursor, suggesting subducted oceanic crust as a common protolith.

In the study of numerous diamonds from an Udachnaya eclogite (U-51; Keller et al., 1999; Taylor L.A. et al., 2000), 15 of the larger ($>30\,\mu m$) clinopyroxene inclusions were analyzed for REEs, and other trace elements (Fig. 24). Eight clinopyroxene inclusions from one diamond (W) are similar in REE concentrations with concave-upward patterns. They are uniformly enriched in REEs compared to clinopyroxene in the host eclogite, as well as in most other diamondiferous eclogites from Udachnaya (Jerde et al., 1993a, b; Snyder et al., 1997, 1998a, b). Similar REE patterns were observed from three clinopyroxene inclusions in another diamond (L); however, from grain to grain, their concentrations are significantly different, with La_n ranging from 1.2 to 26.9 and Yb_n from 1.1 to 4.7 (Fig. 24). These three clinopyroxene composition inclusions from the same diamond (Fig. 24) cover almost the entire range of clinopyroxene composition from all diamondiferous eclogite xenoliths from Udachnaya. Also, except for one sample, the other two have much higher REEs than clinopyroxene in the host eclogite (Fig. 24). Additionally, strong +Eu anomalies were detected in these three inclusions, definite crustal signatures by way of plagioclase involvement. Similar REE features were also observed from clinopyroxene inclusions in other diamonds. As documented above for the major elements, concentrations of REEs in the clinopyroxene inclusions from this single eclogite xenolith vary significantly between diamonds, as well as between multiple inclusions in single diamonds.

In some cases, multiple-inclusions of one phase from a single diamond are chemically identical (Meyer, 1987), which has been utilized to argue that diamonds generally formed in a stable $P-T-X$ environment. However, many exceptions to this have been reported. For example, significant chemical variations among multiple-garnets from single diamonds were discovered in eclogitic inclusions from west Australian diamonds (Griffin et al., 1988). As highlighted above, extreme variations for 35 garnet inclusions from a single diamond were also reported by Sobolev et al., 1998a. Actually, the major- and trace-element compositions of inclusions from this single diamond range over almost the entire compositional range of mantle eclogite clinopyroxenes from Yakutia (Sobolev et al., 1998a) (Fig. 25). Furthermore, in some rare instances, both peridotitic and eclogitic inclusions have been found in single diamond hosts (Prinz et al., 1975; Hall and Smith, 1985; Otter and Gurney, 1989; Moore and Gurney, 1989; Wang, 1998).

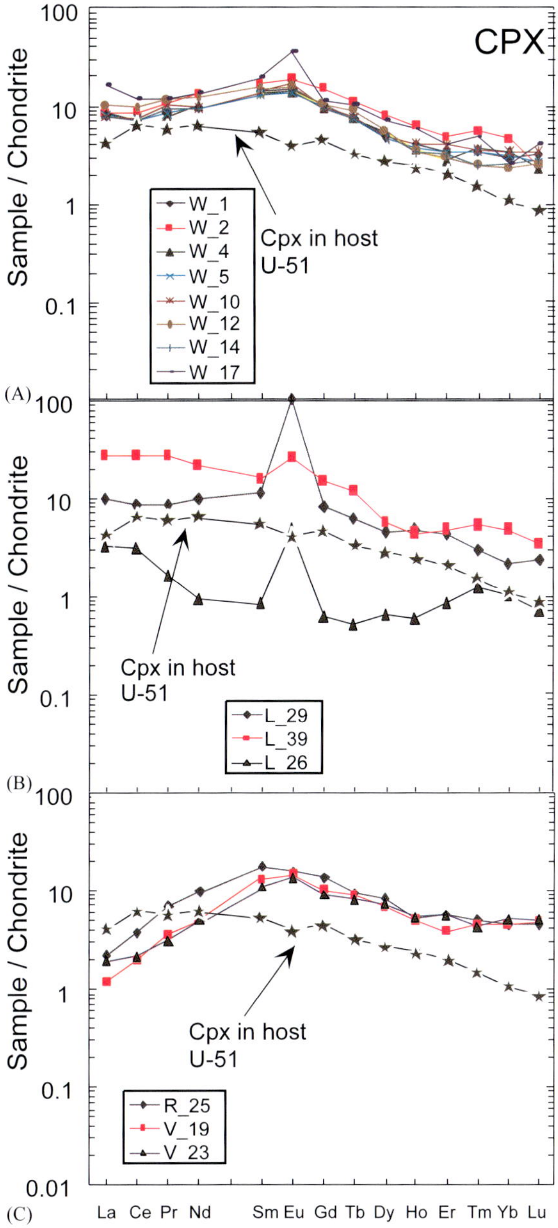

Fig. 24. Rare-earth-element patterns for clinopyroxene inclusions in diamonds from Udachnaya eclogite, U-51, modified after Taylor L.A. et al. (2000). (A,B) are for multiple inclusions in single diamonds, W and L, respectively. Note the +Eu anomalies. (C) is for multiple inclusions in diamond V and a similar inclusion composition in diamond R. Shaded area represents envelope for all clinopyroxenes from Udachnaya eclogites from Jerde et al. (1993b).

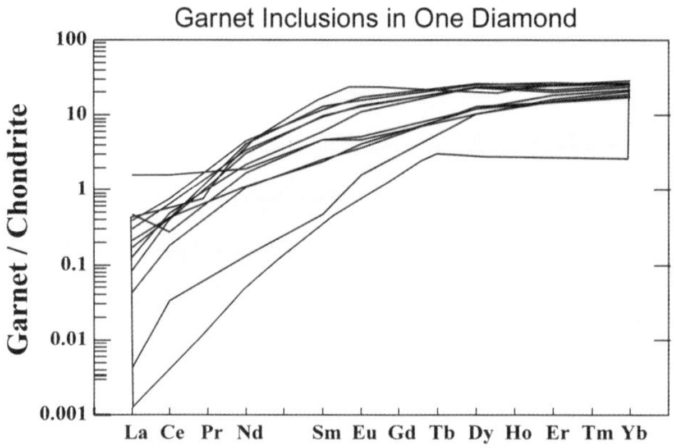

Fig. 25. Rare-earth-element patterns for 10 of the 35 garnets in a single diamond from the Mir Pipe. Note that the spread of patterns from garnets in this one diamond virtually covers the entire region for all garnets from Yakutia. Data from Sobolev et al. (1998a).

5.5.3. Chemical diversity of diamond inclusions in a single eclogite

All mineral inclusions from diamonds in the U-51 xenolith are of eclogitic heritage; no inclusions of ultramafic paragenesis were observed. However, the inclusions do exhibit very complex compositional features. Four diamonds of the 30 from this xenolith contain multiple inclusions. As mentioned earlier, significant compositional variations were observed between clinopyroxene inclusions in single diamonds, including the first reported presence of +Eu anomalies. Almost all clinopyroxene inclusions are richer in K_2O than the clinopyroxene in the host eclogite, a function of re-equilibration of the host clinopyroxene to lower pressures (Sobolev V.S. et al., 1972; Harlow and Veblen, 1991; Harlow, 1997; Taylor L.A. et al., 1998; Sobolev N.V. et al., 1998b, 1999a). Large variations in REE concentrations also occur in the clinopyroxenes in diamond "L". In contrast, clinopyroxenes in other diamonds are homogeneous, in both major and REEs.

A comparison of data for inclusions in the entire population of diamonds in the U-51 eclogite revealed that clinopyroxenes exhibit an exceptionally large range in major- and trace-element contents (Figs. 22–24). The three garnet inclusions have consistently lower Mg# (53.8–63.5), as compared with 64.9 of the host garnet. Even concentrations of REEs in clinopyroxene inclusions in diamonds "W" and "L" are distinctly higher than those in the host clinopyroxene (Fig. 24). However, LREEs in clinopyroxene inclusions from diamonds "R" and "V" are lower than those of the host (Fig. 24).

Inclusions in diamonds from a single eclogite xenolith can have complex compositional features such that it is not obvious as to how they relate to the conditions of the actual diamond formation (Taylor L.A. et al., 2000). Fig. 24 shows that the composition of the clinopyroxene inclusions from this one eclogite xenolith cover more than two-thirds of the compositional range of all eclogite clinopyroxenes in xenoliths from the entire Udachnaya kimberlite pipe.

If it is assumed that diamond inclusions are representative of the host eclogite at different times, the range in the chemical composition of U-51 diamond inclusions is tremendous. In any case, the large variations in compositions of the inclusions cannot be easily explained as a result of simple partial melting and/or metasomatism of the host eclogite. Instead, the amount and degree of metasomatism must have been abnormally great, or varied over time. We suggest a possible scenario whereby episodically a CO_2-rich melt/fluid metasomatically invaded an eclogite lithology, and at suitable physical conditions and oxygen fugacity, crystallization of diamond was initiated. Partial melting of the eclogite lithology may have occurred at the same time, because infiltration of the fluid/melt would lead to a substantial decrease in the solidus temperature. With changes in the $P-T$ and oxygen-fugacity conditions, the chemical compositions of the solidus phases could vary greatly, particularly if one also allows for the occurrence of local disequilibrium. The diamonds could hesitate in their growth, be partially resorbed, only to continue with additional growth later, all due to changes in $P-T-X$ conditions. As a result, minerals captured by growing diamonds at various stages of these processes may exhibit large compositional variations.

Garnet and clinopyroxene inclusions in a single diamond could be incorporated in two absolutely different stages in the crystallization history of a diamond. This scenario, where the inclusions are captured at different periods of the continuing process of chemical change, may explain why the inclusion compositions are not truly representative of the host eclogite at all. Indeed, the inclusions may be "snapshots" of the eclogite during its evolution. Subsequent recrystallization of the eclogite after the formation of the diamonds may have largely erased any chemical heterogeneities in the host eclogite. However, the large chemical diversity of the inclusions was perfectly retained, due to encapsulation in the host diamonds.

5.6. Oxygen isotopes

There are relatively few published $\delta^{18}O$ values on eclogitic diamond inclusions compared to the data on host eclogite garnets and clinopyroxenes. In fact, there are but two studies that have presented $\delta^{18}O$ data on eclogitic mineral inclusions in diamonds (10 eclogitic garnets in Finch diamonds, Lowry et al. (1999); coesite inclusions in Gauniamo diamonds, Schulze et al. (2003)). In both studies, the ranges in $\delta^{18}O$ values are large (5.7–8‰ and 10.2–16.9‰, respectively), presenting evidence for the subduction hypothesis for the origin of eclogitic mineral inclusions as well as the xenoliths (Fig. 26). In addition, Anand et al. (2004) presented oxygen-isotope data on five clinopyroxene and one garnet inclusion in diamonds from the U-51 eclogite. The overall variation in the $\delta^{18}O$ values in clinopyroxene inclusions are between 2.4‰ and 4.4‰, significantly lower than the mantle value of 5.5 ± 0.2‰ (Eiler, 2001). In contrast, the single garnet inclusion has a $\delta^{18}O$ value of 8.6‰, the highest oxygen-isotope value reported for an eclogitic garnet inclusion. In fact, this is only the second time that such a high $\delta^{18}O$ value has been found in any eclogitic garnet. Deines et al. (1991) found $\delta^{18}O$ values up to 9.2‰ in eclogitic garnets from Orapa. Such high $\delta^{18}O$ values in the eclogitic garnets are consistent with the

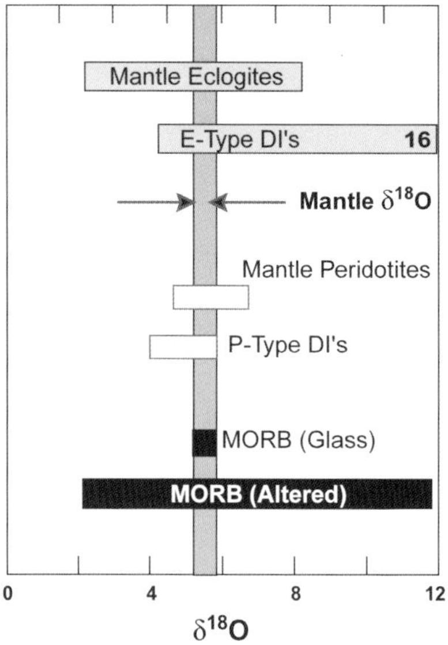

Fig. 26. Oxygen isotopes for eclogitic and peridotite inclusions in diamond from data compiled by Taylor L.A. et al. (2003d). The MORB (Mantle) range (5.4 ± 0.2‰) is from Eiler (2001). Note the non-mantle values for inclusions in both P- and E-type diamonds.

hypothesis that these are derived from recycled oceanic crust. The δ^{18}O ratios in clinopyroxene inclusions, however, define the lower range seen in eclogites. It is even more interesting to note that multiple clinopyroxene inclusions from the same diamonds show different δ^{18}O values (Anand et al., 2004). These observations highlight the variability in oxygen isotopic compositions of eclogitic inclusions, and are consistent with extreme variations in trace-element compositions of multiple inclusions in and between diamonds in diamondiferous eclogites.

5.7. Sulfur isotopes

Studies of sulfur isotopes in diamond inclusions present an attractive means for investigating the possible roles of crust–mantle interaction in the formation of diamonds (Tsai et al., 1979; Deines and Harris, 1995). Sulfur isotope ratios measured in most mantle or meteoritic samples exhibit minimal deviation from the meteoritic standard, whereas the ratios in crustal materials deviate by as much as ±70‰ (Ohmoto and Rye, 1979).

Mantle sulfides are thought to have sulfur-isotopic compositions similar to sulfides in meteorites, with $\delta^{34}S_{CDT}$ values close to zero (Faure, 1986). However, studies of sulfide inclusions in diamonds from African kimberlites revealed large variations in δ^{34}S (−11‰ to +14‰; Fig. 27), which were ascribed to non-uniform mantle sources, and to the incorporation into the mantle of subducted oceanic crust and sedimentary

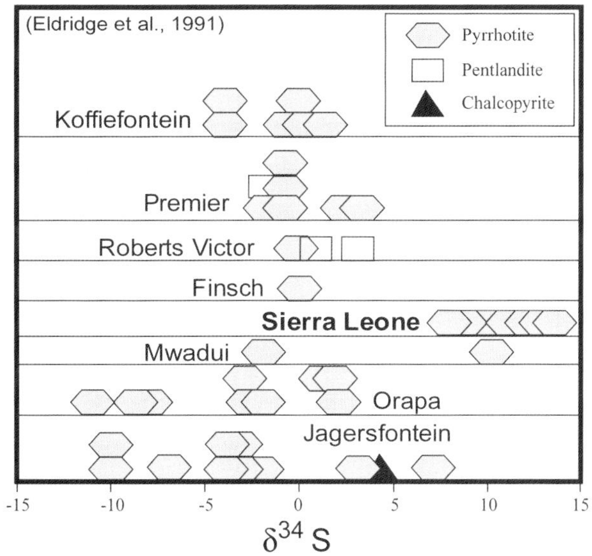

Fig. 27. Range of $\delta^{34}S$ for sulfide mineral inclusions in diamonds from Africa. Data from Eldridge et al. (1991).

rocks (e.g., Chaussidon et al., 1987; Eldridge et al., 1991). In contrast, Rudnick et al. (1993b) found much less variation in $\delta^{34}S$ for sulfide inclusions in Siberian diamonds (−4‰ to +5‰). They also analyzed multiple inclusions within diamonds and found no evidence for differences in sulfur-isotope compositions, although small variations may have escaped detection due to analytical uncertainty.

Ruzicka et al. (1999) studied inclusions from four Siberian diamonds (3 E-type and 1 P-type) and reported a limited range in $\delta^{34}S_{CDT}$ varying from ~−3‰ to +3‰ (Fig. 28), with a mean and standard deviation (among 21 analyses) of 0.0±1.3‰, implying a mantle source for both P- and E-type sulfide phases from a mantle-like reservoir below the Siberian platform. In addition, this entire spread of valves, albeit relatively limited compared to African data, among sulfide inclusions from four diamonds was covered by six sulfide inclusion compositions from one diamond (Mir-1702; Ruzicka et al., 1999). On the assumption that the sulfides had not undergone an "open-system' behavior, these data were interpreted as possibly due to partial isotopic exchange among different sulfide grains during quenching from high temperature. The overall data were also interpreted as indicating that the sulfide phases were possibly incorporated into the diamond from slightly different sulfur-isotopic sources in the mantle, with individual diamonds sampling different isotopic sources during their growth. This, in conjunction with sulfur isotopic studies by Rudnick et al. (1993b), seems to indicate that sulfides incorporated in Siberian diamonds were generally derived from S-bearing sources that were isotopically less variable than those from southern Africa.

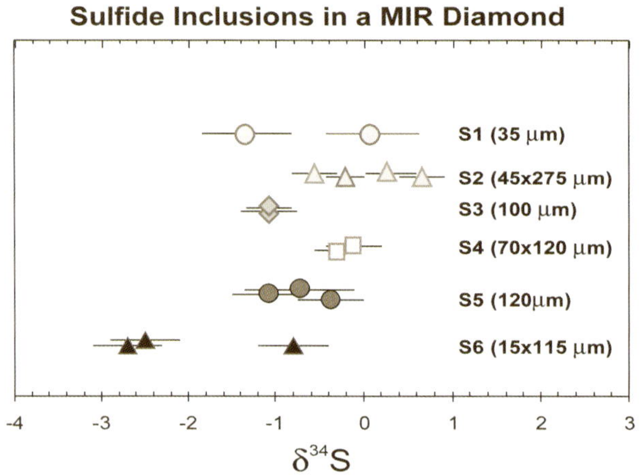

Fig. 28. Range of $\delta^{34}S$ for six sulfide mineral inclusions from a single diamond from the Mir pipe. Note the range in values between inclusions, even within individual sulfides. For a single grain, this range is indicative of "open-system" behavior for the sulfide inclusion. Data from Ruzicka et al. (1999).

5.8. Ages of diamonds

The close similarity of diamond inclusions and minerals in diamondiferous xenoliths strongly suggests that diamonds were formed within such rocks and definitely are not phenocrysts that grew in the erupting kimberlite. However, they may have formed in the mantle just prior to the kimberlite eruption or long before it. Diamonds, consisting of nearly pure carbon, do not provide direct information about their age. The closest age information can be obtained by using different nitrogen aggregation states in the diamond lattice, where nitrogen occurs as impurity and substitutes for carbon. However, the aggregation state of nitrogen is strongly dependent upon temperature and residence time in the mantle. Large uncertainties are associated with the accurate estimation of mantle temperature, and hence, temporal information from N-aggregation studies is not conclusive.

Other age constraints for diamonds are from the radiometric ages derived from mineral inclusions within the diamonds. Kramers (1979) reported the "model Pb" ages for sulfide inclusions in diamonds from Finsch and Kimberly in excess of 2 Ga, much older than the Cretaceous eruption age of the kimberlites. Richardson et al. (1984) reported Nd model ages of 3.3 Ga on harzburgitic garnet diamond inclusions from these two mines that clearly established the antiquity of such inclusions, and by association, of the host diamonds. Similar garnets from Udachnaya yielded an isochron age of 2010 ± 30 Ma (Richardson and Harris, 1997). Another old isochron age of 1930 ± 40 Ma was obtained for lherzolitic garnets and clinopyroxenes from Premier (Richardson et al., 1993). All of these ages are much older than the eruption ages of the host kimberlites. It should be mentioned that these analyses were all performed on composites collected from inclusions liberated from hundreds of diamonds, a necessity for diamond age determinations by Rb–Sr and Sm–Nd isotopic systematics.

Burgess et al. (1989) performed ^{40}Ar–^{39}Ar analyses and age determinations on individual eclogitic clinopyroxenes in diamond, which typically contain 0.1–0.8% K_2O. However, it was suspected that some of the Ar had collected between the clinopyroxene and the diamond interface (Burgess et al., 1992), such that it escaped when the clinopyroxene was exposed by the diamond polishing, prior to irradiation and Ar analysis, thereby yielding young ages. A procedure has been adapted recently (Burgess et al., 2003) wherein the clinopyroxene inclusion in the diamond is not opened prior to analysis, only while in the Ar-detector mass spectrometer. This is accomplished by effectively drilling a hole into the diamond down to the clinopyroxene with a UV laser. This technique holds promise for age determinations on individual clinopyroxene inclusions.

Four sulfides in a Siberian diamond yielded old model ages of 3.1–3.5 Ga, with no isochron relationship (Pearson et al., 1999). Two peridotitic sulfide inclusions in a single diamond from Koffiefontein yielded a very young age of 69 ± 30 Ma (Pearson et al., 1998b). This age is within error of the kimberlite eruption age (89 Ma). Older isochron and model ages were obtained for five eclogitic sulfide inclusions from Koffiefontein diamonds. However, if the bulk of the Re and Os of diamond host rocks is carried by isolated sulfides (e.g., Pearson et al., 1998a, b; Snyder et al., 1998b), then recrystallization and trapping of such sulfides during diamond formation should not affect their isotopic evolution. The system is disturbed only if diamond growth involves new addition of Os and/or Re or partial melting and fractionation of Re and Os between melt and sulfides. Diamond formation related to such recrystallization events may only be associated with minor disturbances of the Re–Os isotopic systems, and hence, in such cases, the Re–Os age of the sulfide inclusions yield a maximum age for the diamond. An important observation to remember with regard to sulfide inclusions is that they are virtually always associated with cracking of the diamond. It seems that the initially trapped Fe–Ni–Cu monosulfide solid solution, upon subsequent cooling, always exsolves to the three minerals, pyrrhotite, pentlandite, and chalcopyrite. This exsolution involves a substantial increase in ΔV, and coupled with the differential thermal coefficients between the diamond and sulfide, the diamond is caused to crack. It is essential that the sulfide inclusions to be used for age determinations be closely examined while in the diamonds, under cathodoluminescent conditions, for cracks to the surface of the diamonds that could have been the avenues for "open-system" behavior.

Shimizu and Sobolev (1995) studied garnet inclusions in diamonds from the Mir pipe and concluded that the Sr inhomogeneities in the garnets indicated a young age for that diamond, approximating the age of the kimberlite. However, because this inclusion was retrieved by burning of the diamond, it is possible that the presence of minute cracks in the diamond creating an avenue to the outside of the diamond was missed—i.e., the Sr zonation may have been a secondary effect due to metasomatism. However, it is just this kind of evidence that must be sought, albeit in garnets measured *in situ* on closely examined polished surfaces of diamond.

Navon (1999) and Pearson and Shirey (1999) provided excellent reviews of the isotopic dating of diamonds based on their mineral inclusions. Navon (1999) addressed the concept of Sm–Nd and Rb–Sr dating of literally hundreds of garnets and or clinopyroxenes in order to arrive at an age for diamonds from a particular pipe. He argued that in order to date a the diamond-formation event, the following criteria must be fulfilled: (1) there was a discrete event of diamond formation, common to all the diamonds that are analyzed; (2) this event reset the isotopic systems equally in all the inclusions; and (3) the analyzed inclusions remained a closed system since diamond formation (encapsulation). This third requirement is easily satisfied if the inclusions have remained encapsulated in the diamonds since their entrapment. The fulfillment of the first two requirements, however, is difficult to prove. If, and only if, all the above three conditions are satisfied, the isotopic age from the inclusions represents diamond formation age. In all other cases, e.g., with only the third condition satisfied, the radiometric ages from mineral inclusions may date some non-geological events, not related to the formation of the diamond itself, and thus, place only maximum age constraints for the age of diamonds. Navon (1999) presented an illustrative scenario for Sm–Nd isotopic dating, wherein the diamonds formed over a large time interval (e.g., 1 Ga). He started with composite samples of silicate inclusions (i.e., garnets from ~180 diamonds) using a random-entrapment model of inclusions in the diamonds. He showed that by combining samples of each mineral from 60 diamonds into a composite for Sm–Nd analysis, the Sm–Nd plot could well indicate an age of ~1.8 Ga, which would only be the average of all the different ages of the diamonds.

In summation, while isotopic dating of diamond inclusions is a very important tool in deciphering the history of diamond formation in the mantle, care must be exercised in interpreting the results. If the analyzed samples are a composite of many inclusions, independent evidence must be sought that the isochrons define distinct events of diamond formation. The alternative possibility that each composite averages many diamonds of different ages must be ruled out. Even when single inclusions are analyzed, it is necessary to verify that the age obtained dates the diamond formation event and not older evolution of the trapped mineral prior to encapsulation in the diamond. It is obvious that only "pristine" inclusions should be used for age dating, requiring close optical examination of the diamonds by photo- and cathodo-luminescence. Our experience has shown that "open-system" behavior of inclusions due to minute cracks may be quite common.

6. Origin of diamonds and their mantle hosts

6.1. Introduction

In the preceding sections, we have reviewed our current understanding of the processes that are involved in the formation of diamonds and their host rocks in the mantle. This knowledge has been gained by studying mineral inclusions in diamonds,

diamonds themselves, and the host rocks to diamonds using a plethora of petrological/geochemical tools. This endeavor, which required experts from almost all fields of petrology/geochemistry, has now made it possible to speculate on the origin of diamonds and rocks in which they formed. We will now recap the most important aspects concerning petrogenetic characteristics of diamonds and their host rocks and conclude with a discussion of future directions in diamond research.

Based on studies of peridotitic diamond inclusions, it was suggested that peridotitic diamonds possibly grew under subsolidus conditions (Boyd and Finnerty, 1980; Hervig et al., 1980). Studies of eclogitic diamond inclusions, however, have yielded enigmatic results. The higher equilibration temperatures of eclogitic inclusions compared to peridotitic diamond inclusions, and the report that multiple inclusions from a single eclogitic diamond follow an igneous fractionation trend from core to rim in the diamond (Bulanova, 1995), suggested that some eclogitic diamonds may form by igneous crystallization. This is speculated to have been from a sulfide–silicate melt (e.g., Bulanova, 1995; Bulanova et al., 1998) or a sulfide-immiscible melt (e.g., Haggerty, 1986). However, variations in compositions of multiple diamond inclusions in many other eclogitic diamonds are not systematic (Sobolev et al., 1998a, b; Taylor L.A. et al., 1998; Keller et al., 1999), and require disequilibrium conditions that are entirely inconsistent with igneous fractionation. An additional complication is the discovery of compositional differences between many eclogitic diamond inclusions and the same minerals in the host eclogite xenoliths (Sobolev et al., 1983; Ireland et al., 1994; Taylor L.A. et al., 1996, 2000; Keller et al., 1999). Thus, either the diamonds and the xenolith were not cogenetic, or the xenolith composition had undergone significant post-diamond chemical modification. Identification of mineral inclusions in diamonds, which are clearly of lower-mantle paragenesis, has further highlighted the complex petrogenetic history of E-type diamond mineral inclusions. Below, we summarize several lines of evidence that have been used to argue in favor of or against the subduction hypothesis for the origin of mantle eclogites. We use similar arguments and pieces of evidence to hypothesize on the origin of mantle peridotites, particularly the diamondiferous ones.

6.2. Origin of diamondiferous eclogites

The long-standing debate about the origin of eclogites from subducted oceanic crust is mainly based on anomalously high and low oxygen isotopic values in eclogite xenoliths (Jagoutz et al., 1984). The accepted mantle oxygen isotopic composition is 5.5 ± 0.4‰ (Mattey et al., 1994). In the case of Udachnaya eclogites, Jacob et al. (1994) showed that $\delta^{18}O$ ratios in garnets and clinopyroxenes range from 5.19‰ to 7.38‰. Snyder et al. (1997) also reported a range of $\delta^{18}O$ values from 4.9‰ to 7‰ on garnets and clinopyroxenes from Udachnaya eclogites. These ranges in Udachnaya eclogites are relatively smaller than those seen in the case of South African eclogites (e.g., Roberts Victor, 2–8‰, Garlick et al., 1971). Nonetheless, it is clear that eclogite xenoliths have both lower and higher $\delta^{18}O$ ratios than typical mantle, and thus the origin of many of the eclogite xenoliths by subduction of oceanic crust is indisputable. Additional pieces of evidence that point to subduction origin of mantle eclogites are:

- Correlation between mineral chemistry and $\delta^{18}O$, inherited from crustal protolith (Jacob et al., 1994; Snyder et al., 1997);
- $\delta^{18}O$ of E-type garnet mineral inclusions in diamonds range from +5.7‰ to 8.6‰, indicative of crustal protoliths (Lowry et al., 1999; Anand et al., 2004);
- $\delta^{34}S$ of sulfides as inclusions in E-type diamonds are −11‰ to +14‰, possibly a signature of crustal (biogenic?) input (Eldridge et al., 1991);
- Garnets and clinopyroxenes with "positive Eu anomalies" in E-type diamond inclusions suggest the involvement of plagioclase in their genesis, i.e., low-pressure fractionation in the case of oceanic crust (Taylor L.A. et al., 2000);
- Extreme values of ε_{Nd} sometimes as high as +250, possibly resulting from interaction between an evolved protolith and crustal fluids during subduction zone processing (e.g., devolatilization; partial melting) (Neal et al., 1990; Snyder et al., 1997);
- High $^{87}Sr/^{86}Sr$ ratios (>0.710) in bulk-rock eclogites point to seawater alteration (Shervais et al., 1988; Neal et al., 1990);
- Pb isotope data in E-type sulfide diamond inclusions indicate the possibility of involvement of an evolved crust in the development of eclogites (Eldridge et al., 1991);
- Origin of the Trondhjamite–Tonalite–Granodiorite (TTG) suite of rocks by partial melting of subducted eclogites; i.e., at least some eclogites are "restites" (Ireland et al., 1994; Rudnick et al., 1998);
- Group A–C eclogites correlate well with the lithologies of an ophiolite sequence (Taylor and Neal, 1989; Beard et al., 1996).

In summary, it appears that there are "no true mantle signatures" that apply to these rocks, only a "lack of crustal signatures" that points to a mantle origin of some eclogites. Without any proof to the contrary, one suspects a mantle origin. However, most eclogites do contain crustal signatures, and the overwhelming evidence is in favor of crustal protoliths for most, if not all, mantle eclogites.

6.3. Origin of diamondiferous peridotites

As opposed to the subduction origin of eclogite xenoliths in kimberlites and E-type mineral inclusions, peridotite xenoliths have always been considered to be of true mantle origin, i.e., no crustal signatures have been identified in the case of peridotite xenoliths. The majority of mineral inclusions in diamonds of peridotitic affinity also do not show any crustal signatures. Thus, it is indisputable at the moment that most of the peridotitic xenoliths in kimberlites and by the same token, the P-type diamond mineral inclusions, are of mantle origin. However, in some cases, there are some crustal signatures that warrant a careful rethinking of the origin of at least some peridotite xenoliths from kimberlites. It may be possible that during subduction of an ophiolite complex, the bottom part, which is mainly made up of indisputable. Additional pieces of evidence that point to subduction origin of mantle eclogites are:

- Correlation between mineral chemistry and $\delta^{18}O$, inherited from crustal protolith (Jacob et al., 1994; Snyder et al., 1997);
- $\delta^{18}O$ of E-type garnet mineral inclusions in diamonds range from +5.7‰ to 8.6‰, indicative of crustal protoliths (Lowry et al., 1999; Anand et al., 2004);
- $\delta^{34}S$ of sulfides as inclusions in E-type diamonds are −11‰ to +14‰, possibly a signature of crustal (biogenic?) input (Eldridge et al., 1991);
- Garnets and clinopyroxenes with "positive Eu anomalies" in E-type diamond inclusions suggest the involvement of plagioclase in their genesis, i.e., low-pressure fractionation in the case of oceanic crust (Taylor L.A. et al., 2000);
- Extreme values of ε_{Nd} sometimes as high as +250, possibly resulting from interaction between an evolved protolith and crustal fluids during subduction zone processing (e.g., devolatilization; partial melting) (Neal et al., 1990; Snyder et al., 1997);
- High $^{87}Sr/^{86}Sr$ ratios (>0.710) in bulk-rock eclogites point to seawater alteration (Shervais et al., 1988; Neal et al., 1990);
- Pb isotope data in E-type sulfide diamond inclusions indicate the possibility of involvement of an evolved crust in the development of eclogites (Eldridge et al., 1991);
- Origin of the Trondhjamite–Tonalite–Granodiorite (TTG) suite of rocks by partial melting of subducted eclogites; i.e., at least some eclogites are "restites" (Ireland et al., 1994; Rudnick et al., 1998);
- Group A–C eclogites correlate well with the lithologies of an ophiolite sequence (Taylor and Neal, 1989; Beard et al., 1996).

In summary, it appears that there are "no true mantle signatures" that apply to these rocks, only a "lack of crustal signatures" that points to a mantle origin of some eclogites. Without any proof to the contrary, one suspects a mantle origin. However, most eclogites do contain crustal signatures, and the overwhelming evidence is in favor of crustal protoliths for most, if not all, mantle eclogites.

6.3. Origin of diamondiferous peridotites

As opposed to the subduction origin of eclogite xenoliths in kimberlites and E-type mineral inclusions, peridotite xenoliths have always been considered to be of true mantle origin, i.e., no crustal signatures have been identified in the case of peridotite xenoliths. The majority of mineral inclusions in diamonds of peridotitic affinity also do not show any crustal signatures. Thus, it is indisputable at the moment that most of the peridotitic xenoliths in kimberlites and by the same token, the P-type diamond mineral inclusions, are of mantle origin. However, in some cases, there are some crustal signatures that warrant a careful rethinking of the origin of at least some peridotite xenoliths from kimberlites. It may be possible that during subduction of an ophiolite complex, the bottom part, which is mainly made up of periods of time so short that they are within the analytical error of isotopic dating techniques.

It is generally agreed that diamonds grew in the mantle during metasomatic events, from a carbon-rich fluid percolating through the mantle, evidenced by the presence of mineral inclusions of magnesite (Bulanova, 1995) and calcite (Sobolev N.V. et al., 1997) in diamonds. Growth from a metasomatic fluid appears to account for the rapidly changing conditions during diamond growth (e.g., Taylor, 1988; Taylor and Green, 1989; Hauri et al., 1993; Rudnick et al., 1993a; Taylor W.R. et al., 1995; Yaxley et al., 1998; Leost et al., 2003). It may also explain the problem of the source of carbon for the diamond formation (Deines and Harris, 1994, 1995; Stachel and Harris, 1997; Spetsius and Griffin, 1998; Taylor L.A. et al., 1998). Numerous experimental studies have reinforced the possibility of carbonate-rich melts being likely sources of diamond nucleation and growth (e.g., Chepurov et al., 1997; Pal'yanov et al., 1997; Arima et al., 2002). Based on exhaustive research on mineral inclusions from both P- and E-type diamonds, the following features can be highlighted with regard to the origin of diamonds and their mineral inclusions:

- $\delta^{13}C$ values of E-type diamonds range from +5‰ to −35‰; i.e., indicate possible biogenic and/or crustal origin for the carbon (Sobolev and Sobolev, 1980; Deines et al., 2001);
- Many inclusions in P- and E-type diamonds have crustal $\delta^{18}O$ values (Lowry et al., 1999; Schulze et al., 2003);
- Multiple garnet inclusions in a single E-type diamond can span a large range in composition (Sobolev et al., 1998a);
- Eclogitic multiple inclusions of clinopyroxene from core and rim zones of a single diamond commonly have different major- and trace-element compositions (Taylor L.A. et al., 1998, 2000), displaying no systematics;
- Clinopyroxene compositions from diamond cores to rims and from diamond inclusions and hosts are not those expected from a differentiating melt (Taylor L.A. et al., 2000);
- Clinopyroxene compositions of diamond inclusions can be grossly different from those of their host eclogites (Taylor L.A. et al., 2000);
- Both eclogitic and peridotitic mineral inclusions can exist within one diamond (Prinz et al., 1975; Wang, 1998);
- Most eclogites have never experienced a magmatic stage, albeit possible partial melting during subduction of their protoliths into the mantle;
- Many harzburgitic garnet inclusions in diamonds existed before incorporation into their diamonds (Taylor L.A. et al., 2003a, c);
- Crystallographic and other criteria for syngenesis are not unambiguous (Taylor L.A. et al., 2003a, c);
- Experimental evidence exists for involvement of a carbonatitic fluid with diamond formation (Pal'yanov et al., 2002; Arima et al., 2002);
- The chemical composition of diamonds and their major inclusions, sulfides, point to the diamond-forming fluid having major C–O–H–N–S components (e.g., Haggerty, 1986; Stachel et al., 1998; Deines et al., 2001);

- ◆ Most evidence would appear to indicate that eclogitic diamonds have nucleated and grown by metasomatic processes (Kopylova et al., 1997; Sobolev et al., 1998a, b; Stachel and Harris, 1997; Stachel et al., 1998; Taylor et al., 1998).

7. Future directions

This review of the petrogenesis of diamonds and their mantle hosts, largely from the Yakutian kimberlites, has addressed several major topics in diamond/mantle research. It is meant mainly as an introduction into the continuing progress and improvement in the scientific understanding of the origin of the diamonds in the mantle. Many avenues of diamond research in the near future will greatly further our understanding of the nature and evolution of the lithospheric mantle through geological time. Some of these rapidly progressing areas of diamond research are discussed below in the form of areas of concern, and warnings of unforeseen and/or unappreciated difficulties.

7.1. Ages of diamonds

Traditionally, the ages of diamonds have been established by radiometric dating of their "syngenetic" mineral inclusions. However, a valid query to raise is if the dating of mineral inclusions indeed gives the formation age of the diamond. In some cases, for sure, but in others it is not that obvious. A careful rethinking of this important problem in diamond genesis is needed. At present, we have very limited information on the duration of diamond growth in the mantle. The instrumentation employed in the radiogenic isotope studies of diamond mineral inclusions has been evolving rapidly, and it is now becoming possible to radiometrically date individual mineral inclusions. However, there is always the possibility that the diamond had minute cracks such that "open-system" behavior has cryptically modified the composition of the mineral inclusion.

With diamond formation in the mantle, it is assumed that there is some event in time that "triggers" the formation of diamond. At present, we do not, however, have any means of predicting such discrete diamond-forming events, and thus, more radiogenic work on diamond mineral inclusions is needed to establish a statistical database from which one may be able to infer possible episodic diamond growth in the mantle. This in turn will help in predicting the diamond potential of a particular region.

7.2. Diamond formation in the mantle

One of the most intriguing topics in diamond research is the nature of the fluids that precipitated the diamonds. Now that we have been largely weaned away from a purely magmatic silicate melt source, there are several alternative sources. A considerable deal can be learnt by some properly designed high-pressure and high-temperature experiments, such as the ones that have already been carried out in carbonate–water–sulfide systems. It seems obvious that, in any scenario, the C–O–H–N–S components must be accounted for in the diamond-precipitating media.

In most cases, the mineral inclusions in diamonds are the only evidence that we have of the type of host in which the diamond actually formed. The biggest advantage of studying mineral inclusions in diamonds is the likelihood that, once encapsulated by the growing diamond, most of the inclusions have remained pristine over time. It is, therefore, imperative that a variety of analytical data be gathered on these mineral inclusions in order to gain further insights into the diamond-forming processes in the mantle. Major advances have been made in this direction, particularly, from the major- and trace-element perspective of mineral inclusions. Only recently, oxygen isotopic compositions of diamond mineral inclusions have begun to accrue in the literature. Similarly, with the invaluable radiogenic age determinations, it is absolutely essential to make sure that these data are obtained only on pristine inclusions. The study of multiple mineral inclusions in diamonds can potentially provide "snapshots" of the geochemical environment present at the time of each inclusion encapsulation.

The non-carbon chemistry of diamond remains somewhat of a mystery as well. The real cause and meaning of the various nitrogen-aggregation states are not well understood. How can CL zones as small as several microns exist with such sharp boundaries over billions of years, when the aggregation of nitrogen is supposed to continue to occur while the diamond is residing in the mantle at high P and T? These zones can each have drastically different total nitrogen contents and aggregation states, much like the oscillatory zoning we see in other minerals. Does this reflect the diamond growth over an extended time period, under constantly changing fluid compositions? The presence of multiple inclusions with different chemical compositions lends support to this idea. The $\delta^{13}C$ and $\delta^{15}N$ of the diamond may hold the clues to the source of the carbon and nitrogen, and may also help resolve the long-standing debate about the dominant contribution from mantle or crustal sources.

Could the eclogite in which a diamond occurs today be but its last "resting place," having moved from its original birthplace into different mineralogic surroundings? How else can one explain the presence of both an eclogite and a peridotitic assemblage in one diamond? It is generally agreed that at least some diamonds have been formed in the asthenosphere, with all its plastic deformation. Is the contrast in strength (e.g., Reynolds number) between the diamond and the hosting silicate minerals such that the diamond effectively can move relative to its local mineralogy? The geophysicists say certainly. But, what about the "reidity" (i.e., deformation versus time) of the lower lithosphere beneath cratons? This is the essence of the outrageous hypothesis put forth by Taylor L.A. et al. (2000).

7.3. In situ analyses of diamond inclusions and high-resolution X-ray tomography

It has become abundantly clear that the best method with which to study mineral inclusions in diamonds is by *in situ* analyses of these minerals as exposed on polished diamond surfaces. The polished surfaces should be examined in detail with high-powered microscopy using CL in order to ascertain the possible presence of healed cracks from the inclusion to the surface of the diamond. Our experience with diamond inclusions, however limited, has shown that in some cases, significant numbers (i.e., $>30\%$) of inclusions are surrounded by minute cracks. Importantly, these cracks are often totally invisible under optical examination, but easily seen with CL.

The HRCXT of diamondiferous eclogites has shown that there is a possible relationship between the secondary formation of diamonds and zones of permeability in the host rocks. Preliminary HRCXT studies of diamondiferous peridotites also appear to indicate some association of diamonds with secondary fabric in the host rock. This effective 3D mapping of the mineralogy of these xenoliths can reveal considerable information about diamond formation, especially when coupled with the systematic dissection of the host and thorough examination of all aspects of the mineralogy, including that of the diamonds.

As we progress in our studies of diamonds and their host xenoliths, it has become apparent that we are narrowing the gap in our understanding of the origin(s) and formation of diamonds. However, in this continuing quest for knowledge, it seems that each investigation solves one problem and uncovers two new ones. But such is the nature of the scientific progress. We hope that this review article will inspire some of the next generation of scientists, who will unravel many more mysteries of diamonds.

Acknowledgements

It was our honor to accept the offer from Professor Klaus Keil to write this Invited Review for "his" journal of long-standing and high esteem. We thank him for encouragement and his review and editing of the manuscript. The very thorough and incisive review by Pin Prompreted was particularly helpful for putting many things into better perspective. To both Klaus and Pin, we are eternally grateful. We would also like to express our sincere appreciation to Dawn Taylor, for her assistance with the figure construction and computer graphics, particularly the tomographic imaging, and whose devotion to our research continues to be invaluable. Special mention is also warranted for Allan Patchen, our "EMP Wizard," whose high standards for quality analyses and microprobe operation have made our numerous studies so successful.

The invigorating discussions, arguments, and constructive dialogue that the senior author has had over the years with numerous colleagues have greatly contributed to his education in the exciting fields of the petrology and geochemistry of kimberlites, their mantle xenoliths, and diamonds. These persons include, especially Nick Sobolev, Nick Pokhilenko, Joe Boyd, Slava Spetsius, Bill Carlson, the late Henry Meyer, Galina Bulanova, Chris Smith, Dan Schulze, John Valley, Christine Floss, Ghislaine Crozaz, Rich Ketchum, Bill Griffin, Kula Misra, and including several Postdoctoral Fellows who have worked with the Tennessee research group, particularly Greg Snyder, Eric Jerde, John Shervais, Clive Neal, Vlad Sobolev, the late Bob Hunter, Alex Ruzicka, Vlad Malkovets, Wuyi Wang, Rene Wiesli, and Brian Beard. This work was supported in part by research grants from the National Science Foundation, for which we are most grateful.

BIBLIOGRAPHY (emphasis mainly on Yakutia)

Afanasiev, V.P., Yefimova, E.S., Zinchuk, N.N., Koptil, V.I., 2000. Atlas of morpholohy of diamonds from Russian sources. Novosibirsk: SPC UIGGM SB RUS, 291.

Agashev, A.M., Fomin, A.S., Watanabe, T., Pokhilenko, N.P., 1998. Preliminary age determination of recently discovered kimberlites of the Siberian kimberlite province. *Extended Abstracts 7th Intern. Kimberlite Conf*, Cape Town, 11-13.

Agashev, A., Watanabe, T., Bydaev, D., Pokhilenko, N., Fomin, A., Maehara, K. & Maeda, J., 2001. Geochemistry of kimberlites from the Nakyn field, Siberia: Evidence for unique source composition. *Geology* **29**, 267-270.

Agashev, A.M., Watanabe, T., Bydaev, D A (Bydayev,D.A.), Pokhilenko, N.P., Fomin, A.S., Maehara, K. & Maeda, J., 2001. Geochemistry of kimberlites from the Nakyn Field, Siberia; evidence for unique source composition. *Geology (Boulder)* **29**, 267-270.

Agashev, A.M., Watanabe, T., Kuligin, S.S., Pokhilenko, N.P. and Orihashi, Y., 2001. Rb-Sr and Sm-Nd isotopes in garnet pyroxenite xenoliths from Siberian kiberlites: an insight into litospheric mantle. *Journ. of Mineralogical and Petrological Sci.*, **96**, 7-18.

Agashev, A.M., Pokhilenko, N.P., Tolstov, A.V., Polyanichko, V.G., Mal'kovets, V.G. & Sobolev, N.V., 2004. New age data on kimberlites from the Yakutian diamondiferous province. *Doklady Earth Sciences* **399**, 1142-1145.

Agashev, A.M., Pokhilenko, N.P., Mal'kovets, V.G. & Sobolev, N.V., 2006. Sm-Nd isotopic system in garnet megacrysts from the Udachnaya kimberlite pipe (Yakutia) and petrogenesis of kimberlites, *Doklady Earth Sciences* **407**, 491-494.

Agee C.B., 1998. Phase transformations and seismic structure in the upper mantle and transition zone. *Rev. Mineral* **37**, 165-203.

Akaogi M. and Akimoto A., 1977. Pyroxene-garnet solid-solution equilibria in the systems $Mg_4Si_4O_{12}$-$Mg_3Al_2Si_3O_{12}$ and $Fe_4Si_4O_{12}$-$Fe_3Al_2Si_3O_{12}$ at high pressures and temperatures. *Phys. Earth. Planet. Inter.* **15**, 90-106.

Akagi, T. and Masuda, A., 1988. Isotope and elemental evidence for a relationship between kimberlite and Zaire cubic diamonds. *Nature* **336**, 665-667.

Alard, O., Griffin, W.L., Lorand, J.P., Jackson, S.E., and O'Reilly, S.Y., 2000. Non-chondritic distribution of the highly siderophile elements in mantle sulphides. *Nature* **407**, 891-894.

Alard, O., Griffin, W.L., Pearson, N.J., Lorand, J.P., & O'Reilly, S.Y., 2002. New insights into the Re-Os systematics of subcontinetal lithospheric mantle from in situ analysis of sulphides. *Earth and Planetary Science Letters* **203**, 651-663.

Anand, M., Misra, K.C., Taylor, L.A. & Sobolev, N.V., 2001. In-situ chemical analyses of mineral inclusions in diamonds in kimberlitic eclogites from Yakutia (abstract). *AGU Fall Meet.* V12C-1006.

Anand, M., Taylor L.A., Carlson W.D., Misra K.C., and Sobolev N.V., 2003. Diamondiferous eclogite dissection: Anomalous diamond genesis? *8th International Kimberlite Conf.,* Victoria, BC, Extended Abstract.

Anand, M., Taylor, L. A., Carlson, W. D., Taylor, D.-H. & Sobolev, N. V., 2003. Diamond genesis revealed by x-ray tomography of diamondiferous eclogites. *8th International Kimberlite Conference, Canada*, Extended abstract.

Anand M., Taylor L.A., Carlson, W.D., Taylor D.-H., Sobolev, N.V., 2004. Nature of diamonds in Yakutian ecologites: views from eclogite tomography and mineral inclusions in diamonds. *Lithos* **77**, 333-348.

Appleyard, C.M., Viljoen, K.S. and Dobbe, R., 2004. A study of eclogitic diamonds and their inclusions from the Finsch kimberlite pipe, South Africa. *Lithos* **77**, 317-332.

Arima, M., Nakayama, K., Akaishi, M., Yamaoka, S., and Kanda, H., 1993. Crystallization of diamond from a silicate melt of kimberlite composition in high-pressure and high temperature experiments. *Geology* **21**, 968-970.

Arima, M., Kozai, Y., & Akaishi, M., 2002. Diamond nucleation and growth by reduction of carbonate melts under high-pressure and high-temperature conditions. *Geology* **30**, 691-694.

Ashchepkov, I.V., Vladykin, N.V., Rotman, A., Pokhilenko, N.P., Logvinova, A.M., Kuchkin, A.M., Palessky, S.S., Saprykin, A.I., Anoshin, G.N. & Khemel'nikova, O., 2005. Regularities of mantle structure beneath Siberian Craton; Abstracts of the 15th annual V. M. Goldschmidt conference. *Geochimica et Cosmochimica Acta,* **69**, 438.

Ashchepkov, I.V., Pokhilenko, N.P., Logvinova, A.M., Vladykin, N.P., Rotman, A.Y., Palessky, S.V., Alymova, N.V. & Vishnyakova, E.V., 2007. Evolution of kimberlite magmatic sources beneath Siberia; Abstracts of the 17th annual V. M. Goldschmidt conference. *Geochimica et Cosmochimica Acta,* **71**, no. 15S, A39.

Ashchepkov, I.V., Pokhilenko, N.P., Vladykin, N.V., Rotman, A.Y., Afanasiev, V. P., Logvinova, A.M., Kostrovitsky, S. I., Pokhilenko, L.N., Karpenko, M.A., Kuligin, S.S., Malygina, E.V., Stegnitsky, Y. B., Alymova, N.A. & Khmelnikova, O.S., 2008. Reconstruction of mantle sections beneath Yakutian kimberlite pipes using monomineral thermobarometry; Metasomatism in oceanic and continental lithospheric mantle. *Geological Society Special Publications* **293**, 335-352.

Aulbach, S, Stachel, T., Viljoen, K.S, Brey, G.P., & Harris, J.W., 2002. Eclogitic and websteritic diamond sources beneath the Limpopo Belt - is slab-melting the link? *Contrib. Mineral. Petrol.* **143**, 56-70.

Bakumenko, I.T., Sobolev, N.V., Khokhriakov, A.F., & Chepurov, A.I., 1984. Faceted inclusions in diamond crystals. *Dokl Akad Nauk SSR* **278,** 1461-1465.

Ballhaus, C., Berry, R. F., & Green, D. H., 1991. High pressure experimental calibration of the olivine-orthopyroxene-spinel oxygen barometer: Implications for the oxidation state of the upper mantle. *Contributions to Mineralogy and Petrology* **107**, 27-40.

Banas A., Stachel, T., Muehlenbachs, K. and McCandless, T.E., 2007. Diamonds from the Buffalo Head Hills, Alberta: Formation in a non-conventional setting. *Lithos* **93**, 199-213.

Barashkov, Y.P. and Beskrovanov, V.V., 1994. Model of evolution of growth conditions of natural diamond; International Mineralogical Association, 16th general meeting; abstracts, *Abstracts of the ...General Meeting of the International Mineralogical Association* **16**, 29.

Barnes, S. J., and Roeder, P. L., 2001. The range of spinel compositions in terrestrial mafic and ultramafic rocks. *Journal of Petrology* **42**, 2279-2302.

Barren, L. M., David R. M., Pattison, M. R. St-Onge, and Normand J. B., 2005. A linear model and topology for the host-inclusion mineral system involving diamond; truth and beauty in metamorphism; a tribute to dugald carmichael. *The Canadian Mineralogist* **43**, 203-224.

Barth, M.G., Rudnick R.L., Horn I., McDonough W.F., Spicuzza M.J., Valley J.W. & Haggerty

S.E., 2001. Geochemistry of xenolithic eclogites from West Africa, part I: A link between low MgO eclogites and Archean crust formation. *Geochimica et Cosmochimica Acta* **65**, 1499-1527.

Bartoshinsky, Z.V., 1961. Comparison of diamonds from different diamondiferous regions of Western Yakutia. *Sov. Geol. Geophys.* **N6**, 40-50 (in Russian).

Beard, B. L., Fraracci, K. N., Taylor, L.A., Snyder, G.A., Clayton, R.N., Mayeda, T., & Sobolev, N.V., 1996. Petrography and geochemistry of eclogites from the Mir kimberlite, Yakutia, Russia. *Contributions to Mineralogy and Petrology* **125**, 293-310.

Becker M., and Le Roex A.P., 2006. Geochemistry of South African On- and Off-craton, Group I and Group II Kimberlites: Petrogenesis and Source Region Evolution. *J. Petrology*, **47**,673 - 703.

Bell, D. R., Gregoire, M., Grove, T. L., Chatterjee, N., Carlson, R.W., Buseck, P.R., 2005. Silica and volatile-element metasomatism of Archean mantle: a xenolith-scale example from the Kaapvaal Craton. *Contrib. Mineral. Petrol.* **150**, 251–267.

Berg, J.H., 1977. Morphodynamic development and preservation of physical sedimentary structures in two prograding Recent ridge and runnel beaches along the Dutch coast. *Geologie en Mijnbouw,* **56**(3), 185-202.

Berg, J.H., 1980. Evidence of picritic magmatism in the Nain Complex, Labrador; Sixth annual Midwest AGU meeting. *EOS, Transactions, American Geophysical Union,* **61**(48), pp. 1193.

Berg, J.H. and Docka, J.A., 1983. Geothermometry in the Kiglapait contact aureole, Labrador. *American Journal of Science,* **283**(5), pp. 414-434.

Berg, J.H. and Wiebe, R.A., 1985. Petrology of a xenolith of ferro-aluminous gneiss from the Nain Complex. *Contributions to Mineralogy and Petrology,* **90**, 226-235.

Beskrovanov, V.V., 1986. Natural diamonds with orange photoluminescence in Physical properties and mineralogy of natural diamond; collection of scientific papers, ed. V.V. Koval'skiy, Akad. Nauk SSSR, Sib. Otd., Yakutsk. Fil., Yakutsk, USSR (SUN), USSR (SUN).

Beskrovanov, V.V., 1986. Principles for the classification of diamond materials" in Physical properties and mineralogy of natural diamond; collection of scientific papers, ed. V.V. Koval'skiy, Akad. Nauk SSSR, Sib. Otd., Yakutsk. Fil., Yakutsk, USSR (SUN), USSR (SUN).

Beskrovanov, V.V. & Smith, C.B., 1986. Growth history of natural diamond; Fourth international kimberlite conference, *Abstracts - Geological Society of Australia* **16**, 368-370.

Bibby, D.M., 1982. Impurities in natural diamond. In: Thrower, P.A. (ed.). Chemistry and Physics of Carbon **18**, 1-91. Marcel Dekker, New York.

Bizzarro M., and Stevenson, R.K., 2003. Major element composition of the lithospheric mantle under the North Atlantic craton: Evidence from peridotite xenoliths of the Sarfartoq area, southwestern Greenland. *Contributions to Mineralogy and Petrology* **146**, 223-240.

Blichert-Toft, J., Ionov, D.A., and Albarède, F., 2000. The nature of the sub-continental lithospheric mantle: Hf isotope evidence from garnet peridotite xenoliths from Siberia. *Journal of Conference Abstracts* **5**, 217.

Bobrievich, A.P., Bondarenko, M.N., Gnevushev, M.A, Krasov, A.M., Smirnov, G.I., and Yurkevich, R.K., 1959. The Diamond Deposits of Yakutia. Moscow: Gosgeoltekhizdat, 527 (in Russian).

Bobrievich, A.P., Smirnov, G.I., Sobolev, V.S., 1960. On the mineralogy of xenoliths of grossular-pyroxene-diasthene (grospydite) from Yakutian kimberlites. *Geologiya i Geofizika*, **3**, 18-24.

Bokil, G.B., Bezrukov, G.N., Klyuev, Y.A., Nafetov, A.M., & Nepsha, V.I., 1986. Natural and synthetic diamond crystals, Nauka Press, Moscow, Russia (in Russian).

Bonney, T.G., 1899. The original rock of the south-African diamond. *Natural Science*, London **XV**, 173–182.

Boyd, F. R., 1970. Garnet-peridotite and the system $CaSiO_3$-Mg SiO_3-Al_2O_3, *Miner. Soc. Amer. Spes. Pap.***3**, 63-75.

Boyd, F.R., 1984. Siberian geotherm based on lherzolite xenoliths from Udachnaya kimberlite, USSR. *Geology* **12**, 528-530.

Boyd, F.R., 1989. Compositional distinction between oceanic and cratonic lithosphere. *Earth and Planetary Science Letters* **96**, 15-26.

Boyd, F.R., Finnerty, A.A., 1980. Conditions of origin of natural diamonds of peridotitic based on lherzolite xenoliths from Udachnaya kimberlite, USSR. *Geology* **12**, 528-530.

Boyd F.R., Dawson J.B., and Smith J.V., 1984. Granny Smith diopside megacrysts from the kimberlites of the Kimberley area and Jagersfontein, South Africa. *Geochimica et Cosmochimica Acta* **48**, 381-384.

Boyd F.R., Nixon P. H., and Boctor N. Z., 1984. Rapidly crystallized garnet pyroxenite xenoliths possibly related to discrete nodules. *Contributions to Mineralogy and Petrology* **86**, 119-130.

Boyd, F.R., Mattey, D.P., Pillinger, C.T., Milledge, H.J., Mendelsson, M., & Seal, M.J., 1987. Multiple growth events during diamond genesis: An integrated study of carbon and nitrogen isotopes and nitrogen aggregation state in coated stones. *Earth Planet. Sci. Lett.* **86**, 341-353.

Boyd, F.R., Pillinger, C.T., Milledge, H.J., & Seal, M.J., 1992. C and N isotopic composition and Infrared absorption spectra of coated diamonds: evidence for the regional uniformity of CO_2-H_2O rich fluids in lithospheric mantle. *Earth Planet. Sci. Lett.* **108**, 139-150.

Boyd, F.R., Pearson, D.G., Nixon, P.H., & Mertzman, S. A., 1993. Low-calcium garnet harzburgites from southern Africa: Their relation to craton structure and diamond crystallization. *Contributions to Mineralogy and Petrology* **113**, 352-366.

Boyd, F.R., Pineau, F., & Javoy, M., 1994. Modeling the growth of natural diamonds. *Chemical Geology* **116**, 29-42.

Boyd, F.R., Kiflawi, I., & Woods, G.S., 1994. The relationship between infrared absorption and A-defect concentration in diamond. *Phil. Mag.* **B69**, 1149-1153.

Boyd, F.R., Pokhilenko, N. P., Pearson, D.G., Mertzman, S. A., Sobolev, N. V., and Finger, L. W., 1997. Composition of the Siberian cratonic mantle: evidence from Udachnaya peridotite xenoliths. *Contributions to Mineralogy and Petrology* **128**, 228-246.

Boyd, F.R., Pearson, D.G., & Mertzman, S.A., 1999. Spinel-facies peridotites from the Kaapvaal root. In: Gurney, J.J., Gurney, J.L., Pascoe, M.D., and Richardson, S.H. (eds.). *Proceedings Volume of 7th International Kimberlite Conference*, *1*. Cape Town: Red Roof Design, 40-48.

Brakhfogel, F.F., 1984. Geological aspects of kimberlite magmatism in the northeastern Siberian platform, Yakutian Institute of Geology Publication, Yakutsk (in Russian).

Brenan, J.M., Shaw, H.F., Ryerson, F.J., & Phinney, D.L., 1995. Mineral-aqueous fluid partitioning of trace elements at 900°C and 2.0 GPa: Constraints on the trace element chemistry of mantle and deep crustal fluids. *Geochimica et Cosmochimica Acta* **55**, 2203-2214.

Brenker, F.E., Stachel, T., & Harris, J.W., 2002. Exhumation of lower mantle inclusions in diamond: A TEM investigation of retrograde phase transitions, reactions and exsolution. *Earth and Planetary Science Letters* **198**, 1-9.

Brey, G., and Köhler, T., 1990. Geothermobarometry in four-phase lherzolites II. New thermobarometers and practical assessment of existing thermobarometers. *Journal of Petrology* **31**, 1353-1378.

Brey, G., Köhler, T., and Nickel, K. G., 1990. Geothermobarometry in four-phase lherzolites I. Experimental results from 10 to 60 kb. *Journal of Petrology* **31**, 1313-1352.

Bulanova, G.P., 1995. The formation of diamond. *Journal of Geochemical Exploration* **53**, 1-23.

Bulanova, G.P., Varshavsky, A.V., Leskova, N.V., & Nikishova, L.V., 1979. About central inclusions in natural diamonds. *Doklady Akademy Nauk SSSR* **244**, 704-706 (in Russian).

Bulanova, G.P., Spetsius, Z.V., and Leskova, N.V., 1990. Sulfides in Diamonds and Xenoliths from Kimberlite Pipes of Yakutia. Novosibirsk: Nauka, 117 pp. (in Russian).

Bulanova, G.P., Milledge, H.J., 1995. Origin and history of growth of macrodiamonds from Yakutian kimberlites. *6th Int. Kimb. Conf., Novosibirsk*, Extended Abstract, 77–79.

Bulanova, G.P., Griffin, W.L., and Ryan, C.G., 1996. Trace elements in sulfide inclusions from Yakutian diamonds. *Contributions to Mineralogy and Petrology* **124**, 111-125.

Bulanova, G.P., Griffin, W.L., and Ryan, C.G., 1998. Nucleation environment of diamonds from Yakutian kimberlites. *Min. Mag.* **62**, 409– 419.

Bulanova, G. P., Shelkov, d., Milledge, H.J., Haury, E.H., & Smith, S.B., 1999. Nature of eclogitic diamonds from Yakutian kimberlites: Evidence from isotopic composition and chemistry of inclusions, in *Proceedings of 7th Intern. Kimberlite Conf.* **1**, 57-65.

Bulanova, G.P., Griffin, W.L., Kaminsky, F.V., Davies, R., Spetsius, Z.V., Ryan, C.G., Andrew, A., & Zahkarchenko, O.D., 1999. Diamonds from Zarnitsa and Dalnaya kimberlites, Yakutia; their nature and lithospheric mantle source. In: Gurney, J.J., Gurney, J.L., Pascoe, M.D., and Richardson, S.H. (eds). *Proceedings Volume of 7th International Kimberlite Conference, I.* Cape Town: Red Roof Design, 49-56.

Bulanova, G. P., Shelkov, D., Milledge, H. J., Hauri, E. H., & Smith, S. B., 1999. Nature of eclogitic diamonds from Yakutian kimberlites: Evidence from isotopic composition and chemistry of inclusions. In: Gurney, J.J., Gurney, J.L., Pascoe, M.D, and Richardson, S.H. (eds). *Proceedings Volume of 7th International Kimberlite Conference, I.* Cape Town: Red Roof Design, 57-65.

Bulanova, G.P., Pearson, D.G., Hauri, E.H., and Griffin, B.J., 2002. Carbon and nitrogen isotope systematics within a sector-growth diamond from the Mir kimberlite, Yakutia. *Chemical Geology* **188**, 105-123.

Bulanova, G.P., Muchemwa, E., Pearson, D.G., Griffin, B.J., Kelley, S.P., Klemme, S., & Smith, C.B., 2004. Syngenetic inclusions of yimengite in diamond from Sese kimberlite (Zimbabwe): evidence for metasomatic conditions of growth. *Lithos* **77**, 181– 192.

Bulanova, G.P., Varshavsky, A.V. and Kotegov, V.A., 2005. A venture into the interior of natural diamond; genetic information and implications for the gem industry. *Journal of Gemmology* **29**, 377-386.

Burgess, R. & Turner, G., 1995. Halogen composition of mantle fluids in diamond; Sixth international kimberlite conference, extended abstracts, *Proceedings of the International Kimberlite Conference* **6**, 89-91.

Burgess, S., and Harte, B., 1999. Tracing lithosphere evolution through the analysis of heterogeneous G9/G10 garnets in peridotite xenoliths I: Major element analysis. In: Gurney, J.J., Gurney, J.L., Pascoe, M.D., & Richardson, S.H. (eds.). *Proceedings Volume of 7th International Kimberlite Conference, 1*. Cape Town: Red Roof Design, 66-80.

Burgess, R., Turner, G., Laurenzi, M. & Harris, J. W., 1989. ^{40}Ar-^{39}Ar laser probe dating of individual clinopyroxene inclusions in Premier eclogitic diamonds. *Earth and Planetary Science Letters* **94**, 22-28.

Burgess, R., Turner, G., Laurenzi, M., & Harris, J.W., 1992. ^{40}Ar-^{39}Ar laser probe studies of clinopyroxene inclusions in eclogitic diamonds. *Geochim Cosmochim Acta* **56**, 389-402.

Burgess, R., Turner, G. & Mattey, D.P., 1995. Helium, argon and carbon isotope constraints on the formation of cubic and polycrystalline diamond; Sixth international kimberlite conference; extended abstracts, *Proceedings of the International Kimberlite Conference* **6**, 92-94.

Burgess, R., Johnson, L.H., Mattey, D.P., Harris, J.W. & Turner, G., 1998. He, Ar and C isotopes in coated and polycrystalline diamond. *Chemical Geology* **146**, 205-217.

Burgess, R., Layzelle, E., Turner, G. & Harris, J.W., 2002. Constraints on the age and halogen composition of mantle fluids in Siberian coated diamonds. *Earth and Planetary Science Letters* **197**, 193-203.

Burgess, R., Kiviets, G., & Harris, J.W., 2003. Different age populations of eclogitic diamonds in the Venetia kimberlite: Evidence from Ar-Ar dating of syngenetic clinopyroxene inclusions. *8th Internat Kimb Conf*, Victoria, BC, Extended Abstract.

Burgess, S.R. & Harte, B. 2004. Tracing lithosphere evolution through the analysis of heterogeneous G9-G10 garnets in peridotite xenoliths; II, REE chemistry. *Journal of Petrology* **45**, 609-634.

Canil, D., 1999. The Ni-in-garnet geothermometer: Calibration at natural abundances. *Contributions to Mineralogy and Petrology* **136**, 240-246.

Canil, D., 2002. Vanadium in peridotites, mantle redox and tectonic environments: Archean to present. *Earth and Planetary Science Letters* **195**, 75-90.

Canil, D., 2004. Mildly imcompatible elements in peridotites and the origins of mantle lithosphere, *Lithos* **77**, 375-394.

Canil, D., and Bellis, A.J., 2007. Ferric iron in $CaTiO_3$ perovskite as an oxygen barometer for kimberlite magmas II: Applications. *J. Petrology* **48**, 231 - 252.

Carbino, G.B. and Canil, D., 2002. Mantle structure beneath the SW Slave Craton, Canada: Constraints from garnet geochemistry in the Drybones Bay kimberlite. *Journal of Petrology* **43**, 129–142.

Carlson, W.D., and Denison, C., 1992. Mechanisms of porphyroblast crystallization: Results from high-resolution computed x-ray tomography. *Science* **257**, 1236–1239.

Carlson, R.W., Irving, A.J. & Hearn, B.C.,Jr., 1999. Chemical and isotopic systematics of peridotite xenoliths from the Williams Kimberlite, Montana; clues to processes of lithosphere formation, modification and destruction, Pproceedings of the 7th international kimberlite conference, *Proceedings of the International Kimberlite Conference* **7**, 90-98.

Carlson, R.W., Pearson, D.G., Boyd, F.R., Shirey, S.B., Irvine, G., Menzies, A.H. & Gurney, J.J. 1999. Re-Os systematics of lithosphere peridotites; implications for lithosphere formation and preservation, Proceedings of the 7Ith international kimberlite conference, *Proceedings of the International Kimberlite Conference* **7**, 99-108.

Carlson, W.D., Denison, C., & Ketcham, R.A., 1995. Contribols on the nucleation and growth of porphyroblasts: kinetics from natural textures and numerical models. *Geol. Jour.* **30**, 207–225.

Caro, G., Bennett, V.C., Bourdon, B., Harrison, T.M., von Quadt, A., Mojzsis, S.J. & Harris, J.W., 2008. Application of precise $^{142}Nd/^{144}Nd$ analysis of small samples to inclusions in diamonds (Finsch, South Africa) and Hadean Zircons (Jack Hills, Western Australia). *Chemical Geology* **247**, 253-265.

Carswell, D.A., 1991. The garnet-orthopyroxene Al barometer: problematic application to natural garnet lherzolite assemblages. *Mineralogical Magazine* **55**, 19-31.

Carswell, D. A. and Herman L. M. R., 2005. On multi-phase mineral inclusions associated with microdlamond formation in mantle-derived perldotite lens at bardane on fjortoft, west norway. *European Journal of Mineralogy* **17**, 31-42.

Cartigny, P., 2005. Stable isotopes and the origin of diamond. *Elements* 1, 79-84.

Cartigny, P., Boyd, S.R., Harris, J.W., & Javoy, M., 1997. Nitrogen isotopes in peridotitic diamonds from Fuxian, China: the mantle signature. *Terra Nova* **9**, 175-179.

Cartigny, P., Harris, J.W., & Javoy, M., 1998. Eclogitic diamond formation at Jwaneng: No room for a recycled component. *Nature* **280**, 421-1423.

Cartigny, P., Harris, J.W., Phillips, D., Girard, M., & Javoy, M., 1998. Subduction-related diamonds? - The evidence for a mantle-derived origin from coupled $\delta^{13}C$-$\delta^{15}N$ determinations. *Chem. Geol.* **147**, 147-159.

Cartigny, P., Harris, J.W., & Javoy, M., 1999. Eclogitic, peridotitic and metamorphic diamonds and the problems of carbon rcycling the case of Orapa (Botswana). *Proceedings of the 7th International Kimberlite Conference* **V1**, 117-124.

Cartigny, P., Harris, J.W., & Javoy, M., 2001. Diamond genesis, mantle fractionations and mantle nitrogen content: A study of $\delta^{13}C$-N concentrations in diamonds. *Earth and Planetary Science Letters* **185**, 85-98.

Cartigny, P., Stachel, T., Harris, J.W. & Javoy, M., 2004. Constraining diamond metasomatic growth using C- and N-stable isotopes: examples from Namibia. *Lithos* **77**, 359-373.

Chalapathi Rao, N.V., Gibson S.A., Pyle, D.M., & Dickin, A.P., 2004. Petrogenesis of Proterozoic Lamproites and Kimberlites from the Cuddapah Basin and Dharwar Craton, Southern India. *J. Petrology* **45**, 907 - 948.

Chaussidon, M., Albarède, F., & Sheppard, S. M. F., 1989. Sulfur isotope variations in the mantle from ion-microprobe analyses of micro-sulfide inclusions. *Earth and Planetary Science Letters* **92**, 144-156.

Chepurov, A.I., Khokhryakov, A.F., Sonin, V.M., & Pal'yanov, Y.N., 1985. Forms of dissolution of diamond crystals in silicate melts at high pressure. *Doklady Akademii Nauk SSSR* **285**, 212–216 (in Russian).

Chepurov, A.I., Fedorov, I.I., & Sonin, V.M., 1997. Experimental modelling of the diamond formation processes. Novosibirsk: Siberian Branch RAS, 197 (in Russian).

Cherny, S.D., Fomin, A.S., Yanygin, Ju.T., Banzeruk, V.I., Kornilova, V.P., 1998. Geology and composition of the Nakyn field kimberlite pipes and diamond properties (Yakutia). *Extended Abstracts of 7th International Kimberlite Conference*, Cape Town, South Africa, 147-148.

Chertl, Y.-P., Neuser, R.D., Sobolev, N.V., & Shatsky, V.S., 2004. UHP-metamorphic rocks from Dora Maira/Western Alps and Kokchetav/Kazakhstan: new insights using cathodoluminescence petrography. *European Journal of Mineralogy* **16**, 49-57.

Chinn, I.L., Gurney, J.J., Milledge, H.J., Taylor, W.R., & Mccallum, M.E., 1995. CO_2 – bearing diamonds from the George Creek K1 kimberlite dyke of the Colorado-Wyoming State line district. *Ext. Abstr. 6th Internat. Kimb. Conf., Novosibirsk,* 113-115.

Chinner, J.A., and Cornell, D.H., 1974. Evidence of kimberlite-grospydite reaction, *Contrib. Mineral and Petrol* **45**, 153-160.

Chrenko, R.M., McDonald, R.S., & Darrow, K.A., 1967. Infra-red spectra of diamond coat. *Nature* **213**, (5075), 474-476.

Coleman, R.G., 1977. Ophiolites—Ancient Oceanic Lithosphere? Springer, Berlin, 229.

Coleman, R.G., Lee, D.E., Beatty, L.B., & Brannock, W.W., 1965. Eclogites and eclogites: their differences and similarities. *Geological Society of America Bulletin* **76**, 483–508.

Coltorti, M., Beccaluva, L., Bonadiman, C., Salvini, L., & Siena, F., 2000. Glasses in mantle xenoliths as geochemical indicators of metasomatic agents. *Earth and Planetary Science Letters* **183**, 303-320.

Craig, J.R. and Scott, S.D., 1974. Sulfide phase equilibria. In: P.H., Ribbe (ed.). *Sulfide Mineralogy*, Mineralogical Society of America, Washington, D.C., 1–110.

Daniels, L. R. M., and Gurney, J. J., 1999. Dokolwayo diamond carbon isotopes. In: *The J. B. Dawson volume; proceedings of the VIIth international kimberlite conference;* **1**, ed. J. J. Gurney, J. L., Gurney, M. D., Pascoe, and S. H., Richardson, Red Roof Designs, Cape Town, 143-147.

Davies, G., 1976. The nitrogen aggregate in diamond: its symmetry and possible structure. *Jour. Phys.* **C9**, L537-L542.

Davies, G., 1984. Diamond. Adam Hilger Ltd., Bristol, UK, 255.

Davis, G.L., Sobolev, N.V., & Kharkiv, A.D., 1980. New data on the age of Yakutian kimberlites obtained by U-Pb method on zircon. *Dokl. Akad. Nauk SSSR* **254**, 175-179.

Davies, R.M., O'Reilly, S.Y., & Griffin, W.L., 1999. Growth structures and nitrogen characteristics of group B alluvial diamond crystals from Bingara and Wellington, Eastern Australia. *Proceedings of the 7th International Kimberlite Conference*, **1**, 156-163.

Davies, R.M. , Griffin, W.L., O'Reilly, S.Y. & Doyle, B.J., 2004. Mineral inclusions and geochemical characteristics of microdiamonds from the DO27, A154, A21, A418, DO18, DD17 and Ranch Lake kimberlites at Lac de Gras, Slave Craton, Canada. *Lithos* **77**, 39-55.

Davies, R.M., Griffin, W.L., O'Reilly, S.Y. & McCandless, T.E., 2004. Inclusions in diamonds from the K14 and K10 kimberlites, Buffalo Hills, Alberta, Canada: diamond growth in a plume? *Lithos* **77**, 99-111.

Dawson, J.B. and Smith, J.V., 1975. Occurrence of diamond in a mica-garnet lherzolite xenolith from kimberlite. *Nature* **254**, 580-58.

Dawson, J.B., Smith, J.V. & Delaney, J.S., 1978. Multiple spinel–garnet peridotite transitions in upper mantle: evidence from a harzburgite xenolith. *Nature* **273**, 741–743.

Dawson, J.B., 1980. Kimberlites and their Xenoliths. Berlin: Springer-Verlag, 252.

Dawson, J.B., 1987. Metasomatized harzburgites in kimberlite and alkaline magmas: enriched restites and "flushed" lherzolites. In: Menzies, M.A., and Hawkesworth, C.J. (eds.). *Mantle Metasomatism*. London: Academic Press, 125-144.

Dawson, J.B., 2002. Metasomatism and partial melting in upper-mantle peridotite xenoliths from the Lashaine volcano, Northern Tanzania. *Journal of Petrology* **43**, 1749-1777.

De Corte, K., Cartigny, P., Shatsky, V.S., Sobolev, N.V., & Javoy, M., 1998. Evidence of fluid inclusions in metamorphic diamonds from the Kokchetav Massif, Northern Kazakhstan. *Geochimica et Cosmochimica Acta* **62**, 3765-3773.

De Corte, K., Cartigny, P., Shatsky, V.S., De Paepe, P., Sobolev, N.V., & Javoy, M., 1999. Characteristics of microdiamonds from UHPM rocks of the Kokchetav Massif, Kazakhstan. In: Gurney, J.J., Gurney, J.L., Pascoe, M.D. and Richardson, S.H. (eds.). *Proceedings Volume of 7th International Kimberlite Conference*, **1**. Cape Town: Red Roof Design, 174-182.

De Stefano, A., Lefebvre, N. & Kopylov, M., 2006. Enigmatic diamonds in Archean calc-alkaline lamprophyres of Wawa, southern Ontario, Canada. *Contributions to Mineralogy and Petrology* **151**, 2: 158-173.

Deines, P., 1980. The carbon isotopic composition of diamonds: Relationship to diamond shape, color, occurrences and vapor composition. *Geochimica et Cosmochimica Acta* **44**, 943-961.

Deines, P., 1992. Mantle carbon: concentration, mode of occurrence and isotopic composition. In: Schidlowski, M., Golubic, S., Kimberley, M.M., McKirdy, D.M., Trudinger, E.A. (eds.). *Early Organic Evolution*. Springer, Berlin, 133-146.

Deines, P., 2002. The carbon isotope geochemistry of mantle xenoliths. *Earth-Science Review* **58**, 247-278.

Deines P. and Harris, J.W., 1994. On the importance of fluids for diamond growth (ext. abs.). *Goldschmidt Conference*, Edinburgh, 219–220.

Deines, P. and Harris, J.W., 1995. Sulfide inclusion chemistry and carbon isotopes of Africa diamonds. *Geochimica et Cosmochimica Acta* **59**, 3173-3188.

Deines, P., Harris, J.W., & Gurney, J.J., 1977. Carbon isotope ratios, nitrogen content and aggregation state, and inclusion chemistry of diamonds from Jwaneng, Botswana. *Geochim. Cosmochim. Acta* **61**, 3993–4005.

Deines, P., Gurney, J.J., & Harris, J.W., 1984. Associated chemical and carbon isotopic composition variations in diamonds from Finsch and Premier kimberlite, South Africa. *Geochimica et Cosmochimica Acta* **48**, 325-342.

Deines, P., Harris, J.W., Spear, P.M., & Gurney, J.J., 1989. Nitrogen and $\delta^{13}C$ content of Finch and Premier diamonds and their implications. *Geochimica et Cosmochimica Acta* **53**, 1367-1378.

Deines, P., Harris, J.W., Robinson, D., Gurney, J.J., & Shee, S.R., 1991. Carbon and oxygen isotope variations in diamond and graphic eclogites from Orapa. *Geochimica et Cosmochimica Acta* **55**, 515-524.

Deines, P., Harris, J.W., & Gurney, J.J., 1993. Depth-related isotope and nitrogen concentration variability in the mantle below the Orapa kimberlite, Bostwana. *Geochimica et*

Cosmochimica Acta **57**, 2781-2796.

Deines, P., Viljoen, F., & Harris, J.W., 2001. Implications of the carbon isotope and mineral inclusion record for the formation of diamonds in the mantle underlying a mobile belt: Venetia, South Africa. *Geochimica et Cosmochimica Acta*, **65**, 813-838.

Deines P. and Harris J.W., 2004, New insights into the occurrence of ^{13}C-depleted carbon in the mantle from two closely associated kimberlites: Letlhakane and Orapa, Botswana. *Lithos* **77**, 125-142.

Denison, C., Carlson, W.D., & Ketcham, R.A., 1997. Three-dimensional quantitative textural analysis of metamorphic rocks using high-resolution computed X-ray tomography: Part I. Methods and techniques. *Jour. Metamorphic Geol.* **15**, 29–44.

Denison, C., Carlson, W.D., & Ketcham, R.A., 1997. Three-dimensional quantitative textural analysis of metamorphic rocks using high-resolution computed X-ray tomography: Part II. Application to natural samples.. *Jour. Metamorphic Geol.* **15**, 45-57..

Dobrzhinetskaya, L.F. & Green, H.W.,II 2007. Diamond synthesis from graphite in the presence of water and SiO (sub 2) ; implications for diamond formation in quartzites from Kazakhstan. *International Geology Review,* **49**, 389-400.

Dobrzhinetskaya, L.F., Wirth, R., & Green II., H.W., 2006. Nanometric inclusions of carbonates in Kokchetav diamonds from Kazakhstan: A new constraint for the depth of metamorphic diamond crystallization. *Earth and Planetary Science Letters 243,* 85-93.

Dong, Z., Taylor, L.A., & Taylor, D.-H., 1998. Pyropes of China: Peridotite xenoliths from kimberlites versus megacrysts in basalts. *International Geology Review* **39**, 141-150.

Donnelly, C.L., Stachel, T., Creighton, S., Muehlenbachs, K., & Whiteford, S., 2007. Diamonds and their mineral inclusions from the A154 South Pipe, Diavik Diamond Mine, Northwest Territories, Canada. *Lithos* 98, 160-176.

Doroshev, A.M., Brey, G.P., Girnis, A.V., Turkin, A.I. & Kogarko, L.N., 1997. Pyrope–knorringite garnets in the Earth's upper mantle: experiments in the $MgO–Al_2O_3–SiO_2–Cr_2O_3$ system. *Russian Geology and Geophysics* **38**, 559–586.

Doroshkevich A.G., Wall, F., & Ripp, G. S., 2007. Magmatic graphite in dolomite carbonatite at Pogranichnoe, North Transbaikalia, Russia. *Contributions to Mineralogy and Petrology* **153**, 339-353.

Downes, H., 2001. Formation and modification of the shallow sub-continental lithospheric mantle: a review of geochemical evidence from ultramafic xenolith suites and tectonically emplaced ultramafic massifs of western and central Europe. *Journal of Petrology* **42**, 233-250.

Downes P.J., Wartho J.-A., & Griffin B.J., 2006. Magmatic Evolution and Ascent History of the Aries Micaceous Kimberlite, Central Kimberley Basin, Western Australia: Evidence from Zoned Phlogopite Phenocrysts, and UV Laser 40Ar/39Ar Analysis of Phlogopite–Biotite. *J. Petrology* **47**, 1751 - 1783.

Ducea, M.N., Saleeby, J., Morrison, J., & Valencia, V.A., 2005. Subducted carbonates, metasomatism of mantle wedges, and possible connections to diamond formation: An example from California, *American Mineralogist* **90**, 864 - 870.

Durazzo, A., and L.A. Taylor, 1982, Exsolution in the Mss-pentlandite system: Textural and genetic implications for Ni-sulfide ores. Mineral. Deposita 17, 313-332.

Durazzo, A., L.A. Taylor, and J.W. Shervais, 1984, Ultramafic lamprophyre in a carbonate platform environment, Mt. Queglia, Abruzzo, Italy. Neues Jahrbuch Mineral. 150, 199-217.

Eggler, D.H., 1987. Solubility of major and trace elements in mantle metasomatic fluids: experimental constraints, In: Menzies, M.A., Hawkesworth, C.J. (eds.). Mantle Metasomatism, Academic Press, London, 21-41.

Eiler, J.M., McInnes, B. Valley, J.W., Graham, C.M., & Stolper, E.M., 1998. Oxygen isotope evidence for slab-derived fluids in the sub-arc mantle. *Nature* **393**, 777-781.

Eiler, J.M., 2001. Oxygen isotope variations of basaltic lavas and upper mantle rocks. In: Valley, J.W., Cole D.R. (eds.), *Reviews in Mineralogy and Geochemistry* ***43****: Stable Isotope Geochemistry*, 319-364.

Eldridge, C.S., Compston, W., Williams, I.S., Harris, J.W., & Bristow, J.W., 1991. Isotope evidence for the involvement of recycled sediments in diamond formation. *Nature* **353**, 649-653.

Ellis, D.I., and Green, D.N., 1979. An experimental study of the effect of Ca upon garnet - clinopyroxene Fe-Mg exchange equilibria. *Contributions to Mineralogy and Petrology* **71**, 13-22.

Erlank, A.J., Water, F. G., Hawkesworth, C. J., Haggerty, S. E., Allsopp, H. L., Rickard, R. S., & Menzies, M., 1987. Evidence for mantle metasomatism in peridotite nodules from kimberlite pipes, South Africa. In: Menzies, M.A., and Hawkesworth, C.J. (eds.). *Mantle Metasomatism*. London: Academic Press, 221-311.

Evans, T., Qi, Z., 1982. The kinetics of aggregation of nitrogen atoms in diamond. *Proceedings of the Royal Soc. London* **A381**, 159-178.

Fadili, S. El., & Demaiffe, D., 1999. Petrology of eclogite and granulite nodules from the Mbuji Mayi Kimberlites, Kasai, Congo: Significance of kyanite-omphacite intergrowths. In: Gurney, J.J., Gurney, J.L., Pascoe, M.D., and Richardson, S.H. (eds.). *Proceedings Volume of 7th International Kimberlite Conference*, *1*. Cape Town: Red Roof Design, 205-213.

Farquhar, J., Wing, B. A., McKeegan, K. D., Harris, J. W., Cartigny, P., & Thiemens, M. H., 2002. Mass-Independent Sulfur of Inclusions in Diamond and Sulfur Recycling on Early Earth. *Science* **298,** 2369-2372.

Faure, G., 1986. Principles of Isotope Geology. 2nd Edition. John Wiley & Sons, New York. 589.

Fedortchouk, Y.and Canil D., 2004. Intensive Variables in Kimberlite Magmas, Lac de Gras, Canada and Implications for Diamond Survival. *J. Petrology*, **45**, 1725 - 1745.

Fedortchouk, Y., Canil, D., & Carlson, J.A., 2005. Dissolution forms in Lac de Gras diamonds and their relationship to the temperature and redox state of kimberlite magma. *Contributions to Mineralogy and Petrology* **150**, 54-69.

Finnerty, A. A., 1989. Xenolith-derived mantle geotherms: Whither the inflection? *Contrib. Mineral. Petrol.* **102**, 367-375.

Finnerty, A.A., and Boyd, F.R., 1984. Evaluation of thermobarometers for garnet peridotites. *Geochimica et Cosmochimica Acta* **48**, 15-27.

Finnerty, A. A., and Boyd, F.R., 1987. Thermobarometry for garnet peridotite xenoliths: a basis for upper mantle stratigraphy. In: Nixon, P.H. (ed.). *Mantle Xenoliths*. Wiley, 381-402.

Finnie, K.S., Fisher, D., Griffin, W.L., Harris, J.W. & Sobolev, N.V., 1994. Nitrogen aggregation in metamorphic diamonds from Kazakhstan. *Geochimica et Cosmochimica Acta,* **58**, 5173-5177.

Fipke, C.E., Gurney, J.J., & Moore, R.O., 1995. Diamond exploration techniques emphasizing indicator mineral geochemistry and Canadian examples. *Geol. Surv. Canada, Bulletin* **423**, 1-86.

Fitzsimons, I.C.W., Harte, B., Chinn, J.J., Gurney, J.J., & Taylor, W.R., 1999. Extreme chemical variation in complex diamonds from George Creek, Colorado: a SIMS study of carbon isotope composition and nitrogen abundance. *Min. Mag.* **63**, 857-878.

Franz, L., Brey, G.P., & Okrusch, M., 1995. Metasomatic reequilibration of mantle xenoliths from the Gibeon kimberlite province, Namibia., *Extended Abstracts of 6th International Kimberlite Conference*, Novosibirsk (Russia), 169-171.

Fesq, H.W., Bibby, D.M., Erasmus, C.S., Kable, E.J.D. & Sellschop, J.P.F., 1975. A comparative trace element study of diamonds from Premier, Finsch and Jagersfontein mines, South Africa. *Physics and Chemistry of the Earth* **9**, 817-836.

Fung, A. T., and Haggerty, S. E., 1995. Petrography and mineral composition of eclogites from the Koidu kimberlite complex, Sierra Leone. *J. Geophys. Res. B.* **100**, 20451-20473.

Galimov, E.M., 1984. The relationship between formation conditions and variations in isotope composition of diamonds. *Geokhimiya* **8**, 1091-1118 (in Russian).

Galimov, E.M., 1985. The relation between formation conditions and variations in isotopic composition of diamonds. *Geochem. Int.* **22**, 118-142.

Galimov, E.M., 1991. Isotope fractionation related to kimberlite magmatism and diamond formation. *Geochim. Cosmochim. Acta* **55**, 1697-1708.

Garlick, G. D., MacGregor, I. D. & Vogel, D. E., 1971. Oxygen isotope ratios in eclogites from kimberlites. *Science* **172**, 1025-1027.

Gautheron, C., Cartigny, P., Moreira, M., Harris, J.W. & Allègre, C.J., 2005. Evidence for a mantle component shown by rare gases, C and N isotopes in polycrystalline diamonds from Orapa (Botswana). *Earth and Planetary Science Letters* **240**, 559-572.

Gillet, P., Sautter, V., Harris, J., Reynard, B., Harte, B. & Kunz, M. 2002, "Raman spectroscopic study of garnet inclusions in diamonds from the mantle transition zone", *American Mineralogist,* **87**, 312-317.

Girnis, A. V., and Brey, G. P., 1999. Garnet-spinel-olivine-orthopyroxene equilibria in the FeO-MgO-Al_2O_3-SiO_2-Cr_2O_3 system: II. Thermodynamic analysis. *Eur. J. Mineral.* **11**, 619-636.

Golovin, A.V., Sharygin, V.V., Pokhilenko, N.P., Malkovets, V.G., Kolesov, B.A., & Sobolev, N.V., 2003. Secondary melt inclusions in olivine from unaltered kimberlite of the Udachnaya-East pipe, Yakutia. *Doklady Earth Sciences* **388**, 93-96.

Golovin, A.V., Sharygin, V.V. & Pokhilenko, N.P., 2007. Melt inclusions in olivine phenocrysts from the unaltered kimberlites of East Udachnaya Pipe, Yakutia; aspects of the kimberlitic magma evolution at the late stages of crystallization. *Petrologiya* **15**, 178-195.

Golubeva, Y.Y., Pervov, V.A., and Kononova, V.A., 2006. Petrogenesis of autoliths from kimberlitic breccias in the V. grib pipe (arkhangelsk district). *Doklady Earth Sciences* **411**, 1257-1262.

Gornova, M.A., Solovjeva, L.V., Glasunov, O.M., and Belozerova, O.Y., 2001. The problems of Arhaean lithospheric mantle formation – geochemical analysis of coarsegrained peridotites from kimberlites of the Siberian platform. *Alkaline magmatism and the problem of mantle sources*. Irkutsk: Russian Academy of Sciences, 226-245 (in Russian).

Gorshkov, A. I., Titkov, S.V., Nan, B.Y., Ryabchlkov, I.D., and Magazina, L.O., 2006. Mikrovklyuchenlya v almazakh oktaedricheskogo gabitusa iz kimberlltov provintsii shandu vostochnogo kitaya. microlnclusions In diamonds of octahedral habit from kimberlites of the shandong province, Eastern China. *Geologiya Rudnykh Mestorozhdeniy* **48**, 375-384.

Gounelle, M., Young, D.E., Shahar, A., Tonui, E., and Kearsley, A., 2007. Magnesium Isotopic constraints on the origin of CB sub(b) chondrites. *Earth and Planetary Science Letters* **256**, 521-533.

Green, D. H., and Ringwood, A. E., 1970. Mineralogy of peridotitic compositions under upper mantle conditions. *Phys. Earth Planet. Int.* **3**, 359-371.

Gregory, R.T., and Taylor, H.P. Jr., 1981. An oxygen isotope profile in a section of Cretaceous oceanic crust, Samail ophiolite, Oman: Evidence for $\delta^{18}O$ buffering of the oceans by deep (>5 km) seawater-hydrothermal circulation at mid-ocean ridges. *Jour. Geophys. Res.* **86**, 2737-2755.

Griffin, B.J., Rissanen, J., Pooley, G.D., Lee, D.C., Macdonald, I. & Kinny, P.D., 1995. A new diamondiferous eclogite-bearing kimberlitic occurrence from Finland. *Ext. Abs. Sixth Int. Kimb. Conf., Novosibirsk,* 198-200.

Griffin, W.L. and Ryan, C.G., 1995. Trace elements in indicator minerals: area selection and target evaluation in diamond exploration. *Journal of Geochemical Exploration* **53**, 311–337.

Griffin, W.L, Jagues, A.L., Sie, S.H., Ryan, C.G., Coursens, D.R., & Suter, G.F., 1988. Conditions of diamond growth: a proton microprobe study of inclusions in West Australian diamonds. *Contribution to Mineralogy and Petrology* **88**, 143-158.

Griffin, W.L., Cousens, D. R., Ryan, C. G., & Suter, G. F., 1989. Ni in chrome garnet: A new geothermometer. *Contrib. Mineral. Petrol.* **103**, 199-202.

Griffin, W. L., Smith, D., Boyd, F. R., Cousens, D. R., Ryan, C. G., Sie, S. H., & Suter, G. F., 1989. Trace-element zoning in garnets from sheared mantle xenoliths. *Geochimica et Cosmochimica Acta* **53**, 561-567.

Griffin, W.L., Gurney, J.J., and Ryan, C.G., 1992. Variations in trapping temperatures and trace elements in peridotite-suite inclusions from African diamonds: evidence for two inclusion suites, and implications for lithosphere stratigraphy. *Contributions to Mineralogy and Petrology* **110**, 1-15.

Griffin, W.L., Sobolev, N.V., Ryan, C.G., Pokhilenko, N.P., Win, T.T., & Yefimova E.S., 1993. Trace elements in garnets and chromites: diamond formation in the Siberian lithosphere. *Lithos* **29**, 235-256.

Griffin, W.L., Ryan, C.G., Gurney, J.J., Sobolev, N.V. & Win, T.T., 1994. "Chromite macrocrysts in kimberlites and lamproites; geochemistry and origin; Kimberlites, related rocks and mantle xenoliths, *CPRM - Special Publication,* vol. 1A, pp. 366-377.

Griffin, W.L., Kaminsky, F.V., Ryan, C.G., O'Reilly, S.Y., Win, T.T., & Ilupin, I.P., 1996. Thermal state and composition of the lithospheric mantle beneath the Daldyn kimberlite field, Yakutia. *Tectonophysics* **262**, 19-33.

Griffin, W.L., Fisher, N. I., Friedman, J., Ryan, C. G., & O'Reilly, S. Y., 1999. Cr-pyrope garnets in the lithospheric mantle. I. Compositional systematics and relations to tectonic setting. *Journal of Petrology* **40**, 679-704.

Griffin, W.L, Ryan, C.G., Kaminsky, F.V., O'Reilly, S.Y., Natapov, L.M., Win, T.T., Kinny, P.D.,

& Ilupin, I.P., 1999. The Siberian lithosphere traverse: Mantle terranes and the assembly of the Siberian craton. *Tectonophysics* **310**, 1-35.

Griffin, W.L., O'Reilly, S. Y., & Ryan, C. G., 1999. The composition and origin of sub-continental lithospheric mantle. In: Fei, Y., Bertka, C. M., and Mysen, B. O. (eds.), Mantle petrology: Field observations and high pressure experimentation, Geochemical Society Special Publication 6, The Geochemical Society, 332

Griffin, W.L., Shee, S.R., Ryan, C.G., Win, T.T., & Wyatt, B.A., 1999. Harzburgite to lherzolite and back again: metasomatic processes in ultramafic xenoliths from the Wesselton kimbelite, Kimberley, South Africa. *Contributions to Mineralogy and Petrology* **134**, 232-250.

Griffin, W.L., Spetsius, Z.V., Pearson, N.J., & O'Reilly, S.Y., 2002. In-situ Re-Os analysis of sulfide inclusions in kimberlitic olivine: new constraints on depletion events in the Siberian lithospheric mantle. *Geochemistry, Geophysics, Geosystems* **3**, 1069.

Griffin, W.L., Natapov, L.M., O'Reilly, S.Y., Achterbergh, E., Cherenkova, A.F., and Cherenkov. V.G., 2005. The kharamai kimberlite field, Siberia; modification of the lithospheric mantle by the Siberian trap event. *Lithos* **81**, 167-187.

Grütter, H.S., Gurney, J.J., Menzies, A.H. & Winter, F., 2004. An updated classification scheme for mantle-derived garnet, for use by diamond explorers. *Lithos* **77**, 841–857.

Grütter, H., Latti, D., & Menzies, A., 2006. Cr-Saturation Arrays in Concentrate Garnet Compositions from Kimberlite and their Use in Mantle Barometry. *J. Petrology* 47, 801 - 820.

Guiraud, M. and Powell, R., 2006. P-V-T relationships and mineral equilibria in inclusions In minerals. *Earth and Planetary Science Letters* **244**, 683-694.

Gurney, J. J. 1984. A correlation between garnets and diamonds. In: Glover, J. E. & Harris, P. G. (eds) Kimberlite Occurrence and Origin: a Basis for Conceptual Models in Exploration. Geology Department and University Extension, University of Western Australia Publication 8, 143–166.

Gurney, J.J., 1989. Diamonds. In: Ross, J. et al. (eds.). *Diamonds, Kimberlites and Related Rocks, Vol 2. Geol. Soc. Aust. Spec. Publ.* **14**, Blackwell Scientific Publications, Melbourne, 935-965.

Gurney, J. J., 1990. The diamondiferous roots of our wandering continent. *South African Journal of Geology* **93**, 423-437.

Gurney, J.J. and Switzer, G.S., 1973. The discovery of garnets closely related to diamonds in the Finsch pipe, South Africa. *Contributions to Mineralogy and Petrology* **39**, 103–116.

Gurney, J.J. and Harte, B., 1980. Chemical variations in upper mantle nodules from southern African kimberlites. *Phil. Trans. Roy. Soc. Lond.* **A 297**, 273-293.

Gurney, J.J., and Hatton, C.J., 1989. Diamondiferous minerals from the Star Mine, South Africa. In: Ross, J. et al. (eds.). *Diamonds, Kimberlites and Related Rocks, Vol 2. Geol. Soc. Aust. Spec. Publ.* **14**, Blackwell Scientific Publications, Melbourne, 1022-1028.

Gurney, J.J., and Zweistra, P., 1995. The interpretation of the major element compositions of mantle minerals in diamond exploration. *Journal of Geochemical Exploration* **53**, 293-309.

Gurney, J.J., Harris J.W., and Rickard R.S., 1984. Minerals associated with diamonds from the Roberts Victor mine. In: Kornprobst, J. (ed.). Kimberlites II: The Mantle and Crust - Mantle Relationships. Elsevier, Amsterdam, 3-10.

Gurney, J. J., Moore, R. O., Otter, M. L., Kirkley, M. B., Hops, J. J., & McCandless, T. E. 1991. Southern African kimberlites and their xenoliths. In: Kampunzu, A. B., and Lubala, R. T. (eds.). *Magmatism in Extensional Structural Settings*. Springer-Verlag, 495-536.

Gurney J.J., Hildebrand P.R., Carlson J.A., Fedortchouk Y. and Dyck D.R. , 2004, The morphological characteristics of diamonds from the Ekati property, Northwest Territories, Canada. *Lithos* **77**, 21-38.

Haggerty, S.E., 1986. Diamond genesis in a multiply-constrained model. *Nature*, **320**, 34-38.

Haggerty, S.E., 1987. Metasomatic mineral titanates in upper mantle xenoliths. In: Nixon, P. H. (ed.). Mantle Xenoliths. Wiley, 671-690.

Haggerty, S. E., 1989. Upper mantle opaque mineral stratigraphy and the genesis of metasomites and alkali-rich melts. In: Ross, J. et al. (eds.). Diamonds, Kimberlites and Related Rocks, Volume 2. Geol. Soc. Aust. Spec. Publ. **14**, Blackwell Scientific Publications, Melbourne, 687-699.

Haggerty, S. E., and Sautter, V., 1990. Ultradeep, greater than 300 kilometers, ultramafic upper mantle xenoliths. *Science* **248**, 993-996.

Haggerty, S.E., and Birkett, T., 2004, Geological setting and chemistry of kimberlite clan rocks in the Dharwar Craton, India. *Lithos* **76**, 535-549.

Hall, A.E., and Smith, C.B., 1984. Lamproite diamonds - are they different? In: Glover, J.E. and Harris, P.G. (eds.). Kimberlite Occurrence and Origin. University Extension Publication **8**. The University of Western Australia, 167-212.

Hammer, P.T.C., Clowes, R.M. & Ramachandran, K., 2004. Seismic reflection imaging of thin, kimberlite dykes and sills: exploration and deposit characterization of the Snap Lake dyke, Canada. *Lithos* **76**, 359-367.

Harley, S. L., 1984. An experimental study of the partitioning of Fe and Mg between garnet and orthopyroxene. *Contrib. Mineral. Petrol.* **86**, 359-373.

Harlow, G.E., 1997. K in clinopyroxene at high pressure and temperature: An experimental study. *American Mineralogist* **82**, 259-269.

Harlow, G.E., 1998. What are diamonds? In: Harlow, G.E. (ed.). The Nature of Diamonds. Cambridge University Press, Cambridge, UK, 5-22.

Harlow, G.E. and Veblen, D.R., 1991. Potassium in clinopyroxene inclusions from diamonds. *Science* **251**, 652-655.

Harris, J. W., 1968. The recognition of diamond inclusions. Pt. 1: Syngenetic inclusions. *Industrial Diamond Review* **28**, 402-410.

Harris, J.W., 1987. Recent physical, chemical, and isotopic research of diamond in Mantle xenoliths, ed. P.H. Nixon, John Wiley & Sons, Chichester, United Kingdom (GBR)

Harris, J. W., 1987. Recent physical, chemical and isotopic research of diamond. In: Nixon, P. H. (ed.). Mantle Xenoliths. John Wiley and Sons, England, 477-500.

Harris, J.W., 1992. Diamond geology. In: Field, J.E. (ed.). The properties of natural and synthetic diamond, London, UK, Academic Press, 345-393.

Harris, J.W. 2006. Mineral inclusions in diamond; a Scouser's view. Abstracts of the 16th annual V. M. Goldschmidt conference, *Geochimica et Cosmochimica Acta,* **70**, A233.

Harris, J. W. and Gurney, J. J., 1979. Inclusions in diamond. Academic Press, London, 555-591.

Harris, J.W., Hawthorne, J.B., Oosterveld, M.M. & Smith, C.B., 1986. A comparison of characteristics of diamonds from the Orapa and Jwaneng kimberlite pipes in Botswana. Fourth international kimberlite conference, *Abstracts - Geological Society of Australia* 16, 395-397.

Harris, J.W., Spear, P.M. & Smith, C.B., 1986. Systematic studies of nitrogen in diamonds from known sources; Fourth international kimberlite conference. *Abstracts - Geological Society of Australia,* vol. 16, pp. 398-400.

Harris, J.W., Duncan, D.J., Zhang, F., Mia, Q. & Zhu, Y., 1991. The physical characteristics and syngenetic inclusion geochemistry of diamonds from pipe 50, Liaoning Province, People's Republic of China; Volume 2; Diamonds; characterization, genesis and exploration", *Proceedings of the International Kimberlite Conference* **5**, 106-115.

Harris, J.W., Hutchison, M.T., Hursthouse, M., Light, M., & Harte, B., 1997**.** A new tetragonal silicate mineral occurring as inclusions in lower-mantle diamonds. *Nature* **387**, 486-488.

Harris, J.W., Stachel, T., Léost, I., & Brey, G.P., 2004. Peridotitic diamonds from Namibia: constraints on the composition and evolution of their mantle source. *Lithos* **77**, 209-223.

Harte, B., Gurney, J.J., & Harris, J.W., 1980. The formation of peridotitic suite inclusions in diamonds. *Contributions to Mineralogy and Petrology*, **72**, 181-190.

Harte, B., Winternburn, P.A., and Gurney, J.J., 1987. Metasomatic and enrichment phenomena in garnet peridotite facies mantle xenoliths from the Matsoku kimberlite pipe, Lesotho. In: Menzies, M.A., and Hawkesworth, C.J. (eds.). Mantle Metasomatism. London: Academic Press, pp. 145-220.

Harte, B., Fitzsimons, C.W., Harris, J.W., & Otter, M.L., 1999. Carbon isotope ratios and nitrogen abundances in relation to cathodoluminescence characteristics for some diamonds from Kaapvaal provinces, South Africa. *Min. Mag.* **63**, 829-856.

Harte, B., Hutchison, M.T., Lee, M., & Harris, J.W., 1999. Inclusions of (Mg, Fe)O in mantle diamonds. *Extended Abstracts of 7th International Kimberlite Conference*, Cape Town, South Africa, 308-310.

Harte, B., Harris, J.W., Wilding, M., Sautter, V., McCammon, C. & Parsons, I., 2002. Eclogite-garnetite inclusions in diamonds from the Sao Luiz area, Brazil; 18th general meeting of the International Mineralogical Association; mineralogy for the millennium; programme with abstracts", *Abstracts of the General Meeting of the International Mineralogical Association,* **18**, 74.

Hatton, C. J., and Gurney, J. J., 1987. Roberts Victor eclogites and their relation to the mantle. In: Nixon, P. H. (ed.). Mantle Xenoliths. Wiley, 453-463.

Hauri, E.H., Shimizu, N., Dieu, J.J., & Hart, S.R., 1993. Evidence for hotspot-related carbonatite metasomatism in the oceanic upper mantle. *Nature* **365**, 221-227.

Hauri, E., Pearson, G., Bulanova, G.P., and Milledge, H.J., 1999. Microscopic variations in C and N isotopes within mantle diamonds revealed by SIMS. In: Gurney, J.J., Gurney, J.L., Pascoe, M.D., and Richardson, S.H. (eds.), *Proceedings Volume of 7th International Kimberlite Conference, 1*. Cape Town: Red Roof Design, 341-347.

Hauri, E. H., Wang, J., Pearson, D.G., & Bulanova, G.P., 2002. Microanalysis of □^{13}C, □ ^{15}N, and N abundances in diamonds by secondary ion mass spectrometry. *Chemical Geology* **185**, 149–163.

Heaman, L. M., Kjarsgaard, B. A. & Creaser R. A., 2003. The timing of kimberlite Magmatism in North America: implications for global kimberlite genesis and diamond exploration. *Lithos* **71**, 153-184.

Heaman, L.M., Creaser R.A., Cookenboo H.O., & Chacko T., 2006. Multi-Stage Modification of the Northern Slave Mantle Lithosphere: Evidence from Zircon- and Diamond-Bearing Eclogite Xenoliths Entrained in Jericho Kimberlite, Canada. *J. Petrology* **47**,821 - 858.

Helmstaedt, H.H., 1993. Natural diamond occurrences and tectonic setting of "primary" diamond deposits. *Prospectors and Developers Association of Canada*, 3-72.

Hervig, R.L., Smith, J.V., Steele, I.M., Gurney, J.J,, Meyer, H.O.A., Harris, J.W., 1980. Diamonds— minor elements in silicate inclusions: Pressure-temperature implications. *Jour. Geophys. Res.* **85**, 6919–6929.

Hills, D. V., and Haggerty, S. E., 1989. Petrochemistry of eclogites from the Koidu Kimberlite Complex, Sierra Leone. *Contrib. Mineral. Petrol.* **103**, 397-422.

Hoal, K.E.O., Hoal, B.G., Erlank, A.J., and Shimizu, N., 1994. Metasomatism of the mantle lithosphere recorded by rare earth elements in garnets. *Earth and Planetary Science Letters* **126**, 303-313.

Hunter, R. H., and Taylor, L. A., 1982. Instability of garnet from the mantle: Glass as evidence of metasomatic melting. *Geology* **10**, 617-620.

Hunter, R.H., and Taylor, L.A., 1984. Magma mixing in the low velocity zone: Kimberlitic megacrysts from Fayette County, Pennsylvania. *Amer. Mineral*. **69**, 30-40.

Hunter, R.H., Kissling, R.D., & Taylor, L.A., 1984. Mid-to late-state kimberlitic melt evolution: Phlogopites and oxides from the Fayette County kimberlite, Pennsylvania. *Amer. Mineral*. **69**, 16-29.

Hutchison M., Hursthouse M., & Light M., 2001. Mineral inclusions in diamonds: associations and chemical distinctions around the 670-km discontinuity. *Contributions to Mineralogy and Petrology* **142**,119-126.

Hutchison, M. T., Nixon, P. H. & Harley, S. L., 2004. Corundum inclusions in diamonds—discriminatory criteria and a corundum compositional dataset. *Lithos* **77**, 273-286.

Hwang, S.-L., Chu, H.-T., Yui, T.-F., Shen, P., Schertl, H.-P., Liou, J.G., Sobolev, N.V., 2006. Nanometer-size P/K-rich silica glass (former melt) inclusions in microdiamond from the gneisses of Kokchetav and Erzgebirge massifs: Diversified characteristics of the formation media of metamorphic microdiamond in UHP rocks due to host-rock buffering. *Earth and Planetary Science Letters* **243**, 94-106.

Hwang, S.-L., Shen, P., Chu, H.-T., Yui, T.-F., Liou, J.G., Sobolev, N.V., Shatsky, V.S., 2005. Crust-derived potassic fluid in metamorphic microdiamond. *Earth and Planetary Science Letters* **231**, 295-306.

Ionov, D.A., Ashchepkov, I.V., Stosch, H.-G., Witt-Eickschen, G. & Seck, H.A., 1993. Garnet peridotite xenoliths from the Vitim volcanic field, Baikal region: the nature of the garnet–spinel peridotite transition zone in the continental mantle. *Journal of Petrology* **34**, 1141–1175.

Ionov, D. A., Chanefo, I., and Bodinler, J-L., 2005. Origin of Fe-rich Iherzolites and wehrlites from Tok, SE Siberia by reactive melt percolation In refractory mantle peridotites. *Contributions to Mineralogy and Petrology* **150**, 335-353.

Ionov, D. A., Griffin, W. L., and O'Reilly, S. Y., 1997. Volatile-bearing minerals and lithophile trace elements in the upper mantle. *Chemical Geology* **141**, 153-184.

Ireland, T. R., Rudnick, R. L, & Spetsius, Z. V., 1994. Trace elements in diamond inclusions from eclogites reveal link to Archean granites. *Earth and Planetary Science Letters* **128**, 199-213.

Irifune, T., Ohtani, E. & Kumazawa, M., 1982. Stability field of knorringite $Mg_3Cr_2Si_3O_{12}$ at high pressure and its implication to the occurrence of Cr-rich pyrope in the upper mantle. *Physics of the Earth and Planetary Interiors* **27**, 263–272.

Irifune, T., 1987. An experimental investigation o the pyroxene-garnet transformation in a pyrolite composition and its bearing on the composition of the mantle. *Earth Planet. Sci. Lett.* **45**, 324-336.

Irvine, G. J., Carslon, R. W., Kopylova, M. G., Pearson, D. G., Shirey, S. B., and Kjarsgaard, B. A., 1999. Age of the lithospheric mantle beneath and around the Slave craton: A rhenium-osmium isotopic study of peridotite xenoliths from the Jericho and Somerset Island kimberlites. *Abst. Ninth Annual V.M. Goldschmidt Conference,* LPI contribution no. 971, Lunar and Planetary Institute, Houston, 134-135.

Irvine, G. J., 2002. Time constraints on the formation of lithospheric mantle beneath cratons: A Re-Os isotope and Platinum Group Element study of peridotite xenoliths from Northern Canada and Lesotho, Unpubl. Ph.D. Thesis, University of Durham.

Izraeli, E.S., Harris, J.W. & Navon, O. 1999. Raman barometry of diamond formation. *Earth and Planetary Science Letter,* **173**, 351-360.

Izraeli, E.S., Harris, J.W., & Navon, O., 2001. Brine inclusions in diamonds: a new upper mantle fluid. *Earth and Planetary Science Letters* **187**, 323-332.

Izraeli, E.S., Harris, J.W., & Navon O., 2004. Fluid and mineral inclusions in cloudy diamonds from Koffiefontein, South Africa. *Geochimica et Cosmochimica Acta* **68**, Issue 11, 2561-2575.

Jackson, D.E., R.H. Hunter, and L.A. Taylor, 1983, Petrogenesis of shallow-level kimberlite from New York State. Jour. Geol., 361-373.

Jacob D.E., 2004. Nature and origin of eclogite xenoliths from kimberlites. *Lithos* **77,** 295–316.

Jacob, D., and Jagoutz, E., 1994. A diamond-graphite bearing eclogite xenolith from Roberts Victor, South Africa: Implications for petrogenesis from Pb-, Nd-, and Sr-isotopes. In: Meyer, H. O. A., and Leonardos, O. H. (ed.). Kimberlites, related rocks and mantle xenoliths. 1, CPRM Spec. Publ., 304-317.

Jacob, D. E., and Foley, S. F., 1999. Evidence for Archean ocean crust with low high field strength element signature from diamondiferous eclogite xenoliths. *Lithos* **48**, 317-336.

Jacob, D.E., Jagoutz, E., Lowry, D., Mattey, D., & Kudrjavtseva, G.P., 1994. Diamondiferous eclogites from Siberia: Remnants of Archean oceanic crust. *Geochimica et Cosmochimica Acta* **58**, 5191-5207.

Jacob, D., Jagoutz, E., Lowry, D. & Zinngrebe, E., 1998. Comment on 'The origins of Yakutian eclogite xenoliths' by Snyder, G.A., Taylor, L.A., Crozaz, G., Halliday, A.N., Beard, B.L., Sobolev, V.N. & Sobolev, N.V., *J. Petrol.* **39**, 1527-1533.

Jacob E., Kronz, A., and Viljoen K.S., 2004. Cohenite, native iron and troilite inclusions in garnets from polycrystalline diamond aggregates. *Contributions to Mineralogy and Petrology* **146**,566-576.

Jacob, D. E., Bizimis, M., and Salters, V.I.M., 2005. Lu/Hf and geochemlcal systematics of recycled ancient oceanic crust; evidence from Roberts Victor eclogites. *Contributions to Mineralogy and Petrology* **148**, 707-720.

Janse, A. J. A., 1992. New ideas in subdividing cratonic areas. *Russian Geology and Geophysics* **33**, 9-25.

Jaques, A.L, Lewis, J.D., & Smith, C.B., 1986. The kimberlites and lamproites of Western Australia. *Geol. Surv. West Australia Bulletin* **132**,1-268.

Jaques, A.L., Hall, A.E., Sheraton, J.W., Smith, C.B., Sun, S.S., Drew, R.M., Foudoulis, C., & Ellingsen, K., 1989. Composition of crystalline inclusions and C-isotopic composition of Argyle and Ellendale diamonds. In: Ross, J. et al. (eds.). Diamonds, Kimberlites and Related Rocks, Volume 2. *Geol. Soc. Aust. Spec. Publ.* **14**, Blackwell Scientific Publications, Melbourne, 966-989.

Jaques, A.L., O'Neill, H. St. C. O., Smith, C.B., Moon, J., & Chappell, B.W., 1990. Diamondiferous peridotite xenoliths from the Argyle (AK1) lamproite pipe, Western Australia. *Contributions to Mineralogy and Petrology* **104**, 255-276.

Jaques A. L. and Milligan P. R., 2004. Patterns and controls on the distribution of diamondiferous intrusions in Australia. *Lithos* **77**, 783-802.

Javoy, M, Pineau, F., Demaiffe, D., 1984. Nitrogen and carbon isotopic compositions in the diamonds of Mbuji Maya (Zaire). *Earth Planet. Sci. Lett.* **68**, 399-412.

Javoy, M., Pineau, F., & Delorme, E., 1986. Carbon and nitrogen isotopes in the mantle. *Chem. Geol.* **57**, 41-62.

Jerde, E.A., Taylor, L.A., Crozaz, G., Sobolev, N.V., & Sobolev, V.N., 1993a. Diamondiferous eclogites from Yakutia Siberia: Evidence for a diversity of protoliths. *Contribib. Mineral. Petrol.* **114**, 189-202.

Jerde, E.A., Taylor, L.A., Crozaz, G., Sobolev, & N.V., 1993b. Exsolution of garnet within clinopyroxene of mantle eclogites: Major- and trace-element chemistry. *Contribib. Mineral. Petrol.* **114**, 148-159.

Johnson, L. H., Burgess, R., Turner, G., Milledge, H. J., & Harris, J. W., 2000. Noble gas and halogen geochemistry of mantle fluids: comparison of African and Canadian diamonds. *Geochimica et Cosmochimica Acta* **64**, 717-732.

Joswig, W., Stachel, T., Harris, J.W., Baur, W.H., & Brey, G., 1999. New Ca-silicate inclusions in diamonds – tracers from the lower mantle. *Earth Planet. Sci. Lett.* **173**, 1-6.

Kagi, H., Sato, S., Akagi,T., & Kanda, H., 2007. Generation history of carbonado inferred from photoluminescence spectra, cathodoluminescence imaging, and carbon-isotopic composition. *American Mineralogist* **92**, 217 - 224.

Kamenetsky, M.B., Sobolev, A.V., Sobolev, N.V. & Pokhilenko, N.P., 2002. Kimberlite parental melts; new insights from inclusions in olivine; Abstracts of the 12th annual V. M. Goldschmidt conference. *Geochimica et Cosmochimica Acta,* **66**, 380.

Kamenetsky, M.B., Sobolev, A.V., Kamenetsky, V.S., Maas, R., Danyushevsky, L.V., Thomas, R., Pokhilenko, N.P., Sobolev, N.V., 2004. Kimberlite melts rich in alkali chlorides and carbonates: a potent metasomatic agent in the mantle. *Geology* **32**, 845-848.

Kaminsky, F.V., Zakharchenko, O.D., Davies, R., Griffin, W.L., Khachatryan-Blinova, G.K., & Shiryaev, A.A., 2001. Superdeep diamonds from the Juina area, Mato Grosso State, Brazil.

Contribution to Mineralogy and Petrology **140**, 734-753.

Kaminsky, F. V., Khachatryan, G. K., 2004. The relationship between the distribution of nitrogen impurity centres in diamond crystals and their internal structure and mechanism of growth. *Lithos* **77**, 255– 271.

Kaminsky, F.V., Sablukov, S.M., Sablukova, L.I., & Channer, D.M. DeR., 2004. Neoproterozoic 'anomalous' kimberlites of Guaniamo, Venezuela: Mica kimberlites of 'isotopic transitional' type. *Lithos* **76**, 565-590.

Kaminsky, F.V., Zakharchenko, O.D., Khachatryan, G.K., Griffin, W.L, and Derenzy Channer, D.M., 2006. Diamond from the los coquitos area, Bolivar state, Venezuela. *The Canadian Mineralogist* **44**.

Kamiya, Y., Lang, A.R., 1965. On the structure of coated diamonds. *Phil. Mag.* **11**, 347-357.

Katayama, I., Nakashima, S., and Yurimoto, H.**,** 2006, Water content in natural eclogite and implication for water transport into the deep upper mantle. *Lithos* **86**, 245-259.

Keleman, P. B., Hart, S. R., and Bernstein, S., 1998. Silica enrichment in the continental upper mantle via melt/rock reaction. *Earth and Planetary Science Letters* **164**, 387-406.

Keller, R.A., Taylor, L.A., Snyder, G.A., Sobolev, V.N, Carlson, W.D. Sobolev, N.V., & Pokhilenko, N.P., 1998. 3-D petrography of diamondiferous eclogites from Udachnaya, Siberia. *Ext. Abstr. 7th Internat. Kimb. Conf.*, Novosibirsk, 405-407.

Keller, R.A., Taylor, L. A., Snyder, G.A., Sobolev, V. N., Carlson, W.D., Bezborodov, S.M., and Sobolev, N. V., 1999. Detailed pull-apart of a diamondiferous eclogite xenoliths: Implications for mantle processes during diamond genesis. In: Gurney, J.J., Gurney, J.L., Pascoe, M.D., and Richardson, S.H. (eds.). *Proceedings Volume of 7th International Kimberlite Conference, 1*. Cape Town: Red Roof Design,397-402.

Kennedy, C.S., and Kennedy, G.C., 1976. The equilibrium boundary between graphite and diamond. *J. Geophys. Res.*, **81**, 2467-2470.

Keshav, S., Corgne, A., Gudfinnsson, G.H., Bizimis, M., McDonough, W.F. & Fei, Y., 2005, Kimberlite petrogenesis: Insights from clinopyroxene-melt partitioning experiments at 6 GPa in the CaO-MgO-Al_2O_3-SiO_2-CO_2 system. *Geochim.et Cosmochim. Acta* **9**, 2829-2845.

Kessel, R., Ulmer, P., Pettke, T., Schmidt, M.W., & Thompson, A.B., 2004. A novel approach to determine high-pressure high-temperature fluid and melt compositions using diamond-trap experiments. *American Mineralogist* **89**, 1078 - 1086.

Kesson, S.E., Ringwood, A.E., 1989. Slab-mantle interactions: 2. The formation of diamonds. *Chem. Geol.* **78**, 97-118.

Kharkiv, A.D., 1995. Classification of original sources and concentric haloes of the indicator minerals of kimberlites of the East-European Platform. *Russian Geology and Geophysics,* **36**(1), 44-53.

Kharkiv, A.D., 1995. Two series of megacrysts from kimberlites of the Verkhnemunskoye Field (Republic of Sakha); Sixth international kimberlite conference; extended abstracts. *Proceedings of the International Kimberlite Conference,* **6**, 271-272.

Kharkiv, A.D. and Safronov, A.F., 1995. Composition inhomogeneity of kimberlitic garnet nodules as an indicator of their metasomatism and disintegration; Sixth international kimberlite conference; extended abstracts. *Proceedings of the International Kimberlite Conference,* **6**, 273-274.

Kharkiv, A.D., Zinchuk, N.N., & Zuev, V.M., 1997. History of Diamonds. Moscow: Nedra,

600 (in Russian).

Khokhryakov, A.F., Palyanov, Y.N., 1990. Morphology of diamond crystals dissolved in water-containing silicate melts. *Mineralogicheskiy Zhurnal* **12**, 14–23 (in Russian).

King, J M., Shigley, J.E., Gelb, T,H., Guhin, S.S., Hall, M., and Wang. W., 2005. Characterization and grading of natural-color yellow diamonds. *Gems and Gemology* 41, 88-115.

Kinny, P.D., Black, L.P., & Sheraton J.W., 1993. Zircon ages and the distribution of Archaean and Proterozoic rocks in the Rauer Islands. *Antarctic Science* **5**, 193-206.

Kinny, P.D., Griffin, B.J., Heaman, L.M., Brakhfogel, F.F., & Spetsius, Z.V., 1997. Shrimp U-Pb ages of perovskite from Yakutian kimberlites. In: Dobretsov, N.L. (ed.). Kimberlites, Related Rocks and Mantle Xenoliths, Volume 1. *Sov. Geol. Geophys.* **38**, 97-105.

Kinny, P.D., Trautman, R.L., Griffin, B.J., Fitzsimons, I.C.W., & Harte, B., 1999. Carbon isotopic composition of microdiamonds. In: Gurney, J. J., Gurney, J.L., Pascoe, M. D., and Richardson, S.H. (eds.). *Proceedings Volume of 7th International Kimberlite Conference, 1*. Cape Town: Red Roof Design, 429-436.

Kirkley, M.B., Gurney, J.J., Otter, M.L., Hill, S.J., & Daniels, L.R., 1991. The application of C in diamonds: a review. *Appl. Geochem.* **6**, 447-494.

Kirkley, M. B., Gurney, J. J., & Levinson, A. A., 1992. Age, origin and emplacement of diamonds: a review of scientific advances in the last decade. *CIM Bulletin,* **84**, 48-57.

Klein-BenDavid, O., Logvinova, A.M., Izraeli, E.S., Sobolev, N.V., & Navon, O., 2003. Sulfide melt inclusions in Yubileinayan (Yakutia) diamonds. *8th Internat. Kimb. Conf., Ext Abstr.,* 111.

Klein-BenDavid, O., Izraeli, E.S., Hauri, E. & Navon, O., 2004. Mantle fluid evolution—a tale of one diamond. *Lithos* **77**, 243-253.

Klein-BenDavid, O., Wirth, R., & Navon, O., 2006. TEM imaging and analysis of microinclusions in diamonds: A close look at diamond-growing fluids. *American Mineralogist* **91**, 353 - 365.

Klein-BenDavid, O., Izraeli, E.S., Hauri, E. & Navon, O., 2007. Fluid inclusions in diamonds from the Diavik mine, Canada and the evolution of diamond-forming fluids. *Geochimica et Cosmochimica Acta* **71**, 723-744.

Koch-Müller M., Matsyuk S.S., and Wirth R., 2004, Hydroxyl in omphacites and omphacitic clinopyroxenes of upper mantle to lower crustal origin beneath the Siberian platform. *American Mineralogist* **89**, 921 - 931.

Köhler, T., and Brey, G. P., 1990. Calcium exchange between olivine and clinopyroxene calibrated as a geothermobarometer for natural peridotites from 2-60 kb with applications. *Geochim. Cosmochim. Acta* **54**, 2375-2388.

Kopylova, M. G., and Russell, J. K., 2000. Chemical stratification of cratonic lithosphere: Constraints from the Northern Slave craton, Canada. *Earth Planet. Sci. Lett.* **181**, 71-87.

Kopylova, M.G., Gurney, J. J., & Daniels, L.R.M., 1997. Mineral inclusions in diamonds from the River Ranch kimberlite, Zimbabwe: *Contrib. Mineral. Petrol.* **129**, 366-384.

Kopylova, M.G., Rickard, R.S., Kleyenstueber, A., Taylor, W.R., Gurney, J.J., & Daniels, L.R., 1997. First occurrence of strontian K-Cr loparite and Cr chevkinite in diamonds. *The 6th*

IKC Proceeding, 2, *Russian Geology and Geophysics* **38**, 405-420.

Kopylova, M.G., Russell, J.K. & Cookenboo, H., 1999. Petrology of peridotite and pyroxenite xenoliths from the Jericho kimberlite: implications for the thermal state of the mantle beneath the Slave craton, Northern Canada. *Journal of Petrology* **40**, 79–104.

Kornilova, V.P., Fomin, A.S., Zaitsev, A.I., 2001. New type of diamondiferous kimberlite rocks on the Siberian platform. *Regional Geology and Metallogeny* **13**, 105-117 (in Russian).

Korsakov, A.V. and Hermann, J., 2006. Silicate and carbonate melt inclusions associated with diamonds in deeply subducted carbonate rocks. *Earth and Planetary Science Letters* **241**, 104-118.

Kozai, Y. and Arima, M., 2005. Experimental study on diamond dissolution in kimberlitic and lamproitic melts at 1300–1420 °C and 1 GPa with controlled oxygen partial pressure. *American Mineralogist* **90**, 1759 - 1766.

Kramers, J. D., 1977. Lead and strontium isotopes in Cretaceous kimberlites and mantle-derived xenoliths from Southern Africa. *Earth and Planetary Science Letters* **34**, 419-431.

Kramers, J.D., 1979. Lead, uranium, strontium, potassium, and rubidium in inclusionbearing diamonds and mantle-derived xenoliths from Southern Africa. *Earth and Planetary Science Letters* **42**, 58-70.

Kryuchkov, A.I., Kharkiv, A.D. & Pokhilenko, N.P., 1995. Identification of kimberlite bodies subjected to dynamic effect of traps (by example of systems of bodies in the area of the Yubileinaya Pipe, Yakutia). *Russian Geology and Geophysics,* **36**(5), 61-71.

Kvasnitsya, V. M., Taran, M.M., VIrt, R., Videnbek, M., Tomas, R., Il'chenko, K.O., and Lupashko, T.N., 2005. Novi dani pro ukraylns'ki almazi. new data on Ukrainian diamonds. *Mineralogicheskiy Zhurnal* **27**, 47-58 (in Russian).

Kyser, T.K., O'Neil, J.R., & Carmichael, I.S.E., 1982. Genetic relations among basic lavas and mantle nodules: Evidence from oxygen isotope compositions. *Contribu. Mineral. Petrology* **81**, 88-102.

Lappin, M.A., 1978. The evolution of a grospydite from Roberts Victor Mine, South Africa. *Contribution to Mineralogy and Petrology* **66**, 229-241.

Lavrent'ev, Yu G., Usova, L.V., Korolyuk, V.N. & Logvinova, A.M., 2005. Electron probe microanalysis of Cr-spinel for zinc and nickel traces as applied to study of the geothermometry of peridotites. *Russian Geology and Geophysic* **46**, 741-745.

Lavrent'ev, Yu G., Korolyuk, V.N., Usova, L.V. & Logvinova, A.M., 2006. Electron probe microanalysis of pyrope for nickel traces as applied to study of the geothermometry of peridotite. *Russian Geology and Geophysics* **47**, 1086-1089.

Le Roex, A.P., Bell, D.R., & Davis P., 2003. Petrogenesis of Group I Kimberlites from Kimberley, South Africa: Evidence from Bulk-rock Geochemistry. *J. Petrology* **44**, 2261 - 2286.

Lee, H.-Y., and Ganguly, J., 1988. Equilibrium composition of coexisting garnet and orthopyroxene: Experimental determinations in the system FeO-MgO-Al_2O_3-SiO_2 and applications. *Journal of Petrology* **29**, 93-113.

Leost, I., Stachel, T., Brey, G.P., Harris, J.W., Ryabchikov, .ID., 2003. Diamond formation and source carbonation: Mineral association in diamonds from Namibia. *Contribib. Mineral. Petrol.* **145**, 15-24.

Leung, I.S., Taylor, L.A., Tsao, C.S., & Han, Z., 1996. SiC in diamond and kimberlites:

Implications for nucleation and growth of diamond. *International Geology Review,* **37**, 483-496.

Levinson, A. A., Gurney J. J., & Kirkley, M. B., 1992. Diamond sources and production; past, present, and future. *Gems and Gemology* **28**, 234-254.

Lewis, R. S., Ming, T., Wacker, J. F., Anders, E., & Steel, E., 1987. Interstellar diamonds in meteorites. *Nature,* **326**, 6109, 160-162.

Lipschutz, M. E., 1964. Origin of diamonds in the ureilites. *Science* **143**, 1431-1434.

Litvin, Yu.A., Kurat, G., and Dobosi, G., 2005. Experimental study of diamondite formation in carbonate-silicate melts: a model approach to natural processes. *Russian Geology and Geophysics* **46**, 1285-1299.

Liu, L.G., 1976. The high pressure phases of $MgSiO_3$. *Earth Planet. Sci. Lett.* **31**, 200-208.

Liu, L.G., 1999. Genesis of diamonds in the lower mantle. *Contribution to Mineralogy and Petrology* **134**, 170-173.

Liu, L.G., 2002. An alternative interpretation of lower mantle mineral associations in diamonds. *Contributions to Mineralogy and Petrology* **144**, 16–21.

Liu, Y., Taylor, L.A., Floss, C., Logvinova, A.M. & Sobolev, N.V., 2005. In-situ study of multiple chromite inclusions in Udachnaya diamonds, Siberia; AGU 2005 fall meeting, *EOS, Transactions, American Geophysical Union* **86**, @Abstract V41E-1518.

Liu, Y., Taylor, L.A., Ketchum, R., Carlson, W., Stepanov, A., Hill, E., Shatsky, V., & Sobolev, N.V., 2006. The Metasomatic Origin of Diamond: Evidence from the World's Largest Diamondiferous Eclogite. AGU 2006 fall meeting, *EOS, Transactions, American Geophysical Union* **87,** @Abstract U14B-05.

Logvinova, A.M. and Sobolev, N.V., 1991. Crystalline inclusions in chromites from kimberlites and lamproites; Fifth international kimberlite conference; extended abstracts, *Proceedings of the International Kimberlite Conference* **5**, 240.

Logvinova, A.M. and Sobolev, N.V., 1995. Morphology and composition of mineral inclusions in chromite macrocrysts from kimberlites and lamproites; Sixth international kimberlite conference; extended abstracts, *Proceedings of the International Kimberlite Conference* **6**, 331-332.

Logvinova, A.M. and Sobolev, N.V., 2006. Coesite-bearing assemblages as the direct evidence for the involvement of subducted crust in the deep mantle; AGU 2006 fall meeting, *EOS, Transactions, American Geophysical Union* **87**, @Abstract U14B-03.

Logvlnova, A.M., Taylor, L.A., Floss, C., and Sobolev, N.V., 2005. Geochemistry of multiple diamond Inclusions of harzburgltic garnets as examined in situ. *International Geology Review* **47**, 1223-1233.

Logvinova, A., Taylor, L.A., Sovolev, N.V. & Floss, C., 2005. Eclogitic diamond inclusions from the Komsomolskaya Pipe, Yakutia; AGU 2005 fall meeting, *EOS, Transactions, American Geophysical Union* **86**, @Abstract V41E-1515.

Logvinova, A.M., Ashchepkov, I.V. & Palessky, S.V., 2007. LAM ICP study of cloudy diamonds; implications for diamond formation; Abstracts of the 17th annual V. M. Goldschmidt conference, *Geochimica et Cosmochimica Acta* **71**, A593.

Logvinova, A.M., Wirth, R., Sobolev, N.V., Seryotkin, Y.V., Yefimova, E.S., Floss, C., & Taylor, L.A., 2008. Eskolaite associated with diamond from the Udachnaya Kimberlite pipe, Yakutia, Russia. *American Mineralogist* **93**, 685-690.

Lowry, D., Mattey, D. P., & Harris, J. W., 1999. Oxygen isotope composition of syngenetic inclusions in diamond from the Finsch Mine, RSA. *Geochimica et Cosmochimica Acta* **63**, 1825-1836.

Lu, F., Molan, E., Jinfu, D., and Taylor, L.A., 1991. The mineral re-equilibration in mantle xenoliths from Xilong area, Zhejiang. In: Characteristics and Dynamics of the Upper Mantle of China(ed). Mineral., Petrol., & Geochim. Assoc. China, Earthquake Publ. House, Beijing, 23-33.

Luguet, A., Pearson, D.G., Jaques, A.L., Bulanova, G.P., Smith, C.B., Roffey, S. and Rayner, M.J., 2005. Archean mantle beneath the Halls Creek mobile zone, W. Australia revealed by Re-Os isotopes; Abstracts of the 15th annual V. M. Goldschmidt conference. *Geochimica et Cosmochimica Acta,* **69**, 285.

Luth, R. W., and Canil, D., 1993. Ferric iron in mantle-derived pyroxenes and a new oxybarometer for the upper mantle. *Contributions to Mineralogy and Petrology* **113**, 236-248.

Maas, R., Kamenetsky, M.B., Sobolev, A.V., Kamenetsky , V.S., & Sobolev, N.V., 2005. Sr, Nd and Pb isotope evidence for a mantle origin of alkali chlorides and carbonates in the Udachnaya kimberlite, Siberia. *Geology* **33**, 549-552.

MacGregor, I. D., 1970. The effect of CaO, Cr_2O_3, Fe_2O_3 and Al_2O_3 on the stability of spinel and garnet peridotites. *Physics of the Earth and Planetary Interiors* **3**, 372–377.

MacGregor, I. D., 1974. The system MgO-Al_2O_3-SiO_2: solubility of Al_2O_3 in enstatite for spinel and garnet peridotite compositions. *American Mineralogist* **59**, 110-119.

MacGregor, I.D., and Carter, J.L., 1970. The chemistry of clinopyroxenes and garnets of eclogite and peridotite xenoliths from the Roberts Victor mine, South Africa. *Phys. Earth Planet. Int.* **3**, 391-397.

MacGregor, I. D., and Manton, W. I., 1986. Roberts Victor Eclogites: Ancient oceanic crust. *J. Geophys. Res.* **91**, 14063-14079.

Mahotkin, I.L., Gibson, S.A., Thompson, R.N., Zhuravlev, D.Z., & Zhuerdev, P.U., 2000. Late Devonian diamondiferous kimberlite and alkaline picrite (Proto-kimberlite?) magmatism in the Arkhangelsk region, NW Russia. *Journal of Petrology* **41**, 201-227.

Mainkar, D. & Lehmann, B. 2007. The diamondiferous Behradih kimberlite pipe, Mainpur kimberlite field, Chhattisgarh, India; reconnaissance petrography and geochemistry; Kimberlite and related rocks of India. *Journal of the Geological Society of India,* **69**, 547-552.

Malinovsky, I.Y. & Doroshev, A.M., 1975. Effect of temperature and pressure on the composition of Cr-bearing phases of the garnet+enstatite+spinel+forsterite associations. In: Experimental Research on Mineralogy. Bulletin Institute Geology and Geophysics, Novosibirsk, pp. 121–125 (in Russian).

Malkovets, V., Griffin, W., O'Reilly, S., & Wood, B., 2007. Diamond, subcalcic garnet, and mantle metasomatism: Kimberlite sampling patterns define the link. *Geology* **35**, 339-342.

Manakov, A.V., 2001. Material models of the upper mantle of the Yakutian kimberlite province. *Voronegsky University Vestnik* **11**, 270-277 (in Russian).

Marakushev, A.A., 1973. Acidic-alkalic composition of minerals and some petrochemical calculations. *Vestnik Moskovskogo Universiteta, Seriya 4, Geologiya.Moscow* **28**, 3-25.

Marakushev, A.A., 1973. Petrological-geochemical systematics of igneous and metamorphic rocks. *Izvestiya Vysshikh Uchebnykh Zavedeniy.Geologiya i Razvedka* **8**, 3-16.

Marakushev, A.A., 1974. Metamorphic formations in connection with magmatism. in Problemy magmaticheskikh formatsiy ed. D.S. Shteynberg, Izd. Nauka, Moscow, USSR (SUN), USSR (SUN).

Marakushev, A.A. and Bobrov, A.V., 2005. Problems of primary magmas and the depths of diamondiferous magmatism origin. *Doklady Earth Sciences* **403**, 901-904.

Marakushev, A.A., San Lonkan, Bobrov, A.V., Paneyakh, N.A., Pertsev, N.N. & Zotov, I.A., 2000. Asian eclogite diamond-bearing belts. *Tikhookeanskaya Geologiya = Pacific Geology* **19**, 3-36.

Marakushev, A.A., Bobrov, A.V., Paneyakh, N.A., Garanin, V.K., Lu, F., Sang, L., & Chen, M., 2007. Genetic features of peridotite-eclogite complexes in fold-belts and diamondiferous pipes in cratons. Special issue on "Earth material science" (researches on mantle petrology and geochemistry). *Dixue Qianyuan = Earth Science Frontiers* **14**, 109-128.

Marty, B., and Zimmerman, L., 1999. Volatiles (He, C, N, Ar) in mid-ocean ridge basalts: Assessment of shallow level fractionation and characterization of source composition. *Geochimica et Cosmochimica Acta* **63**, 3619–3633.

Maruoka, T., Kurat, G., Dobosi, G. & Koeberl C., 2004. Isotopic composition of carbon in diamonds of diamondites: record of mass fractionation in the upper mantle. *Geochimica et Cosmochimica Acta* **68**, 1635-1644.

Massone, H.J., 2005. Genesis of diamonds and diamondiferous rocks from the Saxonian Ezebrige, central Europe. *7th International Eclogite Conference, Seggau, Austria, Mitt. Osterr. Mineral. Ges.* **150**, 104.

Mattey, D. P., Lowry, D., & Macpherson, C., 1994. Oxygen isotope composition of mantle peridotite. *Earth and Planetary Science Letters* **128**, 231-241.

Mazonne, P., and Haggerty, S. E., 1989. Peraluminous xenoliths in kimberlite: metamorphosed restites produced by partial melting of pelites. *Geochimica et Cosmochimica Acta* **53**, 1551-1561.

McCallum, M.E., 1994. Lamproitic(?) diatremes in the Golden area of the Rocky Mountain fold and thrust belt, British Columbia, Canada, Kimberlites, related rocks and mantle xenoliths", *CPRM - Special Publication,* 195-210.

McCammon, C., 2001. Geophysics - Deep diamond mysteries. *Science* **293**, 813-814.

McCallum, M.E., and Eggler, D.H., 1976. Diamonds in an upper mantle peridotite nodule from kimberlite in southern Wyoming. *Science* **192**, 253-256.

McCallum, M.E., Huntley, P.M., Falk, R.W. & Otter, M.L., 1991. Morphological, resorption and etch feature trends of diamonds from kimberlites within the Colorado-Wyoming state line district, USA, *Proceedings of the International Kimberlite Conference* **5**, 261-263.

McCammon, C.A., Griffin, W.L., Shee, S.R., & O'Neil, H.S.C., 2001. Oxidation during metasomatism in ultramafic xenoliths from Wesselton kimberlite, South Africa: Implications for the survival of diamond. *Contribution to Mineralogy and Petrology,* **141**, 287-296.

McCammon, C.A., Stachel, T. & Harris, J.W., 2004. Iron oxidation state in lower mantle mineral assemblages; II, Inclusions in diamonds from Kankan, Guinea. *Earth and Planetary Science Letters* **222**, 423-434.

McCandless, T.E., Gurney, J.J., 1986. Sodium in garnet and potassium in clinopyroxene: Criteria for classifying mantle eclogites. *Geol. Soc. Australia Spec. Pub.* **14**, 827-832.

McCandless, T. E., and Gurney, J. J., 1989. Sodium in garnet and potassium in clinopyroxene: criteria for classifying mantle eclogites. In: Ross, J. et al. (eds.). *Diamonds, Kimberlites and Related Rocks, Volume 2. Geol. Soc. Aust. Spec. Publ.* **14**, Blackwell Scientific Publications, Melbourne, 827-832.

McCandless, T.E. and Gurney, J.J., 1997. Diamond eclogites: Comparison with carbonaceous chondrites, carbonaceous shales and microbial the lower mantle. Science **278**, 434-436.

McCormick, T. C., Smyth, J. R., & Caporuscio F. A., 1994. Chemical systematics of secondary phases in mantle eclogites. In: Meyer, H.O.A., and Leonardos, O.H. (eds). Kimberlites, related rocks and mantle xenoliths. Rio de Janeiro: Spec. Publ., Companhia de Pesquisa de Recursos Minerais, 405-419.

McCulloch, M. T., 1989. Sm-Nd systematics in eclogite and garnet peridotite nodules from kimberlites: Implications for the early differentiation of the earth. In: Ross, J. et al. (eds.). Diamonds, Kimberlites and Related Rocks, Volume 2. *Geol. Soc. Aust. Spec. Publ.* **14**, Blackwell Scientific Publications, Melbourne, 864-876.

McCulloch, M.T., Gregory, R.T., Wasserburg, G.J., & Taylor, H.P. Jr., 1981. Sm-Nd, Rb-Sr, and $^{18}O/^{16}O$ isotopic systematics in an oceanic crustal section: evidence from the Samail ophiolite. In: Coleman, R.G., and Hopson, C.A. (eds.). *Journal of Geophysical Research* **B 86**, 2721-2735.

McDade, P. and Harris, J. W., 1999. Syngenetic inclusion bearing diamonds from Letseng-la-Terai, Lesotho. In: Gurney, J.J., Gurney, J.L., Pascoe, M.D., and Rickardson, S.H. (eds.). *Proceedings Volume of 7th International Kimberlite Conference, 1*. Cape Town: Red Roof Design, 557-565

McDade, P. and Harte, B., 2000. Roberts Victor eclogites with a spinel-facies mantle signature. *J. Conf. Abs.* **5**, 690.

McDonough, W. F., and Sun, S. S., 1995. The composition of the Earth. *Chemical Geology* **120**, 223-253.

McKenna, N., Gurney, J. J., Klump, J. & Davidson, J. M., 2004. Aspects of diamond mineralisation and distribution at the Helam Mine, South Africa. *Lithos* **77**, 193-208.

McKenzie, D., 1989. Some remarks on the movement of small melt fractions in the mantle. *Earth and Planetary Science Letters* **95**, 53-72.

Meisel, T., Walker, R. J., Irving, A. J., & Lorand, J. P., 2001. Osmium isotopic compositions of mantle xenoliths: A global perspective. *Geochimica et Cosmochimica Acta* **65**, 1311-1323.

Melton, C.E., Salotti, C.A., and Giardini, A.A., 1972. The observation of nitrogen, water, carbon dioxide, methane and argon as impurities in natural diamonds. *Am. Mineral.* **57**, 1518-1523.

Mendelssohn, M.J., and Milledge, H.J., 1995. Geologically significant information from routine analysis of the mid-infrared spectra of diamonds. *International Geology Review* **37**, 95-110.

Mendelssohn, M. J., and Milledge, H. J., 1995. Morphological characteristics of diamond populations in relation to temperature-dependent growth and dissolution rates. *International Geology Review* **37**, 285-312.

Menzies, A. H., 2001. A detailed investigation into diamond-bearing xenoliths from Newlands kimberlite, South Africa. Ph.D. thesis, University of Cape Town, 253 pp.

Menzies, A. H., Shirey, S. B., Carlson, R. W., & Gurney, J. J., 1999. Re-Os systematics of Newlands peridotite xenoliths: Implications for diamond and lithosphere formation. In: Gurney, J.J., Gurney, J.L., Pascoe, M.D., and Richardson, S.H. (eds.) *Proceedings Volume of 7th International Kimberlite Conference, 2*. Cape Town: Red Roof Design, 566-583.

Menzies, M.A., 1990. Petrology and geochemistry of the continental mantle: an historical perspective. In: Menzies, M.A. (ed.). Continental mantle. Oxford: Science Publication, 50-54.

Menzies, M.A. and Hawkesworth, C. J., 1987. Mantle Metasomatism, New York, Academic Press, 472.

Meyer, H.O.A., 1968. Chrome pyrope: an inclusion in natural diamond. Science **160**, 1446-1447.

Meyer, H.O.A., 1985. Genesis of diamond: a mantle saga. *American Mineralogist* **70**, 344-355.

Meyer, H.O.A., 1987. Inclusions in diamond. In: Nixon, P.H. (ed.). Mantle Xenoliths, John Wiley., and Sons, England, 501-523.

Meyer, H.O.A. and Boyd, F.R., 1968. Inclusions in diamonds: Yrbk Carnegie Inst. Washington **66**, 446-450.

Meyer, H.O.A. and Boyd, F.R., 1969. Mineral inclusions in diamonds. *Carnegie Institute of Washington Year Book*, Washington, D.C., 130-135.

Meyer, H.O.A., and Boyd, F.R., 1972. Composition and origin of crystalline inclusions in natural diamonds. Geochim Cosmochim Acta **36**, 1255-1273.

Mikheenko, V.N., Viadimtrov, B.M., Nenashev, N.T., & Seldiseeva, E.B., 1970. A cobble of diamond-bearing ecologites from a kimberlite in the Mir pipe. *Dokl. Akad. Nauk SSSR* **190**, 1440-1443.

Milledge H.J., Mendelssohn, M. J., Boyd, S.R., Pillinger, C.T., Seal, M., 1989, Infrared topography and carbon and nitrogen isotope distribution in natural and synthetic diamonds in relation to mantle processes. In: *Extended Abs., Workshop in diamonds, 28th Internat. Geol. Congress*, Washington, D.C., 55-60.

Milledge, H.J., Woods, P.A., Taylor, W.R., Bulanova, G.P., & Kaminsky, F.V., 1993. Cathodoluminescence and infrared studies of Russian diamonds. In: *Abstr. IAVCEI General Assembly*, Canberra, 74.

Misra, K., Anand, M., Taylor, L. A., and Sobolev, N. V., 2004. Multi-stage metasomatism of diamondiferous eclogite xenoliths from the Udachnaya kimberlite pipe, Yakutia, Siberia, *Contr. Mineral. Petrol.* **146**, 696-714.

Mitchell, R.H., 1986. Kimberlites: Mineralogy, Geochemistry and Petrology. Plenum NY.

Mitchel, R.H., 1991. Kimberlites and lamproites primary sources of diamond. *Geoscience Canada* **18**, 1-16.

Mitchell, R. M., 1995. *Kimberlites, Orangeites and Related Rocks*. Plenum Press, New York, 410.

Mitchell, R.H., 2006. Potassic magmas derived from metasomatized lithospheric mantle; nomenclature and relevance to exploration for diamond-bearing rocks. *Journal of the Geological Society of India* **67**, 317-327.

Mitchell, R.S. and Giardini, A.A., 1953. Oriented olivine inclusions in diamond. *Amer. Mineral.*

38, 136-138.

Mityukhin, S.I., and Spetsius, Z.V., 2005. Paragenesis of inclusions in diamonds from the Botuobinckaya pipe (Nakyn field, Yakutia). *Russian Geology and Geophysics* **46**, 1246-1258.

Miyazaki, T., and Okushi, H., 2002. Theoretical modeling of sulfur and hydrogen complexes in diamond. *Diamond and Related Materials* **11,** 323-327.

Moore, R.O., and Gurney, J.J., 1985. Pyroxene solid solution in garnets included in diamonds. *Nature* **318**, 553-555.

Moore, R.O., and Gurney, J.J., 1989. Mineral inclusions in diamond from the Monastery kimberlite, South Africa: In: Ross, J. et al. (eds.). Diamonds, Kimberlites and Related Rocks, Volume 2. *Geol. Soc. Aust. Spec. Publ.* **14**, Blackwell Scientific Publications, Melbourne, pp. 1029-1041.

Morten, L., L.A. Taylor, and A. Durazzo, 1989, Spinels in hrzburgite and lherzolite inclusions from the San Giovanni Ilarione Quarry, Lessini Mountains, Veneto Region, Italy. Mineral. Petrol. 40, 73-89.

Mukhopadhyay, B., 1991. Garnet-clinopyroxene geobarometry: The problems, a prospect and an approximate solution. *Am. Mineral.* **76**, 512-529.

Nadolinnyi, V.A., Afanasyev, V.P., Pokhilenko, N.P., Zuev, V. M., Yuryeva, O.P., Bezborodov, S.M. & Mityukhin, S.I., 1994. Diamond paragenesis diagnostics based on optical characteristics of structural nickel; International Mineralogical Association, 16th general meeting; abstracts, *Abstracts of the ...General Meeting of the International Mineralogical Association,* **16**, 293.

Nadolinnyi, V.A., Yur'eva, O.P., Yelisseyev, A.P., Pokhilenko, N.P. & Chepurov, A.A., 2004. Disruption of B1 nitrogen defects in 1aB natural diamonds by plastic deformation and behavior of the defects created on the P-T treatment. *Doklady Earth Sciences* **399**, 1268-1272.

Nadolinnyi, V.A., Shatsky, V.S., Kozmenko, O.A., Stepanov, A.S., Palyanov, Y., Kupriyanov, I.N., 2006. Study of local concentration of single substitutional nitrogen atoms in microdiamonds from the Kokchetav massif. *European Journal of Mineralogy* **18**, 739-743.

Nasdala, L, Hofmeister, W., Harris, J.W., Glinnemann, J., Williams, M.L., and Tracy, R.J., 2005. Growth zoning and strain patterns inside diamond crystals as revealed by Raman maps; monazite microprobe geochronology. *American Mineralogist* **90**, 745-748.

Navon, O., 1999. Diamond formation in the Earth's mantle. In: Gurney, J.J., Gurney, J.L., Pascoe, M.D., and Richardson, S.H. (eds.). *Proceedings Volume of 7th International Kimberlite Conference, 2.* Cape Town: Red Roof Design, 584-605.

Navon, O., Hutcheon, I.D., Rossman, G.R, Wasserburg, G.J., 1988. Mantle-derived fluids in diamonds micro-inclusions. *Nature* **335**, 784-789.

Navon, O., Spettel, B., Hutcheon, I.H., Rossman, G.R., Wasserburg, G.J., 1989. Micro-inclusions in diamonds from Zaire and Botswana (Extended Abstract). *Workshop on Diamonds*, Geophysical Laboratory, Washington D.C., 69-72.

Neal, C.R., and Taylor, L.A., 1989. A negative Ce anomaly in a peridotite xenolith: Evidence for crustal recycling into the mantle. *Geochim. et Cosmochim. Acta* **53**, 1035-1040.

Neal, C.R. and Taylor, L.A., 1990. Comment on "Mantle Eclogites: Evidence of Igneous Fractionation in the Mantle" by Smyth, J.R., Caporuscio, F.A., and McCormick, T.C. *Earth*

Planet. Sci. Lett. **101**, 112-119

Neal, C.R., Taylor, L.A., Davidson, J.P., Holden, P., Halliday, A.N., Paces, J.B., Clayton, R.N., & Mayeda, T.K., 1990. Eclogites with oceanic crustal and mantle signatures from the Bellsbank kimberlite, South Africa, Part 2: Sr, Nd, and O isotope chemistry. *Earth Planet. Sci. Lett.* **99**, 362-379.

Nestola, F, Longo, M., McCammon, C., & Boffa-Ballaran, T., 2007. Crystal-structure refinement of Na-bearing clinopyroxenes from mantle-derived eclogite xenoliths. *American Mineralogist* **92**, 1242-1245.

Nickel, K.G., 1989. Garnet–pyroxene equilibria in the system SMACCR (SiO_2–MgO–Al_2O_3–CaO–Cr_2O_3): the Cr-geobarometer. In: Ross, J. (ed.) Kimberlites and Related Rocks. Geological Society of Australia Special Publication 14, 901–912.

Nickel, K. G., and Green, D. H., 1985. Empirical geothermobarometry for garnet peridotites and implications for the nature of the lithosphere, kimberlites and diamonds. *Earth Planet. Sci. Lett.* **73**, 158-170.

Nimis, P., 1998. Evaluation of diamond potential from the composition of peridotitic chromian diopside. *European Journal of Mineralogy* **10**, 505–519.

Nimis, P., and Taylor, W. R., 2000. Single clinopyroxene thermobarometry for garnet peridotites. Part 1. Calibration and testing of a Cr-in-cpx barometer and enstatite-in-cpx thermometer. *Contrib. Mineral. Petrol.* **139**, 541-554.

Nisbet, D.E.G., Mattey, D.P., & Lowry, D., 1994. Can diamonds be dead bacteria? *Nature* **367**, 694.

Nixon, P. H., 1987. Mantle Xenoliths. John Wiley and Sons, Chichester, UK, 844 .

Nixon, P. H., 1993. Diamonds. *Geoscientist* **3**, 17-19.

Nixon, P. H., 1995. A review of mantle xenoliths and their role in diamond exploration. *Journal of Geodynamics* **20**, 305-329.

Nixon, P. H., Chapman, N. A., & Gurney, J. J., 1978. Pyrope-Spinel, Alkremite xenoliths from kimberlite. *Contributions to Mineralogy and Petrology* **65**, 314-346.

Nixon, P. H., Rogers, N. W., Gibson, I. L., & Grey, A., 1981. Depleted and fertile mantle xenoliths from southern African kimberlites. *Ann. Rev. Earth Planet. Sci.* **9**, 285-309.

Nowell, G. M., Pearson, D. G., Bell, D. R., Dowall, D. P., Carlson, R. W., Kempton, P. D., Noble, S. R., Smith, C. B., & Zartman, R. E., 2004. Hf isotopic systematics of kimberlites and their megacrysts: New constraints on kimberlite source regions and implications for mantle evolution. *Journal of Petrology* **45**, 1583-1612.

Nowicki, T., Crawford, B., Dyck, D., Carlson, J., McElroy ,R., Oshust, P. & Helmstaedt, H., 2004, The geology of kimberlite pipes of the Ekati property, Northwest Territories, Canada. *Lithos* **76**, 1-27.

O'Reilly, S.Y., and Griffin, W.L., 1995. Trace-element partitioning between garnet and clinopyroxene in mantle-derived pyroxenites and eclogites: P-T-X controls. *Chemical Geology* **121**, 105-130.

Ogasawara, Y., 2005. Microdiamonds in ultrahigh-pressure metamorphic rocks. *Elements* **1**, 91-96.

Ogasawara, Y., Ohta, M., Fukusawa, K., Katayama, I., & Maruyama, S., 2000. Diamond-bearing and diamond-free metacarbonate rocks from Kumdy-Kol in the Kokchetav massif, northern Kazakhstan. *The Island Arc* **9**, 400-416.

Ohmoto, H., Rye, R.O., 1979. Isotopes of sulfur and carbon. In: Barnes, H.L. (ed.). Geochemistry of Hydrothermal Ore Deposits, Wiley, New York, 509–567.

Okamoto, K, Katayama, I., Maruyama, S., and Llou, J.G., 2006. Zircon-Inclusion mineralogy of a diamond-grade eclogite from the Kokchetav massif, northern Kazakhstan. *International Geology Review* **48**, 882-891.

O'Neill, H. S. C., and Wood, B. J., 1979. An experimental study of Fe-Mg partitioning between garnet and olivine and its calibration as a geothermometer. *Contributions to Mineralogy and Petrology* **70**, 59-70.

O'Neill, H. S. C., Rubie, D. C., Canil, D., Geiger, C. A., Ross, C. R. I., Seifert, F., & Woodland, A. B., 1993. Ferric iron in the upper mantle and in transition zone assemblages: Implications for relative oxygen fugacities in the mantle. In: Takahashi, E., Jeanloz, R., Rubie, D. (eds.). Evolution of the Earth and Planets, Geophysical Monograph **74**, American Geophysical Union, 73-87.

Ongley, J. S., Basu, A. R., and Kyser, T. K., 1987. Oxygen isotopes in coexisting garnets, clinopyroxenes and phlogopites of Roberts Victor eclogites: Implications for petrogenesis and mantle metasomatism. *Earth and Planetary Science Letters* **83**, 80-84.

Orlov, Y.L., 1977. The Mineralogy of Diamond,. John Wiley and Sons, New York, 235 pp.

Otter, M.L. and Gurney, J.J., 1989. Mineral inclusions in diamond from the Sloan diatremes, Colorado-Wyoming State line kimberltie district, North America. In: Ross, J. et al. (eds.). Diamonds, Kimberlites and Related Rocks, V 2. *Geol. Soc. Aust. Spec. Publ.* **14**, Blackwell Scientific Publications, Melbourne, 1042-1053.

Otter, M.L., Gurney, J.J., McCandless, T.E., Boyd, F.R., Meyer, H.O.A. & Sobolev, N.V., 1989. The carbon isotope composition of Sloan diamonds; Extended abstracts; Workshop on Diamonds, *28th international geological congress; Workshop on Diamonds, Washington, DC.*Geophys. Lab.

Otter, M.L., Gerneke, D.A., Harte, B., Gurney, J.J., Harris, J.W. & Wilding, M.C., 1991. Diamond growth histories revealed by cathodoluminescene and carbon isotope studies; Fifth international kimberlite conference; extended abstracts, *Proceedings of the International Kimberlite Conference* **5**, 318-319.

Otter, M.L., McCallum, M.E. & Gurney, J.J., 1991. A physical characterization of the Sloan (Colorado) diamonds using a revised diamond description scheme. *Proceedings of the International Kimberlite Conference* **5**(2), 15-31.

Ozima, M., and Zashu, S., 1991. Noble gas state of the ancient mantle as deduced from noble gases in coated diamonds. *Earth Planet. Sci. Lett.* **105**, 13-27.

Pacolo, R., and Gasparik, T., 1990. Reversals of the orthoenstatite-clinoenstatite transition at high pressures and temperatures. *Jour. Geophys. Res.* **95**, 15853-15858.

Page, F.Z., Fu, B., Kita, N.T., Fournelle, J., Spicuzza, M.J., Schulze, D.J., Viljoen, F., Basei, M.A.S. & Valley, J.W. 2007, Zircons from kimberlite; new insights from oxygen isotopes, trace elements, and Ti in zircon thermometry, *Geochimica et Cosmochimica Acta* **71**, 3887-3903.

Palyanov, Y.N, Khokhryakov, A.F., Borzdov, Y.M, Sokol, A.G., Gusev, V., Rylov, G.M., & Sobolev, N.V., 1997. Growth conditions and real structure of synthetic diamond crystals. *Russ. Geol. Geophys.* **38**, 245-254.

Palyanov, Y.N., Khokhryakov, A.F., & Borzdov, Y.M., 1997. Conditions of growth and actual structure of synthetic diamond crystals. *Geology and Geophysics* **N5-P**, 882-906.

Palyanov, Y.N., Sokol, A.G., Borzdov, Y.M. & Khokhryakov, A.F., 2002. Fluid-bearing alkaline carbonate melts as the medium for the formation of diamonds in the Earth's mantle: an experimental study. *Lithos* **60,** 145-159.

Palyanov, Y.N., Sokol, A.G., and Sobolev, N.V., 2005. Experimental modeling of mantle diamond-forming processes. *Russian Geology and Geophysics* **46**, 1252-1270.

Pal'yanov, Y.N., Sokol, A.G., Tomilenko, A.A., and Sobolev, N.V., 2005. Conditions of diamond formation through carbonate-silicate Interaction; EMPG X; experimental mineralogy, petrology and geochemistry. *European Journal of Mineralogy* **17**, 207-214.

Palyanov, Y.N., Shatsky, V.S., Sobolev, N.V, and Sokol, A.G., 2006. The role of mantle ultrapotassic fluids in diamond formation. *Proc. Nat'l. Acad. Sci. U.S.A* **104**, 9122-9127.

Palyanov, Y.N., Borzdov, Y.M., Khokhryakov, A.F., Kupriyanov, I.N. & Sobolev, N.V., 2006. Sulfide melts–graphite interaction at HPHT conditions: Implications for diamond genesis. *Earth and Planetary Science Letters* **250**, 269-280.

Palyanov, Y.N, Borzdov, Y.M., Bataleva, Y.V., Sokol, A.G., Palyanova, G.A., and Kupriyanov, I.N., 2007. Reducing role of sulfides and diamond formation in the earth's mantle. *Earth and Planetary Science Letters* **260**, 242-256.

Patchen, A.D., Taylor, L.A., & Pokhilenko, N., 1997. Ferrous iron freudenbergite in ilmenite megacrysts: A unique paragenesis from the Dalnaya kimberlite, Yakutia. *Amer. Mineral.* **82**., 991-1000.

Pattison, D.R.M., and Levinson, A.A., 1995. Are euhedral microdiamonds formed during ascent and decompression of kimberlite magma? Implications for use of microdiamonds in diamond grade estimation. *App. Geochem.*, **10**, 725-738.

Pearson, D.G., 1999. The age of continental roots. *Lithos* **48**, 171-194.

Pearson, D.G. and Shirey, S.B., 1999. Isotopic dating of diamonds. In: Lambert, D., and Ruiz, J. (eds.). *Econ. Geol. Spec. Publ., SEG Reviews* **12**, 143-171.

Pearson, D. G. and Nowell, G. M., 2002. The continental lithospheric mantle: characteristics and significance as a mantle reservoir. In: Davies, J.H., Brodholt, J.P., and Wood, B.J. (eds.), *Proc. Royal Society Series A.* **360**, 1800, 2383-2410.

Pearson, D. G., Shirey, S. B., Carlson, R. W., & Taylor, L. R., 1992. Os isotope constraints on the origin of eclogite xenoliths. *Eos, Trans. Am. Geophys. Union* **73**, 376.

Pearson, D.G, Davies, G.R & Nixon, P. H., 1993. Geochemical constraints on the petrogenesis of diamond facies pyroxenites from the Beni Bousera peridotite massif, North Morocco. *Journal of Petrology* **34**: 125-172.

Pearson, D.G., Boyd, F.R., Haggerty, S.E., Pasteris, J.D., Field, S.W., Nixon, P.H., & Pokhilenko, N.P., 1994. The characterization and origin of graphite in cratonic lithospheric mantle: a petrological carbon isotope and Raman spectroscopic study. *Contribution to Mineralogy and Petrology* **115**, 444-466.

Pearson, D. G., Shirey, S. B., Carlson, R. W., Boyd, F. R., Pokhilenko, N. P., & Shimizu, N. 1995. Re-Os, Sm-Nd and Rb-Sr isotope evidence for think Archaean lithospheric mantle beneath the Siberia craton modified by multistage metasomatism. *Geochimica et Cosmochimica Acta* **59**, 959-977.

Pearson, D. G., Snyder, G. A., Shirey, S. B., Taylor, L. A., Carlson, R. W., & Sobolev, N. V.,

1995. Archaean Re-Os age for Siberian eclogites and constraints on Archaean tectonics. *Nature* **374**, 711-713.

Pearson, D.G., Carlson, R. W., Shirey, S. B., Boyd, F. R., & Nixon, P. H., 1995. The stabilisation of Archaean lithospheric mantle: A Re-Os isotope study of peridotite xenoliths from the Kaapvaal craton. *Earth Planet. Sci. Lett.* **134**, 341-357.

Pearson, D.G., Kelley, S.P., Pokhilenko, N.P., & Boyd, F.R., 1997. Laser $^{40}Ar/^{39}Ar$ analyses of phlogopites from southern African and Siberian kimberlites and their xenoliths: Constraints on eruption ages, melt degassing, and mantle volatile composition. *Russian Geology and Geophysics* **38**, 106-117.

Pearson, D.G., Davies, R.M., Shirey, S.B., Carlson, R.W., & Griffin, W.L., 1998. The age and origin of eastern Australian diamonds: Re-Os isotope evidence from sulfide inclusions in two diamonds from Wellington, New South Wales. *Extended Abstracts of 7th International Kimberlite Conference*, 664-666.

Pearson, D.G., Shirey, S.B., Harris, J.W., & Carlson, R.W., 1998. Sulphide inclusions in diamonds from the Koffiefontein kimberlite, S. Africa: constraints on diamond ages and mantle Re-Os systematics. *Earth and Planetary Science Letters* **160**, 311-326.

Pearson, D.G., Shirey, S.B., Bulanova, G.P., Carlson, R.W., & Milledge, H.J., 1999. Dating and paragenetic distinction of diamonds using the Re-Os isotope system: application to some Siberian diamonds. *Proceedings of the 7th International Kimberlite Conference* **2**, 637-643.

Pearson, D.G., Shirey, S.B., Bulanova, G.P., Carlson, R.W. & Milledge, H.J., 1999. Re-Os isotopic measurements of single sulfide inclusions in a Siberian diamond and its nitrogen aggregation systematics. *Geochim. Cosmochim. Acta* **63**, 703-711.

Pearson, D. G., Irvine, G. J., Carlson, R. W., Kopylova, M. G., and Ionov, D. A., 2002. The development of lithospheric mantle keels beneath the earliest continents: Time constraints using PGE and Re-Os isotope systematics. In: Fowler, M., Ebinger, C. J., and Hawkesworth, C.J. (eds.). The Early Earth, Geological Society of London Special Publication, **199**, 65-90.

Pearson, D.G., Canil, D., and Shirey, S.B., 2003. Mantle samples included in volcanic rocks: xenoliths and diamonds. *Treatise on Geochemistry*, **2**, 171-275.

Pearson, N. J., Alard, O., Griffin, W.L., Jackson, S. E., & O'Reilly, S.Y., 2002. In situ measurement of Re-Os isotopes in mantle sulfides by Laser Ablation Multi-Collector Inductively-Coupled Mass Spectrometry: analytical methods and preliminary results. *Geochimica et Cosmochimica Acta*, **66**, 1037-1050.

Peltonen, P., Kinnunen, K.A., & Huhma, H., 2002. Petrology of two diamondiferous eclogite xenoliths from the Lahtojoki kimberlite pipe, eastern Finland. *Lithos* **63**, 151– 164.

Peltonen, P. and Bruegmann, G., 2006. Origin of layered continental mantle (Karellan craton, Finland); geochemical and Re-Os Isotope constraints. *Lithos* **89**, 405-423.

Perraki, M., Proyer, A., Mposkos, E., Kaindl, R., & Hoinkes, G., 2006. Raman micro-spectroscopy on diamond, graphite and other carbon polymorphs from the ultrahigh-pressure metamorphic Kimi Complex of the Rhodope Metamorphic Province, NE Greece. *Earth and Planetary Science Letters* **241,** 672–685.

Pertsov, N. V., Kogan, B.S., & Balashov, V.N., 1977. Model of fracture magma intrusions in conditions of absorption effect reveal on decreasing of the rocks strength, *Dokl. Akad. Nauk*

SSSR ***235***, 1375-1378, (in Russian) .

Peslier, A. H., Reisberg, L. R., Ludden, J., & Francis, D., 2000. Os isotope systematics in mantle xenoliths: age constraints on the Canadian Cordillera lithosphere. *Chem. Geol.* **166**, 85-101.

Phillips, D., Harris, J.W. & Kiviets, G.B., 2004. ^{40}Ar / ^{39}Ar analyses of clinopyroxene inclusions in African diamonds; implications for source ages of detrital diamonds. *Geochimica et Cosmochimica Acta,* **68**, 151-165.

Pokhilenko, N.P., Sobolev, N.V, & Yefimova, E.S., 1982. Xenolith of broken down diamondiferous disthen eclogite from pipe Udachnaya, Yakutia. *Rep. of USSR AS* **266**, 212-216.

Pokhilenko, N. P., Sobolev, N. V., Boyd, F. R., Pearson, D. G., & Shimizu N., 1993. Megacrystalline pyrope peridotites in the lithosphere of the Siberian platform: Mineralogy, geochemical peculiarities and the problem of their origin. *Russian Journal of Geology and Geophysics* **34**, 1-12.

Pokhilenko, N.P., McDonald, J.A., Melnyk, W. & McCorquodale, J., 1995. Indicator minerals of Cl-25 kimberlite pipe, Slave Craton, North-West Territories, Canada; Sixth international kimberlite conference; extended abstracts, *Proceedings of the International Kimberlite Conference* **6**, 444-445.

Pokhilenko, N.P. & Sobolev, N.V., 1995. Mineralogical mapping of the north-east section of the Yakutian kimberlite province and its main results; Sixth international kimberlite conference; extended abstracts, *Proceedings of the International Kimberlite Conference* **6**, 446-448.

Pokhilenko, N.P., McDonald, J.A., Melnik, U., McCorquodale, J., Reimers, L.F. & Sobolev, N.V., 1997. Indicator minerals from the CL-25 kimberlite pipe, Slave Craton, Northwest Territories, Canada; Proceedings of the Sixth international kimberlite conference; Volume 2, Diamonds; characterization, genesis, and exploration", *Russian Geology and Geophysics* **38**, 550-558.

Pokhilenko, N.P., Sobolev, N.V., Kuligin, S.S. & Shimizu, N., 1999. Peculiarities of distribution of pyroxenite paragenesis garnets in Yakutian kimberlites and some aspects of the evolution of the Siberian Craton lithospheric mantle; Proceedings of the VIIth international kimberlite conference", *Proceedings of the International Kimberlite Conference,* **7**, 689-698.

Pokhilenko, N.P., Boyd, F.R., Ashchepkov, I.V., Kuligin, S.S. & Malygina, E.V., 2000. Layered mantle beneath Udachnaya kimberlite pipe; OPx thermobarometry; European Geophysical Society, 25th general assembly. *Geophysical Research Abstracts,* vol. 2.

Pokhilenko, N.P., Sobolev, N.V., Chernyi, S.D. & Yanygin, Y.T., 2000. Pyropes and chromites from kimberlites in the Nakyn Field (Yakutia) and Snipe Lake District (Slave River region, Canada); evidence for anomalous structure of the lithosphere. *Doklady Earth Sciences* **372**, 638-642.

Pokhilenko, N.P., Sobolev, N.V., McDonald, J.A., Hall, A.E., Yefimova, E.S., Zedgenizov, D.A., Logvinova, A.M., & Reimers, L.F., 2001. Crystalline inclusions in diamonds from kimberlites of the Snap Lake area (Slave Craton, Canada): New evidences for the anomalous lithospheric structure. *Doklady Earth Science*, **380**, 806-811.

Pokhilenko, N.P., McDonald, J.A. & Turner, R.C., 2003. Snap Lake kimberlite dyke system; history and methods of a new type of large primary diamond deposit discovery; Geological Association of Canada, Mineralogical Association of Canada; joint annual meeting; abstracts--Association geologique du Canada, Association mineralogique du Canada;

congres annuel conjoint; recueil des resumes", *Program with Abstracts - Geological Association of Canada; Mineralogical Association of Canada: Joint Annual Meeting* **28**, @unpaginated.

Pokhilenko, N.P., Sobolev, N.V., Reutsky, V.N., Hall, A.E. & Taylor, L.A., 2004. Crystalline inclusions and C isotope ratios in diamonds from the Snap Lake-King Lake kimberlite dyke system; evidence of ultradeep and enriched lithospheric mantle. *Lithos,* **77**, 57-67.

Pollack, H.N., and Chapman, D.A., 1977. On the regional variation of heat flow geotherms and lithospheric thickness. *Tectonophysics* **38**, 279-296.

Ponomarenko, A. I., 1977. Partial melting of eclogites, natural processes. *Doklady Akademy Nauk SSSR* **235**, 1416-1418 (in Russian).

Ponomarenko, A.I. and Leskova, N.V., 1980. Peculiarities of chemical composition of minerals of alkremites from kimberlite pipe "Udachnaya". *Doklady Akademy Nauk SSSR* **252**, 707-711 (in Russian).

Ponomarenko, A. I., Sobolev, N. V., & Pokhilenko, N. P., 1976. Diamondiferous grospydite and diamondiferous disthen eclogites from kimberlite pipe Udachnaya, Yakutia. *Doklady Akademy Nauk SSSR* **226**, 927-930 (in Russian).

Ponomarenko, A.I., Spetsius, Z.V., & Lybushkin, V.A., 1977. Kyanite eclogite with coesite. *Doklady Akademy Nauk SSSR* **236**, 215-219, (in Russian).

Prinz, M., Manson, D.V., Hlava, P.F., Keil, K., 1975. Inclusions in diamonds: garnet lherzolite and eclogite assemblages. *Phys. Chem. Earth* **9**, 797-815.

Promprated, P., Taylor, L.A., & Snyder, G.A., 1999. Petrochemistry of the Mantle beneath Thailand: Evidence from peridotite xenoliths. *Int'l. Geol. Rev.* **41**, 506-530.

Promprated, P., Taylor, L.A., Floss, C., Anand, M., Misra, K.C., & Sobolev, N.V., 2001, Syngenetic diamond mineral inclusions: Are they really? *AGU Fall Meet.* V12C-1005, Abstract.

Promprated, P., Taylor, L.A., Anand, M., Pokhilenko, N.P., Sobolev, N.V., and Patchen, A., 2002, Diamond Inclusions From Snap Lake Kimberlite, Canada: An Insight From In-Situ Analysis, *EOS Trans. AGU,* **82**, Fall Meet. Suppl., Abstract.

Promprated, P., Taylor, L.A., and Neal, C.R., 2003. Petrochemistry of Mafic Granulite Xenoliths from the Chantaburi Basaltic Field: Implications for the lower crust beneath Thailand. *International Geology Review* **45**, 383-406.

Promprated, P. Taylor, L.A., Anand, M., Floss, C., Sobolev, N.V., & Pokheilenko, N.V., 2004. Multiple-mineral inclusions in diamonds from the Snap Lake/King Lake kimberlite dike, Slave craton, Canada: A trace-element perspective. *8th International Kimberlite Conference*, Victoria City, 57-68.

Pyle, J. M., and Haggerty, S. E., 1994. Silicate-carbonate liquid immiscibility in upper mantle eclogites: evidence of natrosilicic and carbonatitic conjugate melts. *Geochim. Cosmochim. Acta* **58**, 2997-3011.

Pyle, J. M., and Haggerty, S. E., 1998. Eclogites and the metasomatism of eclogites from the Jagersfontein kimberlite: Punctuated transport and implications for alkali magmatism. *Geochim. Cosmochim. Acta* **62**, 1207-1231.

Qi, Q., Beard, B.L., Jin, Y., & Taylor, L.A., 1994. Petrology and geochemistry of Al-augite and Cr-diopside group mantle xenoliths from Tahiti, Society Islands. *International Geology Review* **36**, 152-178.

Qi, Q., Taylor, .LA., Snyder, G.A., & Sovolev N.V., 1994. Eclogites from the Obnazhennaya

Kimberlite Pipe, Yakutia, Russia. *International Geology Review* **36**, 911-924.

Qi, Q., Taylor, L.A., and X. Zhou, X., 1994. Petrology and geochemistry of an unusual tridymite-hercynite xenolith in tholeiite from Southeastern China. *Mineral. Petrol.* **50**, 195-207.

Qi, Q., L.A. Taylor, & X. Zhou, 1995. Petrology and geochemistry of mantle peridotite xenoliths in Cenozoic basalts from Southeastern China. *Journal of Petrology* **36**, 55-79.

Qi, Q., Taylor, L.A., Snyder, G.A., Clayton, R.N., Mayeda, T.K., & Sobolev, N.V., 1997. Detailed petrology and geochemistry of a rare corundum eclogite xenolith from Obnazhënnaya. *Russian Geol. Geophys.* **38**, 247-260.

Ragozin, A.L., Shatsky, V.S., Zetgenizov, D.A., and Mityukhin, S.I., 2006. Evidence for evolution of diamond crystallization medium in eclogite xenolith from the Udachnaya kimberlite pipe, Yakutia. *Doklady Earth Sciences* **407**, 465-468 (in Russian).

Rapp, R. P., and Watson, E. B., 1995. Dehydration melting of metabasalt at 8-32 kbar. Implications for continental growth and crust-mantle recycling. *J. Petrol.* **36**, 891-932.

Rehkämper, M., Halliday, A. N., Barfod, D., Fitton, J. G., and Dawson, J. B., 1997. Platinum-group element abundance patterns in different mantle environments. *Science* **278**, 1595-1598.

Rehkämper, M., Halliday, A. N., Alt, J., Fitton, J. G., Zipfel, J., & Takazawa, E., 1999. Non-chondritic platinum-group element ratios in oceanic mantle lithosphere: Petrogenetic signature or melt percolation? *Earth Planet. Sci. Lett.* **172**, 65-81.

Reisberg, L., and Lorand, J. P., 1995. Longevity of sub-continental mantle lithosphere from osmium isotope systematics in orogenic peridotite massifs. *Nature* **376**, 159-162.

Reutsky, V.N., Logvinova, A.M., & Sobolev, N.V., 1999. Isotopic composition of carbon of polycrystal aggregations of chromite-bearing diamonds from Mir kimberlite pipe (Yakutia), *Geokhimiya* **11,** 1191–1196.

Reutsky, V.N. and Zedgenizov, D.A., 2001. Isotopic composition of carbon of diamond microcrystals from Udachnaya pipe (Yakutia), in: Crystal genesis and mineralogy: Proc. of International Conf., 318–319(in Russian).

Reutsky, V.N., Logvinova, A.M., & Sobolev, N.V., 2002. Carbon isotopic composition of macrodiamonds from Yubileinaya Pipe (Yakutia); Abstracts of the 12th annual V. M. Goldschmidt conference. *Geochimica et Cosmochimica Acta,* **66** (15A), 635.

Richard, R.S., Harris, J.W., Gurney, J.J. & Cardoso, P., 1986. Mineral inclusions in diamonds from Koffiefontein mine. In: Ross, J. (ed.) Kimberlites and Related Rocks. Geological Society of Australia Special Publication 14, 1054–1062.

Richardson, S.H., 1984. Sr, Nd and O isotope variation in an extensive Karoo dolerite sheet, southern Namibia (*in* Petrogenesis of the volcanic rocks of the Karoo Province, Erlank,) *Special Publication - Geological Society of South Africa* **13** 289-293.

Richardson, S.H., 1986. Latter-day origin of diamonds of eclogitic paragenesis. *Nature* **322**, 623 – 626.

Richardson, S.H., and Harris, J.W., 1997. Antiquity of peridotitic diamonds from the Siberian Craton. *Earth and Planetary Science Letters* **151**, 271-277.

Richardson, S.H., Gurney, J.J., Erlank, A.J., & Harris, J.W., 1984. Origin of diamonds in old enriched mantle. *Nature* **310**, 198-202.

Richardson, S.H., Erlank, A. J., and Hart, S. R., 1985. Kimberlite-borne garnet peridotite xenoliths from old enriched subcontinental lithosphere. *Earth Planet. Sci. Lett.* **75**, 116-128.

Richardson, S. H., 1986. Latter-day origin of diamonds of eclogitic paragenesis. *Nature* **322**, 623-626.

Richardson, S. H., Erlank, A. J., Harris, J. W., & Hart, S. R., 1990. Eclogitic diamonds of Proterozoic age from Cretaceous kimberlites. *Nature* **346**, 54-56.

Richardson, S. H., Harris, J. W., and Garney, J. J., 1993. Three generations of diamonds from old continental mantle. *Nature*, **366**, 256-258.

Richardson, S.H., Chinn, I.L., and Harris, J.W., 1998. Age and origin of eclogitic diamonds from the Jwaneng kimberlite, Botswana. *Extended Abstracts of 7th International Kimberlite Conference*, Cape Town, South Africa, 734-736.

Richardson, S. H., Shirey, S. B., Harris, J. W., and Carlson, R. W., 2001. Archean subduction recorded by Re-Os isotopes in eclogitic sulfide inclusions in Kimberley diamonds. *Earth and Planetary Science Letters* **191**, 257-266.

Richardson, S. H., Shirey, S. B., and Harris, J. W., 2004. Episodic diamond genesis at Jwaneng, Botswana, and implications for Kaapvaal craton evolution. *Lithos* **77**, 143-154.

Richardson, S.H., Harris, J.W. & Poeml, P.F., 2006. Antiquity of harzburgitic diamonds from the Venetia kimberlite, Limpopo Belt, Kaapvaal craton, Abstracts of the 16th annual V. M. Goldschmidt conference, *Geochimica et Cosmochimica Acta* **70**, @Abstract A531.

Ringwood, A.E., 1956. The olivine-spinel transition in the Earth's mantle. *Nature* **178**, 1303-1304.

Ringwood, A. E., 1966. The chemical composition and origin of the Earth. In: Hurley, P.M. (ed.). Advances in Earth Science, MIT press, Cambridge, Mass, 287-356.

Ringwood, A.E., 1967. The pyroxene garnet transformation in the earth's mantle. *Earth Planet. Sci. Lett.* **2**, 255-263.

Ringwood, A.E., 1982. Phase transformations and differentiation in subducted lithosphere: implications for mantle dynamics, basalt petrogenesis, and crustal evolution. *J. Geology* ***90***, 611-643.

Ringwood, A. E., 1990. Slab-mantle interactions, 3. Petrogenesis of intraplate magmas and structure of the upper mantle. *Chemical Geology* **82**, 187-207.

Ringwood, A.E., 1991. Phase transformations and their bearing on the constitution and dynamics of the mantle. *Geochimica et Cosmochimica Acta* **55**, 2083-2110.

Ringwood, A.E., and Major, A., 1966. Synthesis of Mg_2SiO_4-Fe_2SiO_4 spinel solid solutions. *Earth Planet. Sci. Lett.* **1**, 241-245.

Ringwood, A.E. and Green, D.H., 1967. An experimental investigation of the gabbro-eclogite transformation and some geophysical implications. *Tectonophysics* **3**, 383-427.

Robinson, D.N., 1979. Diamond and graphite in eclogite xenoliths from kimberlite. *Proceedings of the 2nd International Kimberlite Conference* **2**, 50-58.

Robinson, D. N., Gurney, J. J., & Shee, S. R., 1984. Diamond eclogite and graphite eclogite xenoliths from Orapa, Botswana. In: Kornprobst, J. (ed.). Kimberlites II: The Mantle and Crust-Mantle Relationships, Elsevier, 11-24.

Robinson, D. N., Scott, J. A., Van, Niekerk, A., & Anderson, V. G., 1989. The sequence of events reflected in the diamonds of some southern African kimberlites. In: Ross, J. et al. (eds.). Diamonds, Kimberlites and Related Rocks, Volume 2. *Geol. Soc. Aust. Spec. Publ.* **14**, Blackwell Scientific Publications, Melbourne, 990-1000.

Robinson, J. A. C., Wood, B. J., & Blundy, J. D., 1998. The beginning of melting of fertile and depleted peridotite at 1.5 GPa. *Earth and Planetary Science Letters* **155**, 97-111.

Rosen, O.M., 1995. Metamorphic effects of tectonic movements at the lower crust level, Proterozoic collision zones and terraines of the Anabar shield. *Geotectonics* **29**, 91-101.

Rosen, O.M., Serenko, V.P., Spetsius, Z.V., Manakov, A.V., and Zinchuk, N.N., 2002. Yakutian kimberlite province: position in the structure of the Siberian craton and composition of the upper and lower crust. *Russian Geology and Geophysics* **43**, 3-26 (in Russian).

Rosenbaum, J. M., 1993. Mantle phlogopite - A significant lead repository? *Chem. Geol.* **106**, 475-483.

Rosenbaum, J. M., Zindler, A., & Rubenstone, J. L., 1996. Mantle fluids: Evidence from fluid inclusions. *Geochimica et Cosmochimica Acta* **60**, 3229-3252.

Rowe, T., Kappelman, J., Carlson, W.D., Ketcham, R.A., & Denison, C., 1997. High-resolution computed tomography: A breakthrough technology for earth scientists. *Geotimes* **42**, 23–27.

Rudnick, R. L., and Nyblade, A. A., 1999. The thickness and heat production of Archean lithosphere: Constraints from xenolith thermobarometry and surface heat flow. In: Fei, Y., Bertka, C., and Mysen, B.O. (eds.). Mantle Petrology: Field Observations and High pressure Experimentation. Vol. 6, The Geochemical Society, pp. 3-12.

Rudnick, R.L., McDonough, W.F., Chappell, B.W., 1993. Carbonatite metasomatism in the northern Tanzanian mantle: petrography and geochemical characteristics. *Earth Planet. Sci. Lett.* **114**, 463-475.

Rudnick, R.L., Stewart, E.C., & Bulanova, G.P., 1993. Diamond growth history from in-situ measurement of Pb and S isotopic compositions of sulfide inclusions. *Geology* **21**, 13-16.

Rudnick, R. L., McDonough, W. F., & O'Connell, R. J., 1998. Thermal structure, thickness, and composition of continental lithosphere. *Chemical Geology* **145**, 395-411.

Rudnick, R. L., Barth, M. G., Horn, I., & McDonough, W. F., 2000. Rutile-bearing refractory eclogites: Missing link between continents and depleted mantle. *Science* **287**, 278-281.

Russell, J. K., and Kopylova, M. G., 1999. A steady state conductive geotherm for the north central Slave, Canada: Inversion of petrological data from the Jericho kimberlite pipe. *J. Geophys. Res.* **104**, 7089-7101.

Russell, J. K., Dipple, G. M., & Kopylova, M. G., 2001. Heat production and heat flow in the mantle lithosphere, Slave craton, Canada. *Phys. Earth Planet. Int.* **123**, 27-44.

Ruzicka, A., Riciputi, L.R., Greenwood, J., Taylor, L.A., Snyder, G.A., Keller, R.A., Bulanova, G.P., & Millidge, J.H., 1999. Petrogenesis of mantle-derived sulfide inclusions in Yakutian diamonds: Chemical and Isotopic disequilibrium during quenching from high temperatures, *Proc. 7th International Kimberlite Conference* **2**, 741-749.

Ryabchikov, I.D., Sharkov, E.V., Snyder, G.A., & Taylor, L.A., 1999. The Upper Mantle: Direct Samples of the Mantle. Chapter 2.2. In: Advanced Mineralogy **3**, Springer-Verlag, Berlin, 58.

Rylov, G.M., Fedorova, E.N. & Sobolev, N.V., 2006. Study of the internal structure of imperfect diamond crystals by the Laue-Sr method. *Russian Geology and Geophysics* **47**, 242-249.

Sachtleben, T. H. and Seck, H. A., 1981. Chemical control of Al-solubility in orthopyroxene and its implications on pyroxene geothermometry. *Contrib. Mineral. Petrol.* **78**, 157-165.

Safonov, O.G., Perchuk, L.L. & Litvin, Y.A., 2007. Melting relations in the chloride–carbonate–silicate systems at high-pressure and the model for formation of alkalic diamond–forming liquids in the upper mantle. *Earth and Planetary Science Letters* **253**, 112-128.

Salpas, P.A., L.A. Taylor, and J.W. Shervais, 1987, The Blue Ball, Arkansas kimberlite: Mineralogy, petrology, and geochemistry. Jour. Geol. 94, 891-901.

Salters, V. J. M., and Zindler, A., 1995. Extreme $^{176}Hf/^{177}Hf$ in the sub-oceanic mantle. *Earth Planet. Sci. Lett.* **129**, 13-30.

Saltzer, R. L., Chatterjee, N., & Grove, T. L., 2001. The spatial distribution of garnets and pyroxenes in mantle peridotites: Pressure-temperature history of peridotites from the Kaapvaal craton. *J. Petrol.* **42**, 2215-2229.

Samoilovich, M.I., Sangarlinsky, N.G., & Laptev, V.A., 1987. Investigation of processes of diamond nucleation and growth in system metal-graphite. In: Synthesis of minerals, *1*, Moscow: Nedra, 368-389.

Sangarlinsky, N.G., Bokiy, G.B., & Laptev, V.A., 1976. To mekhanism of diamond formation in system metal-carbon. In: Synthesis of minerals and experimental investigations, Moscow: Nedra, 141-145.

Sautter, V., Haggerty, S. E., and Field, S. W., 1991. Ultradeep, >300 kilometers ultramafic xenoliths: Petrological evidence from the transition zone. *Science* **252**, 827-830.

Schmidberger, S. S., and Francis, D., 2001. Constraints on the trace element composition of the Archean mantle root beneath Somerset Island, Arctic Canada. *J. Petrol.* **42**, 1095-1117.

Schmidberger S. S., Simonetti A., & Francis D., 2001. Sr-Nd-Pb isotope systematics of mantle xenoliths from Somerset Island kimberlites: Evidence for lithosphere stratification beneath Arctic Canada. *Geochimica et Cosmochimica Acta,* **65**, 4243-4255.

Schmidberger, S. S., Simonetti, A., & Francis, D., Gariepy, C., 2002. Probing Archean lithosphere using the Lu-Hf isotope systematics of peridotite xenoliths from Somerset Island kimberlites, Canada. *Earth Planet. Sci. Lett.* **6137**, 1-15.

Schrauder, M., and Navon, O., 1994. Hydrous and carbonatitic mantle fluids in fibrous diamonds from Jwaneng. Botswana, *Geochim. Cosmochim. Acta* **58**, 761-771.

Schulze, D. J., 1989. Constraints on the abundance of eclogite in the upper mantle. *J. Geophys. Res.* **94**, 4205-4212.

Schulze, D. J., 1995. Low-Ca garnet harzburgites from Kimberley, South Africa: abundance and bearing on the structure and evolution of the lithosphere. *Journal of Geophysical Research* **100**, 12513–12526.

Schulze, D. J., 1996. Chromite macrocrysts from southern African kimberlites: mantle xenolith sources and post-diamond re-equilibration. In: Kogbe, C. A. (ed.) Diamond Deposits in Africa. Special Issue Africa Geoscience Review **3**, 203–216.

Schulze, D. J., Wiese, D. & Steude, J., 1996. Abundance and distribution of diamonds in eclogite revealed by volume visualization of CT x-ray scans. *J. Geol.* **104**, 109-113.

Schulze, D. J., Valley, J. W., & Spicuzza, M., 2000. Coesite eclogites from the Roberts Victor kimberlite, South Africa. *Lithos* **54**, 23-34.

Schulze, D.J., Valley, J.R., Bell, D.R. & Spicuzza, M.J. 2001, Oxygen isotope variations in Cr-poor megacrysts from kimber*lite,* ***Geochimica et Cosmochimica Acta*** **65**, 4375-4384.

Schulze, D.J., Valley, J.W., Spicuzza, M.J. & Channer, D.M.D. 2003, Oxygen isotope composition of eclogitic and peridotitic garnet xenocrysts from the La Ceniza Kimberlite, Guaniamo, Venezuela, *International Geology Review* **45**, 968-975.

Schulze, D.J., Harte, B., Valley, J.W., Brenan, J.M., & Channer, D.M., 2003. Extreme crustal oxygen isotope signatures preserved in coesite in diamond. *Nature* **423**, 68-70.

Schulze, D.J., Harte, B., Valley, J.W. & Channer, D.M. DeR., 2004. Evidence of subduction and crust–mantle mixing from a single diamond. *Lithos* **77**, 349-358.

Schulze, D.J., Canil, D., Channer, D. M.DeR. & Kaminsky, F.V., 2006, Layered mantle structure beneath the western Guyana Shield, Venezuela: Evidence from diamonds and xenocrysts in Guaniamo kimberlites. *Geochimica et Cosmochimica Acta*, **70**, 192-205.

Scott, B.H., Danchin, R.V., Harris, J.W., & Stracke, K.J., 1984. Kimberlites near Orroroo, South Australia. In: In: Kornprobst, J. (ed.). Kimberlites I: Kimberlites and Related Rocks. Elsevier, Amsterdam, 121-142.

Sellschop, J.P.F., Madiba, C.C.P., & Annegarn, H.J., 1980. Light volatiles in diamond: physical interpretation and genetic significance. *Nucl. Inst. Meth.* **168**, 529-534.

Sharygin, V.V., Golovin, A.V., Pokhilenko, N.P. & Sobolev, N.V., 2003. Djerfisherite in unaltered kimberlites of the Udachnaya-East Pipe, Yakutia. *Doklady Earth Sciences* **390**, 554-557.

Sharygin, V.V., Kamenetsky, V.S., Kamenetskaya, M.B., Seretkin, Y.V. & Pokhilenko, N.P., 2007. Rasvumite from the Udachnaya east pipe; the first finding in kimberlites, *Doklady Earth Sciences* **415A**, 929-934.

Shatsky, V.S., Pal'yanov, Y.N., Sokol, A.G., Tomilenko, A.A., and Sobolev, N.V., 2005. Diamond formation in UHP dolomite marbles and garnet-pyroxene rocks of the Kokchetav massif (northern Kazakhstan): Natural and experimental evidence. *International Geology Review* **47**, 999-1019.

Shatsky, V.S., Zedgenizov, D.A., Ragozin, A.L., Mityukhin, S.L., and Sobolev, N.V., 2005. Evidence for metasomatic formation of diamond in eclogite xenolith from the Udachnaya kimberlite pipe (Yakutia). *Doklady Earth Sciences* **402**, 587-590 (in Russian).

Shatsky, V.S., Pal'yanov, Y.N., Sokol, A.G., Tomilenko, A.A. & Sobolev, N.V., 2006. Diamond formation in UHP dolomite marbles and garnet-pyroxene rocks of the Kokchetav Massif, northern Kazakhstan; natural and experimental evidence; Phase relations, high-pressure terranes, P-T-ometry, and plate pushing; a tribute to W. G. Ernst, *International Book Series* **9**, 108-119.

Shatsky, V.S., Ragozin, A.L. & Sobolev, N.V., 2006. Some aspects of metamorphic evolution of ultrahigh-pressure calc-silicate rocks of the Kokchetav Massif; Geodynamics and petrology of the lithosphere and upper mantle. *Russian Geology and Geophysics* **47**, 105.

Shee, S.R., Gurney, J.J., & Robinson, D.N., 1982. Two diamond-bearing peridotitic xenoliths from the Finsch kimberlite, South Africa. *Contributions to Mineralogy and Petrology* **81**, 79-87.

Shervais, J.W., Taylor,L.A., & Korotev, R.L., 1985. Petrology and mineral chemistry of some African eclogites and the evolution of sub-continental mantle and continental crust. In Lunar & Planetary Science XVI, 769-770.

Shervais, J.W., Taylor, L.A., Lugmair, G.W., Clayton, R.N., Mayeda, T., & Korotev, R.L., 1986. Evolution of sub-continental mantle and crust: Eclogites from southern Africa. *Fourth Int'l. Kimberlite Conf., Geol. Soc. Australia* **16**, 326-328.

Shervais, J.W., Taylor, L.A., Laul, J.C., 1987. Magma-mixing and kimberlite genesis: Mineralogical, petrological, and trace element evidence from eastern U.S.A. kimberlites. *Geol. Soc. Amer. Spec.* **215**, 101-114.

Shervais, J.W., Taylor, L.A., Lugmair, G.W., Clayton R.N., Mayeda, T.K., Korotev, R.L., 1988. Early Proterozoic oceanic crust and the evolution of subcontinental mantle: Eclogites and related rocks from southern Africa. *Geol. Soc. Amer. Bulletin* **100**, 411-423.

Shimizu, N., 1975. Rare earth elements in garnets and clinopyroxenes from garnet lherzolite nodules in kimberlites. *Earth and Planetary Science Letters* **25**, 26-35.

Shimizu, N., 1999. Young geochemical features in cratonic peridotites from southern Africa and Siberia. In: Fei Y., Bertka, C. M., and Mysen, B. (eds.). Mantle Petrology: Field Observations and High Pressure Experimentation **6**, The Geochemical Society, 47-55.

Shimizu, N. and Richardson, S. H., 1987. Trace element abundance patterns of garnet inclusions in peridotite suite diamonds. *Geochimica et Cosmochimica Acta* **51**, 755-758.

Shimizu, N. and Sobolev, N.V., 1995. Young peridotitic diamonds from the Mir kimberlite pipe. *Nature* **375**, 394-397.

Shimizu, N., Boyd, F.R. & Pokhilenko, N.P., 1993. Trace element zoning patterns of mantle garnets. Geological Society of America, 1993 annual meeting", *Abstracts with Programs - Geological Society of America* **25**, 36.

Shimizu, N., Boyd, F.R., Sobolev, N.V., & Pokhilenko, N.P., 1994. Chemical zoning of garnets in peridotites and diamonds. *Mineralogical Magazine* **58A**, 831-832.

Shimizu, N., Sobolev, N. V., & Yefimova, E. S., 1997. Chemical heterogeneities of inclusion garnets and juvenile character of peridotitic diamonds from Siberia. *Russian Journal of Geology and Geophysics.* **38**, 356-372.

Shimizu, N., Sobolev, N.V., & Yefimova, E.S., 1998. Trace element heterogeneities of in-situ diamond inclusion garnets from Siberia. *7th International Kimberlite Conference*, Extended Abstracts, 807.

Shimizu, N., Pokhilenko, N.P., Boyd, F.R. & Pearson, D.G., 1999. Trace element characteristics of garnet dunites/harzburgites, host rocks for Siberian peridotitic diamonds; Proceedings of the VIIth international kimberlite conference", *Proceedings of the International Kimberlite Conference* **7**, 773-782.

Shin, S.-D., Hwang, N. M., and Kim, D.-Y., 2002. High rate of diamond deposition through graphite etching in a hot filament CVD reactor. *Diamond and Related Materials* **11,** 1337–1343.

Shirey, S.B., and Walker, R.J., 1998. The Re-Os isotope system in cosmochemistry and high-temperature geochemistry. *Annual Review of Earth and Planetary Science* **26**, 423-500.

Shirey, S.B., Carlson, R.W., Richardson, S.H., Menzies, A., Gurney, J., Pearson, D.G., Harris, J.W., & Wiechert, U., 2001. Archean emplacement of eclogitic components into the lithospheric mantle during formation of the Kaapvaal Craton. *Geophysics Research Letters* **28**, 2509-2512.

Shirey S. B., Richardson S. H., & Harris J. W., 2004. Integrated models of diamond formation and craton evolution. *Lithos* **77,** 923– 944.

Shiryaev, A.A., Izraeli, E.S., Hauri, E.H., Zakharchenko, O.D. & Navon, O. 2005, Chemical, optical and isotopic investigation of fibrous diamonds from Brazil; Evolution of continental lithosphere, origin of diamonds and their deposits (to the birthday of Academician Nikolai

Vladimirovich Sobolev), *Russian Geology and Geophysics,* **46**, 1185-1201.

Shushkanova, A.V., and Litvin, Y.A., 2005. Phase relations in diamond-forming carbonate-silicate-sulfide systems on melting, *Russian Geology and Geophysics* **46**, 1317-1326.

Shushkanova, A. V., and Litvin, Y.A., 2006. Formation of diamond polycrystals in pyrrhotite-carbonic melt; experiments at 6.7 GPa. *Doklady Earth Sciences* **409**, 916-920 (in Russian).

Silaev, V.I., Petrovsky, V.A., Sukharev, A.E., and Filippov, V.N., 2006. Inclusions of zircon-based solid solutions in diamonds. *Doklady Earth Sciences* **411**, 1318-1323 (in Russian).

Simakov, S.K. and Vaganov, V.I., 1992. A new petrologic criterion for preliminary estimation of the diamond content of plutonic igneous rocks. *Transactions (Doklady) of the U.S.S.R.Academy of Sciences: Earth Science Sections* **325A**, 170-173.

Simakov, S.K. and Taylor, L.A., 2000. Geobarometry for Mantle Eclogites: Solubility of Ca-Tschermaks in Clinopyroxene, *International Geology Review* **42**, 534-544.

Simakov S.K., 2006. Redox state of eclogites and peridotites from sub-cratonic upper mantle and a connection with diamond genesis. *Contributions to Mineralogy and Petrology* **151**, 282-296.

Simon, N.S.C., Carlson, R. W., Pearson, D.G., & Davies, G.R., 2002. The Lu-Hf isotope composition of cratonic lithosphere: disequilibrium between garnet and clinopyroxene in kimberlite xenoliths. *Geochimica et Cosmochimica Acta* **66** (S1), A717.

Simon, N.S.C., Irvine, G.J., Davies, G. R., Pearson, D.G., & Carlson, R. W., 2003. The origin of garnet and clinopyroxene in 'depleted' Kaapvaal peridotites. *Lithos* **71**, 289-322.

Simon N.S.C., Carlson R.W., Pearson D.G., & Davies G.R., 2007. The origin and evolution of the Kaapvaal cratonic lithospheric mantle. *J. Petrology* **48**, 589-625.

Sitepu, H., Kopylova, M.G., Quirt, D.H., Cutler, J.N., & Kotzer, T.G., 2005. Synchrotron micro-X-ray fluorescence analysis of natural diamonds: First steps in identification of mineral inclusions in situ. *American Mineralogist* **90**, 1740 - 1747.

Slodkevich, V.V., 1982. Paramorphism of graphite in relation to diamond. *Zapiski Vsesoyuznogo Mineralogicheskogo Obshchestva* **111**, 13-33.

Slodkevich, V.V., 1983. Graphite paramorphs after diamond. *International Geology Review* **25**, 497-514.

Slodkevich, V.V., 1987. Lherzolites from ex-diamond-associated phlogopite. *Doklady Akademii nauk SSSR* **297**, 942-945.

Slodkevich, V.V., 1988. Former diamond-bearing phlogopite lherzolite. *Transactions (Doklady) of the U.S.S.R.Academy of Sciences: Earth Science Sections* **297**, 117-120.

Slodkevich, V.V. and Lobkov, V.A., 1983. Carbon isotopic composition of graphites of the stratified pluton with basic-ultrabasic rocks in Beni Bouchera, Morocco. *Doklady Akademii nauk SSSR* **272**, 698-701.

Smith, A., Mainwood, A., & Watkins, M., 2002. The kinetics of the capture of nitrogen by nickel defects in diamond. *Diamond and Related Materials* **11**, 312–315.

Smith, C. B., Gurney, J. J., Harris, J. W., Otter, M. L., Kirkley, M. B., & Jagoutz E., 1991. Neodymium and strontium isotope systematics of eclogite and websterite paragenesis inclusions from single diamonds, Finsch and Kimberley Pool, RSA. *Geochim. Cosmochim. Acta* **55**, 2579-2590.

Smith, D., 1999. Temperatures and pressures of mineral equilibration in peridotite xenoliths:

Review, discussion and implications. In: Fei, Y., Bertka, C., and Mysen, B.O. (eds.). Mantle Petrology: Field Observations and High Pressure Experimentation **6**, The Geochemical Society, 171-188.

Smith, D., and Barron, B. R., 1991. Pyroxene-garnet equilibration during cooling in the mantle. *American Mineralogist* **76**, 1950-1963.

Smith, D., and Boyd, F. R., 1992. Compositional zonation in garnets of peridotite xenoliths. *Contributions to Mineralogy and Petrology* **112**, 134-147.

Smyth, J.R. and Hatton, C.J., 1977. A coesite-sanidine grospydite from the Roberts Victor kimberlite. *Earth and Planetary Science Letters* **34**, 284-290.

Snyder, G. A., Jerde, E. A., Taylor, L. A., Halliday, A. N., Sobolev, V. N., and Sobolev, N. V. 1993. Nd and Sr isotopes from diamondiferous eclogites, Udachnaya kimberlite pipe, Yakutia, Siberia: Evidence of differentiation in the early Earth? *Earth and Planetary Science Letters* **118**, 91-100.

Snyder, G.A., Taylor, L.A., Jerde, Sharkov, E., Las'ko, E., 1994. Petrogenesis of garnet pyroxenite and spinel peridotite xenoliths of the Tell-Danun alkali basalt volcano, Harrat as Shamad, southwestern Syria. *International Geology Review* **35**, 1104-1120.

Snyder, G.A., Taylor, L.A., Beard, B.L., Sobolev, V.N., Crozaz, G. & Sobolev, N.V., 1994. Siberian eclogite xenoliths; generalities, oddities and crustal vs. mantle protoliths, AGU fall meeting, *EOS, Transactions, American Geophysical Union* **75**, 745.

Snyder, G.A., Taylor, L.A., Jerde, E.A., Deines, P., Clayton, R.N., Mayeda,T., Rossman G.R., & Sobolev, N.V., 1995. Archean mantle heterogeneity and the origin of diamondiferous eclogites, Siberia: Evidence from stable isotopes and hydroxyl in garnets. *American Mineralogist* **80**, 799-809.

Snyder, G.A., Taylor, L.A. & Sobolev, N.V., 1996. Diamond genesis in Archean Yakutian eclogites, Siberia, Geological Society of America, 28th annual meeting, *Abstracts with Programs - Geological Society of America 28*, 290.

Snyder, G.A., Taylor, L.A., Crozaz, G., Halliday, A.N., Beard, B.L, Sobolev, V.N., & Sobolev, N.V., 1997. The origin of Yakutian eclogite xenoliths. *Journal of Petrology* **38**, 85-113.

Snyder, G.A., Taylor, L.A., Jin, Y., and D.H. Taylor, 1997. Petrology and mineral chemistry of spinel lherzolite xenoliths from eastern China. *International Geology Review* **39**, 671-687.

Snyder, G.A., Taylor, L.A., Kelle, R.A., McCandless, T.E., Ruiz, J., & Sobolev, N.V., 1998. First internal Re-Os isochron for a diamondiferous eclogite, Yakutia, and evidence of Archean subduction of seafloor hydrothermal vent deposits (abs.). *EOS, AGU Trans.* **79**, F1018.

Snyder, G.A., Taylor, L.A., Beard, B.L., Crozaz, G., Halliday, A.N., Sobolev, V.N., & Sobolev, N.V., 1998. Reply to a comment by Jacob, D. et al. on "The Origins of Yakutian Eclogite Xenoliths." *Journal of Petrology* **39**, 1535-1543.

Snyder, G.A. Keller, R.A., Taylor, L.A. Remley, D., Halliday, A.N., and Sobolev, N.V., 1998, The origin of ultramafic (Group A) Eclogites: Nd & Sr isotopic evidence from the Obnazhennaya kimberlite, Yakutia, *7th International Kimberlite Conference,* Capetown, South Africa, 823-825.

Snyder, G.A., Taylor, L.A., Crozaz, G., Halliday, A.N., Beard, B.L., Sobolev, V.N., & Sobolev, N.V., 1998. The diamond-bearing Mir eclogites, Yakutia: Nd and Sr isotopic evidence for the continental crustal input in an Archean oceanic environment. *Extended*

Abstracts of 7th International Kimberlite Conference, Cape Town, South Africa, 826-828.

Snyder, G.A., Taylor, L.A., Beard, B.L., Halliday, A.N., Sobolev, N.V., & Dimakov, 1999. The diamond-bearing Mir eclogites, Yakutia: Nd and Sr isotopic evidence for a possible early to mid-proterozoic depleted mantle source with arc affinitity, *Proc. 7th Int'l Kimberlite Conf.* **2**, National Printers, So. Afr., 808-815.

Sobolev, A.V., Hofmann, A.W., & Kuzmin, D.V., 2007. The amount of recycled crust in sources of mantle – derived melts. *Science* **316**, 412-417.

Sobolev, N.V., 1968. Eclogite clinopyroxenes from the kimberlite pipes of Yakutia. *Lithos* **1**, 54-57.

Sobolev, N.V., 1974. Deep-seated Inclusions in Kimberlites and the Problem of the Composition of the Upper Mantle. Novosibirsk, Nauka, (in Russian), 264.

Sobolev, N.V., 1977. Deep-seated Inclusions in Kimberlites and the Problem of Upper Mantle Composition. Novosibirsk: Nauka, in Russian. Engl. Translation by Brown, D.A. and Boyd, F.R. (eds.). American Geophysics Union, Washington, D.C., 279.

Sobolev, N.V., 1984. Crystalline inclusions in diamonds from New South Wales, Australia. *Publ. Geol. Dept., Univ. W. Aust.* **8**, 213-219.

Sobolev, N.V. (ed.). 1995. Field guide to the kimberlites of Yakutia., *6th International Kimberlite Conference*. Novosibirsk, Almazy Rossii - Sakha Co. Ltd.

Sobolev, N.V., 2006. Coesite as an indicator of ultrahigh pressures in continental lithosphere. *Russian Geology and Geophysics* **47**, 95-104.

Sobolev, N.V. and Lavrent'yev, Y.G., 1971. Isomorphic sodium admixture in garnets formed at high pressures. *Contribib. Mineral. Petrol.* **31**, 1-12.

Sobolev, N.V. and Shimizu, N., 1993. Trace element variations in diamond inclusion garnets from Siberian kimberlite pipes, AGU 1993 fall meeting, *EOS, Transactions, American Geophysical Union* **74**, 637.

Sobolev, N.V. and Yefimova, E.S., 1998. Compositional variations of chromite inclusions as an indicator of the zonation of diamond crystals, *Doklady Earth Sciences* **359**, 163-166.

Sobolev, N.V. and Logvinova, A.M., 2005. Significance of accessory chrome spinel in identifying serpentinite paragenesis. *International Geology Review 47*, 58-64.

Sobolev, N.V., Kuznetsova, I.K., & Zyuzin, N.I., 1968. The petrology of grospydite xenoliths from the Zagadochnaya kimberlite pipe in Yakutia. *Journal of Petrology* **9**, 253-280.

Sobolev, N.V., Lavrent'ev, Y.G., Pospelova, L.N., & bolev, E.V., 1969. Chrome pyropes from the diamonds of Yakutia. *Dokl. Akad. Nauk SSSR* **189**, 162-165.

Sobolev, N.V., Bartoshinsky, Z.V., Yefimova, E.S., Lavrent'yev, Y.G., & spelova, L.N., 1970. Assocaition of olivine, garnet and chrome-diopside in a Yakutsk diamond. *Dokl. Akad. Nauk SSSR* **192**, 1349-1353.

Sobolev, N.V., Gnevushev, M.A., Shemanina, E.I., Lavrent'ev, Yu.G., & Pospelova, L.N., 1971. The composition of garnet and pyroxene inclusions from the diamonds of the Urals. *Dokl. Akad. Nauk SSSR* **198**, 190-193.

Sobolev, N.V., Botkunov, A.I., Bakumenko, I.T., &Sobolev, V.S., 1972. Crystalline inclusions with octahedral facets in diamonds. *Dokl. Akad. Nauk SSSR* **204**, 192-195.

Sobolev, N. V., Lavrent'yev, Y. G., Pokhilenko, N. P., & Usova, L. V., 1973. Chrome-rich garnets from the kimberlites of Yakutia and their paragenesis. *Contrib. Mineral. Petrol.* **40**, 39-52.

Sobolev, N.V., Botkunov, A.I., Lavrent'ev, Y.G. & Usova, L.V., 1976. New data on mineral compositions as coexisting with diamonds of kimberlite pipe 'Mir' (Yakutia). *Geologiya i Geofizika* **12**, 3–15 (in Russian).

Sobolev, N.V., Pokhilenko, N.P., Lavrentyev, Y.G., & Usova, L.V., 1978. Significance of chromium in garnets from kimberlites. In: V.A.Kuznetsov, ed. Problems of Petrology of the Earth's Crust and Upper Mantle. Publ. House Nauka, Novosibirsk, p. 145-168 (in Russian).

Sobolev, N.V., Galimov, E.M., Ivanovskaya, I.N., & Yefimova, E.S., 1979. The carbon isotope compositions of diamonds containing crystalline inclusions. *U.S.S.R. Academy of Science, Doklady, Earth Science Section,* **249**, 1217-1220.

Sobolev, N.V., Yefimova, E.S., Usova, L.V., 1983. Eclogitic paragenesis of diamonds from the Mir kimberlite pipe: In Mantle Xenoliths and Problems of Ultrabasic Mag*mas*. Dokl Akad. Nauk USSR, Siberian Branch (in Russian).

Sobolev, N.V., Pokhilenko, N. P., & Yefimova, E.S., 1984. Diamond-bearing peridotite xenoliths in kimberlites and the problem of the origin of diamonds. *Geologia Geofizica* **25**, 63-80 (in Russian).

Sobolev, N.V., Galimov, E.M., Smith, C.B., Yefimova, E.S., Maltsev, K.A., Hall, A.E., Usova, L.V., 1989. Comparative study of morphology, inclusions, carbon isotope composition of diamonds in alluvials of the King George River, Argyle Lamproite Mine (Western Australia), of cube microdiamonds from Northern Australia. *Geologia Geofizica, (Sov. Geol. Geophys.)* **30**.

Sobolev, N.V., and Shatsky, V.S., 1990. Diamond inclusions in garnets from metamorphic rocks: a new environment for diamond formation. *Nature* **343,** 742-746.

Sobolev, N.V., Bakumenko, I.T., Yefimova, E.S., & Pokhilenko, N.P., 1991. Morphological features of microdiamonds, sodium in garnets and potassium contents in clinopyroxene of two xenoliths from the Udachnaya kimberlite pipe (Yayutia). *Dokl. Akad. Nauk SSSR* **321**, 585-591 (in Russian).

Sobolev, N.V., Pokhilenko, N.P. & Afanas'ev, V.P., 1993. Kimberlitic pyrope and chromite morphology and chemistry, as indicators of diamond grade in Yakutian and Arkhangelsk provinces. In: Dunne, K. P. E. & Grant, B. (eds) Mid-Continent Diamonds. GAC–MAC Symposium Abstracts, Edmonton, June 1993, pp. 63–69.

Sobolev, N.V., Bakumenko, I.T., Yefimova, E.S., & Pokhilenko, N.P., 1994. Morphological characteristics of microscopic diamonds containing traces of sodium in garnets and of potassium in pyroxenes of two eclogite xenoliths from the Udachnaya kimberlite pipe (Yakutia), *Transactions (Doklady) of the Russian Academy of Sciences.Earth Science Sections* **322**, 38-146.

Sobolev, N.V., Shatsky, V.S. & Vavilov, M.A., 1994. Significance of zircon from ultra high pressure metamorphic rocks as the best high-strength mineral container; AGU 1994 fall meeting, *EOS, Transactions, American Geophysical Union* **75**, 743.

Sobolev, N.V., Yefimova, E.S., Reimers, L.F., Zakharchenko, O.D., Makhin, A.I. & Usova, L.V., 1995. Arkhangelsk diamond inclusions, Sixth international kimberlite conference, extended abstracts, *Proceedings of the International Kimberlite Conference* **6**, 558-560.

Sobolev, N.V., Kaminsky, F.V., Griffin, W.L., Yefimova, E.S., Win, T.T., Ryan, C.G., & Bitarnov, A.I., 1997. Mineral inclusions in diamonds from the Sputnik kimberlite pipe, Yakutia.

Lithos **39**, 135-157.

Sobolev, N.V., Yefimova, E.S., Reimers, L.F., Zakharchenko, O.D., Makhin, A.I. & Usova, L.V., 1997. Mineral inclusions in diamonds of the Arkhangelsk kimberlite province. *Russian Geology and Geophysics* **38**, 379–393.

Sobolev, N.V., Snyder, G.A., Taylor, L.A., Keller, R.A., Yefimova, E.S., Sobolev, V.N., & Shimizu, N., 1998. Extreme chemical diversity in the mantle during eclogite diamond formation: Evidence from 35 garnet and 5 pyroxene inclusions in a single diamond, *International Geology Review* **40**, 567-578.

Sobolev, N.V., Taylor, L.A., Zuev, V.M., Bezborodov, S.M., Snyder, G.A., Sobolev, V.N., & Yefimova, E.S., 1998. The specific features of eclogitic paragenesis of diamonds from Mir and Udachnaya kimberlite pipes, Yakutia. *Russ. Geol. Geophys.* **39**, 1653– 1663.

Sobolev, N.V., Sobolev, V.N., Snyder, G.A., Yefimova, & Taylor, L.A., 1999. Significance of eclogitic and related parageneses of natural diamonds. In: Planetary Petrology and Geochemistry, *The Lawrence A. Taylor 60th Birthday Volume*, Geol. Soc. Amer., 15-26.

Sobolev, N.V., Yefimova, E.S., and Koptil, V.I., 1999. Mineral inclusions in diamonds in the northeast of the Yakutian diamondiferous province. In: Gurney, J.J., Gurney, J.L., Pascoe, M.D., and Richardson, S.H. (eds.), *Proceedings Volume of 7th International Kimberlite Conference*, *2*. Cape Town: Red Roof Design, 816-822.

Sobolev, N.V., Fursenko, B.A., Goryainov, S.V., Shu, J., Hemley, R.J., Mao, H.K,& Boyd, F.R., 2000. Fossilized high pressure from the Earth's deep interior: coesite-in-diamond barometer. *Proceedings of National Academy of Sciences of the USA* **97**, 11875-11879.

Sobolev, N.V., Efimova, E.S., Logvinova, A.M., Sukhodol'skaya, O.V., & Solodova, Y.P., 2001. Abundance and composition of mineral inclusions in large diamonds from Yakutia. *Doklady Earth Sciences* **376**, 34-38.

Sobolev, N.V., Logvinova, A.M., Zedgenizov, D.A., Yefimova, E.S., Taylor, L.A., Promprated, P., Koptil, V.I., & Zinchuk, N.N., 2003. Mineral inclusions in diamonds from Komsomolskaya and Krasnopresnenskaya pipes, Yakutia: evidence for deep lithospheric heterogeineties in Siberian Craton. *8th International Kimberlite Conference, Extented Abstracts* **1**, 0141.

Sobolev, N.V., Logvinova, A.M., Zedgenizov, D.A., Seryotkin, Y.V., Yefimova, E.S., Floss, C., Taylor, L.A., 2004. Mineral inclusions in microdiamonds and macrodiamonds from kimberlites of Yakutia: a comparative study. *Lithos* **77,** 225– 242.

Sobolev, N.V., and Logvinova, A.M., 2004, Pyrope inclusions in chrome spinels from kimberlites and lamproites and their significance for estimation of the paragenetic assemblage and formation depth. *Doklady Earth Sciences* **399**, 1074-1079.

Sobolev, N.V., Schertl, H.-P., and Neuser, R.D., 2006. Composition and paragenesis of garnets from ultrahigh-pressure calc-silicate metamorphic rocks of the Kokchetav Massif (northern Kazakhstan). *Russian Geology and Geophysics* **47**, 521-531.

Sobolev, N.V., Shertl, H.-P., Neuser, R.D., and Shastsky, V., 2007. Relict unsusually low iron pyrope-grossular garnets in UHPM calc-silicate rocks of the Kokchetav Massif, Kazakhstan. *International Geology Review* **49**, 717-731.

Sobolev, V.N., Taylor, L.A., Snyder, G.A., & Sobolev, N.V., 1994. Diamondiferous eclogites from the Udachnaya kimberlite pipe, Yakutia. *Internat. Geol. Rev.* **36**, 42-64.

Sobolev, V.N., Taylor, L.A., Snyder, G.A., Sobolev, N.V., Pokhilenko, N.P., & Kharkiv, A. D., 1997. A unique metasomatized peridotite from the Siberian Platform. *Proceedings Volume of 6th International Kimberlite Conference, 1: Kimberlites, Related Rocks, and Mantle Xenoliths*.; and *Russian Geology and Geophysics* **38**, 218-228.

Sobolev, V.N., Taylor, L.A., Snyder, G.A Jerde, E.A., Neal, C.R., &Sobolev, N.V., 1999. Quantifying the Effects of metasomatism in mantle xenoliths: Constraints from secondary chemistry and mineralogy in Udachnaya eclogites, Yakutia, *International Geology Review* **41**, 391-416.

Sobolev, V.N., McCammon, C.A., Tayor, L.A., Snyder, G.A., & Sobolev, N.V., 1999. Precise Mössbauer milliprobe determination of ferric iron in rock-forming minerals and limitations of electron microprobe. *American Mineralogy* **84**, 78-85.

Sobolev, V.S., and Sobolev, N.V., 1980. New proof of submersion to great depths of eclogitized rocks of the earth's crust. *Trans. Dokl. Akad. Nauk SSSR.* **249**, 1217-1220.

Sobolev, V.S., Sobolev, N.V., & Lavrent'ev, Y., 1972. Inclusions in diamonds from a diamond-bearing eclogite: *Dokl. Akad. Nauk. SSR*, **207**, 164-167.

Solovjova, L.V., Vladimirov, B.M., & Dneprovskaya, L.V., 1994. Upper mantle substance under old platforms. Novosibirsk, Nauka, 256 (in Russian).

Song, S., Zhang, L., Niu, Y., Su, L., Jian, P. & Liu, D., 2005. Geochronology of diamond-bearing zircons from garnet peridotite in the North Qaidam UHPM belt, Northern Tibetan Plateau: A record of complex histories from oceanic lithosphere subduction to continental collision. *Earth and Planetary Science Letters* **234**, 99-118.

Song Shuguang, Zhang Lifei, Niu, Y. & Zhang Guibin 2007. Two types of peridotite in continental orogenic belts; a case study from the North Qaidam UHP metamorphic belt; Special issue on "Earth material science" (researches on mantle petrology and geochemistry). *Dixue Qianyuan = Earth Science Frontiers* **14**, 129-138.

Sonin, V.M., Fedorov, I.I., Pokhilenko, L.N., Pokhilenko, N.P., 2000. Diamond oxidation rate as related to oxygen fugacity. *Geology of Ore Deposits* **42,** 549–556 (in Russian).

Sonin, V.M., Zhimulev, E.I., Afanas'ev, V.P., Chepurov, A.I., 2002. Genetic aspects of the diamond morphology. *Geology of Ore Deposits* **44,** 291–299 (in Russian).

Spetsius, Z.V., 1980. The form, nature and conditions of partial melting of eclogites from the kimberlite pipes, *Vestnik MGU, Geology* **4**, 101-105.

Spetsius, Z.V., 1995. Occurrence of diamond in the mantle: a case study from the Siberian Platform. *Journal of Geochemical Exploration* **53**, 25-39.

Spetsius, Z.V., 1995. Diamondiferous eclogites from Yakutia: Evidence for a late and multistage formation of diamonds. *Extended Abstracts of 6th International Kimberlite Conference*, Novosibirsk, Russia, 572-574.

Spetsius, Z.V., 1999. The high-temperature xenolith of diamondiferous peridotite from the Udachnaya kimberlite pipe. *Doklady Akademy Nauk SSSR* **235**, 1416-1418 (in Russian).

Spetsius, Z.V., 1999. Two generation of diamonds in the eclogite xenoliths. In: Gurney, J.J., Gurney, J.L., Pascoe, M.D., and Richardson, S.H. (eds.). *Proceedings Volume of 7th International Kimberlite Conference, 2*. Cape Town: Red Roof Design, 823-828.

Spetsius, Z.V., 2004. Metasomatism and partial melting in xenoliths from the kimberlites of Yakutia: implication to the origin of diamonds. *Proceedings of 4th Intern. Conference: Deep-seated magmatism, its sources and their relation to plume processes.* Ulan-Ude.

Publishing House of the Institute of Geography SB RAS, 128-159.

Spetsius Z.V., 2004, Petrology of highly aluminous xenoliths from kimberlites of Yakutia. *Lithos* **77**, 525-538.

Spetsius, Z.V. and Ponomarenko, A.I., 1979. The amorphozed eclogites - representatives of asthenosphere of the Earth, *Dokl. Akad. Nauk SSSR* **235**, 1415-1419 (in Russian).

Spetsius, Z.V. and Serenko, V.P., 1990. Composition of continental upper mantle and lower crust beneath the Siberian platform. *Moscow:* Nauka, 272 (in Russian).

Spetsius, Z.V. and Bezborodov, S.M., 1992. Mineralogy of diamondiferous eclogites from the Udachnaya kimberlite pipe, new finding. *Doklady Akademy Nauk SSSR* 326, 717-721(in Russian).

Spetsius, Z.V. and Griffin, W. L., 1997. Trace elements in minerals from eclogites from the Udachnaya kimberlite pipe, Yakutia. *Russian Geology and Geophysics* **38**, 240-246.

Spetsius, Z.V. and Griffin, B.J., 1998. Secondary phases associated with diamonds in eclogites from the Udachnaya kimberlite pipe: Implications for diamond genesis, *Extended Abstracts of 7th International Kimberlite Conference*, Cape Town, South Africa, 850-852.

Spetsius, Z.V. and Griffin, W.L., 1998. Trace element composition of garnet kelyphites in xenoliths from Udachnaya as evidence of their origin. *Extended Abstracts of 7th International Kimberlite Conference*, Cape Town, South Africa, 853-855.

Spetsius, Z.V., and Serenko, V.P., 1999. Lithosphere of the Middle-Marxinsky kimberlite field. In: Geology and Tectonics of Platforms and Orogenic Regions in Northeastern Asia. Siberian Brunch Russian Akademy of Sciences, Yakutsk,. **2**, 104-108 (in Russian).

Spetsius, Z.V. and Taylor, L.A., 2002. Partial melting in mantle eclogite xenoliths: Connections with diamond paragenesis. *International Geology Review* **44**, 973-987.

Spetsius, Z.V., Taylor, L.A., 2003. Kimberlite xenoliths as evidence for subducted oceanic crust in the formation of the Siberian craton. *Proceedings of 3d Intern. Workshop: Plumes and problem of deep sources of alkaline magmatism*. Irkutsk, 3-15.

Spetsius, Z.V. and Spetsius, V. Z., 2005. Exsolution textures and minerals inhomogeneity in xenoliths from Yakutian kimberlites: evidence for the mantle evolution. *Proceedings of Intern. Workshop: Plumes and problems of sources of deep magmatism and plumes*. Petropavlovsk-Kamchatsky, 148-169.

Spetsius, Z.V., Taylor, W.R., & Griffin, B.J., 1998. Major and trace-element partitioning between mineral phases in diamondiferous and non-diamondiferous eclogites from the Udachnaya kimberlite pipe, Yakutia. *Extended Abstracts of 7th International Kimberlite Conference*, Cape Town, South Africa, 856-858.

Spetsius, Z.V., Serenko, V.P., Mityukhin, S.I., 2000. The lithosphere of the Nakynsky kimberlite field, Yakutia. *Abstracts 31th IGC*, Brazilia, Rio de Janairo.

Spetsius, Z.V., Belousova, E.A., Griffin, W. L., O'Reilly, S.Y., & Pearson, N. J., 2002. Archean sulfide inclusions in Paleozoic zircon megacrysts from the Mir kimberlite, Yakutia: implications for the dating of diamonds. *Earth and Planetary Science Letters* **199**, 111-126.

Spetsius, Z.V., Mityukhin, S.I., Ivanov, A.S., 2003. First Discovery of Diamondiferous Xenolith in Kimberlite from the Botuobinskaya Pipe, Nakynsky Field, Yakutia. *Dokl. Akad. Nauk* **390**, 1–4 (in Russian).

Spetsius, Z.V., Mityukhin, S.I., Ivanov, A.S., Banzeruk, S.V., 2005. Paragenesis of inclusions in diamonds from the Botuobinskaya pipe (Nakynsky field, Yakutia). *Dokl. Akad. Nauk*, in press (in Russian).

Spetsius, Z.V. Ivanov, A.S, Mityukhin, S.I., 2006. Diamondiferous xenoliths and megacrysts from the Nyurbinskaya kimberlite pipe (Nakynsky field, Yakutia). *Doklady Earth Sciences* **409**, 779-783 (in Russian).

Srikantappa, C. and Malathl, M.N., 2006. Olivine-hosted melt Inclusions In diamondiferous kimberlites from Wajrakarur, Andhra Pradesh, Indla. *Indian Mineralogist* **40**, *73-76.*

Srivastava, R.K. and Taylor, L.A., 1996. Carbon and oxygen isotope variations in Indian carbonatites. *International Geology Review* **38**, 419-429.

Stachel, T., 1999. Diamonds from the asthenosphere and the transition zone. *European Journal of Mineralogy* **13**, 883-892.

Stachel T., 2001, Diamonds from the asthenosphere and the transition zone. *European Journal of Mineralogy* **13**, 883-892.

Stachel, T., 2007. Diamond; Geology of gem deposits, *Short Course Series - Mineralogical Association of Canada,* 37,1-22.

Stachel, T. and Harris, J. W., 1997. Diamond precipitation and mantle metasomatism: Evidence from the trace element chemistry of silicate inclusions in diamonds from Akwatia, Ghana. *Contributions to Mineralogy and Petrology* **129**, 143-154.

Stachel, T. and Harris, J. W., 1997. Syngenetic inclusions in diamond from the Birim field (Ghana): A deep peridotitic profile with a history of depletion and re-enrichment. *Contributions to Mineralogy and Petrology* **127**, 336-352.

Stachel, T, Harris JW, & Brey G, 1998. Inclusions in diamonds from Mwadui (Tanzania) - chemical mush in the source. *7th International Kimberlite Conference*, Cape Town, 859–861.

Stachel, T., Viljoen, K. S., Brey, G., & Harris, J. W., 1998. Metasomatic processes in lherzolitic and harzburgitic domains of the diamondiferous lithospheric mantle: REE in garnets from xenoliths and inclusions in diamonds. *Earth and Planetary Science Letters* **159**, 1-12.

Stachel, T., Brey, G.P., & Harris, J.W., 2000. Kankan diamonds (Guines) I: From the lithosphere down to the transition zone. *Contrib. Mineral. and Petrol.* **140**, 1-15.

Stachel, T., Harris J.W., Brey, G.P., and Joswig, W., 2000. Kankan diamonds (Guines) II: Lower mantle inclusion paragenesis. *Contributions to Mineralogy and Petrology* **140**, 16-27.

Stachel, T, Harris, J.W., Aulbach, S., & Deines, P., 2002. Kankan diamonds (Guinea) III: $\delta^{13}C$ and nitrogen characteristics of deep diamonds. *Contr. Mineral. Petrol.* **142**, 465-475.

Stachel, T., Aulbach, S., Brey, G.P., Harris, J.W., Leost, I., Tappert, R., & Viljoen, K.S., 2004. The trace element composition of silicate inclusions in diamonds: a review. *Lithos* **77**, 1-19.

Stachel, T., Viljoen, K.S., McDade, P. & Harris, J.W., 2004. Diamondiferous lithospheric roots along the western margin of the Kalahari Craton—the peridotitic inclusion suite in diamonds from Orapa and Jwaneng. *Contrib. Mineral. and Petrol.* **147**, 32-47.

Stachel, T., Brey, G.P., and Harris, J.W., 2005. Inclusions In sublithospheric diamonds; glimpses of deep earth. *Elements* **1**, 73-87.

Stachel, T., Banas, A., Muehlenbachs, K., Kurszlaukis, S., & Walker, E.C., 2006. Archean diamonds from Wawa (Canada): samples from deep cratonic roots predating cratonization of the Superior Province. *Contributions to Mineralogy and Petrology* **151**, 737-750.

Stefano, A. D., Lefebvre, N., Kopylova, M., 2006. Enigmatic diamonds in Archean calc-alkaline lamprophyres of Wawa, southern Ontario, Canada. *Contrib. Mineral. Petrol.* **151,** 158–173.

Streckeisen, A., 1976. To each plutonic rock its proper name. *Earth Science Review* **12**, 1-33.

Sun, S. S., 1982. Chemical composition and origin of the Earth's primitve mantle. *Geochimica et Cosmochimica Acta* **46**, 179-192.

Sunagawa, I., 1981. Characteristics of crystal growth in nature as seen from the morphology of mineral crystals. *Bull. Mineral.* **104**, 81– 87.

Sunagawa, I., 1984. Growth of crystals in nature. In: Sunagawa, I. (ed.). *Materials Science of the Earth's Interior*, Tokyo, 61-103.

Sunagawa, I., 1984. Morphology of natural and synthetic diamond crystals. In: Sunagawa, I. (ed.). Materials Science of the Earth's Interior, Terra Scientific Publishing Company, Tokyo, 303-331.

Sunagawa, I., 1990. Growth and morphology of diamond crystals under stable and metastable conditions. *J. Cryst. Growth* **99**, 1156-1161.

Switzer, G., and Melson, W. G., 1969. Partially melted kyanite eclogite from Roberts Victor mine, South Africa. *Smithsonian Contributions to the Earth Sciences* **1**, 1-9.

Szabo, C., and Taylor, L.A., 1994. Mantle petrology and geochemistry beneath the Nógrád-Gömör volcanic field, Carpathian-Pannonian Region. *Int'l Geology Review* **36**, 328-358.

Talnikova, S.B., 1995. Inclusions in natural diamonds of different habits. *Extended Abstracts of the 6th International Kimberlite Conference, Russia*, 603-605.

Tappert, R., Stachel, T., Harris, J. W., Muehlenbachs, K., Ludwig, T., & Brey, G. P., 2005. Diamonds from Jagersfontein (South Africa): messengers from the sublithospheric mantle. *Contrib. Mineral. Petrol.* **150,** 505–522.

Tappert, R., Stachel, T., Harris, J.W., Shimizu, N., & Brey, G.P., 2005. Mineral inclusions in diamonds from the Panda kimberlite, Slave Province, Canada. *European Journal of Mineralogy* **17**, 423-440.

Tappert, R., Stachel, T., Harris, J.W., Muehlenbachs, K., and Brey, G.P., 2006. Placer diamonds from Brazil; Indicators of the composition of the earth's mantle and the distance to their kimberlitic sources. *Economic Geology and the Bulletin of the Society of Economic Geologists* **101**, 453-470.

Tatsumoto, M., Basu, A. R., Wankang, H., Junwen, W., & Guanhong, X., 1992. Sr, Nd and Pb isotopes of ultramafic xenoliths in volcanic rocks of eastern China: Enriched components, EMI and EMII in sub-continental lithosphere. *Earth Planet. Sci. Lett.* **113**, 107-128.

Taylor, L.A., 1984, Kimberlite magmatism in the eastern United States: Relationships to Mid-Atlantic tectonism. Kimberlites I: Kimberlites and Related Rocks, Kornprobst (Ed.), Elsevier Publ., 417-424.

Taylor, L.A., 1993. Evolution of the subcontinental mantle beneath the Kaapvaal craton: A review of evidence for crustal subduction for Bellsbank eclogites. In: *V.S. Sobolev Memorial Volume, Russian Geol. Geophys., Geologiya I Geofizika* **34**, 25-47.

Taylor, L.A., and Hunter, R.H., 1983. Kimberlites in the Eastern United States: location and depths of origin related to Mid-Atlantic tectonism, *3rd International Kimberlite Conference,* Strasbourg, France.

Taylor, L. A., and Neal, C. R., 1989. Eclogites with oceanic crustal and mantle signatures from the Bellsbank kimberlite, South Africa, part 1: Mineralogy, Petrography, and whole rock chemistry. *J. Geol.* **97**, 551-567.

Taylor, L.A., Neal, C.R., 1993. Comment on "Trace-element crystal chemistry of mantle eclogites" by Caporuscio, F.A., and Smyth, J.R. *Contrib. Mineral. Petrol.* **1131**, 280-284.

Taylor, L.A., and Anand, M., 2004. Diamonds: time capsules from the Siberian Mantle, *Chemie der Erde* **64**, 1-74.

Taylor, L.A., Snyder, G.A., Crozaz, G., Sobolev, V.N., Yefimova, E.S., & Sobolev, N.V., 1996. Eclogitic Inclusions in diamonds: Evidence of complex mantle processes over time. *Earth and Planetary Science Letters* **142**, 535-551.

Taylor, L.A., Milledge, H.J., Bulanova, G.P., Snyder, G.A., and Keller, R.A., 1998. Metasomatic eclogitic diamond growth: evidence from multiple diamond inclusions. *International Geology Review* **40**, 663-676.

Taylor, L.A., Snyder, G.A., & Camacho, A., 1999. Diamond: Just another Metamorphic mineral? *Goldschmidt Ann. Geochem. Conf.*, 66-67.

Taylor, L.A., Keller, R.A., Snyder, G.A., Wang, W., Carlson, W.D., Hauri, E.H., McCandless, T., Kim, K.-R., Sobolev, N.V., & Bezborodov, S.M., 2000. Diamonds and their mineral inclusions, and what they tell us: A detailed "pull-apart" of a diamondiferous eclogite. *International Geology Review* **42**, 959-983.

Taylor, L.A., Carlson, W.D., Anand, M., & Misra, K., 2001. High-Resolution X-Ray Computed Tomography (HRXCT) of Diamondiferous Eclogites and the Origin of Diamond, *EOS Trans. AGU,* **82**, Fall Meeting, Abstract V12C-1001.

Taylor, L. A., Spetsius, Z., Wiesli, R. Anand, M. & Valley, J., 2002. Crustal signatures in mantle peridotites from Yakutian kimberlites, *EOS Trans. AGU* **82**, Fall Meet., Abstract.

Taylor, L.A., Snyder, G.A., Keller, R., Remley, D., Anand, M., Wiesli, R., Valley, J., & Sobolev, N.V., 2003. Petrogenesis of group A eclogites and websterites; evidence from the Obnazhennaya Kimberlite, Yakutia",. *Contrib. Mineral. Petrol.* **145,** 424–443.

Taylor, L.A., Anand, M., & Promprated, P., 2003. Evidence for non-syngenetic inclusions in diamonds. *8th International Kimberlite Conf.*, Victoria, BC, Extended Abstract.

Taylor, L.A., Anand, M., & Promprated, P., 2003. Diamonds and their inclusions: Are the criteria for syngenesis valid? *8th Int,l Kimb. Conf., Ext. Abstr.*, Victoria, Canada.

Taylor, L.A., Anand. M., Promprated, P., Floss, C., & Sobolev, N.V., 2003. The significance of mineral inclusions in large diamonds from Yakutia, Russia. *American Mineralogist* **88**, 912-920.

Taylor, L.A., Taylor, D-H, Anand, M., Ketchum, R., Carlson, W., Pokhilenko, N.P., & Sobolev, N.V., 2003. Diamondiferous peridotites: Tomography as a precursor to dissections. *Amer. Geophys. Union Fall Mtg.*, abstract.

Taylor, L.A., Spetsius, Z.V., Wiesli, R., Anand, M., Promprated, P., Spicuzza, M., & Valley, J.W., 2003. Diamondiferous peridotites from subducted oceanic crust: Lessons from the past and signatures from Yakutian kimberlites. *8th Internat. Kimb. Conf.*, Ext. Abstract.

Taylor, L.A., Spetsius, Z.V., Wiesli, R., Spicuzza M., and Valley J.W., 2005. Diamondiferous peridotites from oceanic protolites: crustal signatures from Yakutian kimberlites. *Russian Geolgy and Geophysics* **46,** 1198-1206.

Taylor, W.R., 1988. A reappraisal of the nature of fluids included by diamond – a window to deep-seated mantle fluids and redox conditions. *Geol. Soc. Aust. Spec. Publ.*, **13**, 333-349.

Taylor, W.R., 1998. An experimental test of some geothermometer and geobarometer formulations for upper mantle peridotites with application to the thermobarometry of fertile lherzolite and garnet websterite. *N. Jahrb. Mineral. Abh.* **172**, 381-408.

Taylor, W.R., and Green, D.H., 1989. The role of reduced C-O-H fluids in mantle partial melting. In: Ross J. et al. (eds.). Diamonds, Kimberlites and Related Rocks **2**. *Geol. Soc. Aust. Spec. Publ.* **14**, Blackwell Scientific Publications. Melbourne, 592-602.

Taylor, W.R., and Milledge, H.J., 1995. Nitrogen aggregation character, thermal history and stable isotope composition of some xenolith-derived diamond crystals from Roberts Victor and Finch. *Extended Abstracts of the 6th International Kimberlite Conference*, 620-622.

Taylor, W.R., Jaques, A.L., & Ridd, M., 1990. Nitrogen-defect aggregation characteristics of some Australasian diamonds: Time-temperature constraints on the source regions of pipe and alluvial diamonds. *Amer. Mineral.* **75**, 1290–1310.

Taylor, W.R., Bulanova, G.P., & Milledge, H.J., 1995. Quantitative nitrogen aggregation study of some Yakutian diamonds: constraints on the growth, thermal and deformation history of peridotitic and eclogitic diamonds. *Extended Abstracts of the 6th International Kimberlite Conference*, 608-610.

Taylor, W.R., Canil, D., & Milledge, H.J., 1996. Kinetics of 1b to 1aA nitrogen aggregation in diamond. *Geochim. Cosmochim. Acta* **60**, 4725-4733.

Thomassot, E., Cartigny, P., Harris, J.W. & (Fanus) Viljoen, K.S. 2007, "Methane-related diamond crystallization in the Earth's mantle: Stable isotope evidences from a single diamond-bearing xenolith", *Earth and Planetary Science Letters* **257**, 362-371.

Titkov, S. V., Gorshkov, A.I., Solodova, Y.P., Ryabchlkov, I.D., Magazina, L.O., Sivtsov, A.V., Gasanov, M.D., Sedova, E.A., and Samosorov, G.G., 2006. Mineral microinclusions in cubic diamonds from the Yakutian deposits based on analytical electron microscopy data. *Doklady Earth Sciences* **410**, 1106-1108 (in Russian).

Tomilenko, A.A., Chepurov, A.I., Pal'yanov, Yu. N., Pokhilenko, N.P., & Shebanin, A.P. 1997. Volatile components in the upper mantle, from data on fluid inclusions. *Russian Geology and Geophysics* **38**, 276-285.

Tomlinson, E., Schrijver, I., Corte, K., Jones, A.P., Moens, L. & Vanhaecke, F., 2005. Trace element compositions of submicroscopic inclusions in coated diamond: A tool for understanding diamond petrogenesis. *Geochimica et Cosmochimica Acta* **69**, 4719-4732.

Tomlinson, E.L., Jones, A.P. & Harris, J.W., 2006. Co-existing fluid and silicate inclusions in mantle diamond. *Earth and Planetary Science Letters* **250**, 581-595.

Tomshin, M.D., Fomin, A.S., Kornilova, V.P., Cherniy, S.D., & Yanygin, Yu.T., 1998. Peculiarities of magmatic formations from the Nakyn kimberlite field of the Yakutian province. *Geologiya I Geofizika* **39**, 1693-1703 (in Russian).

Trautman, R.L., Griffin, B.J., Taylor, W.R., Spetsius, Z.V., Smith, C.B., & Lee D.C., 1997. A comparison of the microdiamonds from kimberlite and lamproite of Yakutia and Australia. **2**, *Proc. Sixth Int. Kimb. Conf., and Russian Geol. and Geophys.* **38**, 341-355.

Trautman, R.L., Griffin, B.J., and Bulanova, G.P., 1998. Growth features and nitrogen aggregation properties of microdiamonds derived from kimberlitic diatremes. *7th Int. Kimb. Conf., Ext. Abst.*, 926-928.

Trofimov, V.S., 1966. Diamond diatreams. *Sov. Geol.* **5**, 3-12.

Trofimov, V.S., 1980. Geology of deposits of natural diamonds. Moscow: Nedra, 304 (in Russian).

Troupis, D. K., Gaudin, O., Whitfield, M. D., & Jackman, R. B., 2002. Ion implantation of sulphur, boron and nitrogen in diamond: a charge-based deep level transient spectroscopic investigation. *Diamond and Related Materials* **11,** 342–346.

Tsai, H-M, Meyer, H.O.A., Moreau, J., & Milledge, H.J., 1979. Mineral inclusions in diamond: Premier, Jagersfontein, and Finch kimberlites, South Africa, and Williamson mine, Tanzania. In: Boyd, F.R., Meyer, H.O.A. (eds.). Kimberlites, Diatrimes, and Diamonds: Their Geology, Petrology, and Chemistry. Proc. 2nd Internat. Kimb. Conf. **1,** Amer. Geophys. Union, Washington, DC, 16-26.

Ukhanov, A.V., Ryabchikov, I. D., & Kharkiv, A.D., 1988. Lithospheric mantle of the Yakutian kimberlite province. Moscow, Nauka, 286 (in Russian).

Ukhanov, A.V. and Kharkiv, A.D., 1991. Upper mantle composition beneath Yakutian kimberlite province; Fifth international kimberlite conference; extended abstracts. *Proceedings of the International Kimberlite Conference,* **5**, pp. 567-568.

Ulmer, P. and Sweeney, R.J., 2002. Generation and differentiation of group II kimberlites: constraints from a high-pressure experimental study to 10 GPa. *Geochimica et Cosmochimica Acta* **66**, 2139-2153.

Urusovskaya, A.A. and Orlov, Y.L., 1964. Nature of plastic deformation of diamond crystals. *Doklady Akademii Nauk SSSR* **154,** 112–115 (in Russian).

Ustinov, V.I., Ukhanov, A.V., Grinenko, V.A., & Gavrilov, Y.Y., 1987. $\delta^{18}O$ in eclogites from the Udachnaya and Obnazhennaya kimberlite pipes. *Geokhimiya* **11**, 1637–1641.

Vaganov, V.I., Varlamov, V.A., Feldman, A.A., Olofinsky, L.N., Prusakova, N.A., Boyko, A.N. & Golubev, Y.K., 1995. Diamondiferous magmatism; mineragenetic taxons and prediction-prospecting models, Sixth international kimberlite conference, extended abstracts, *Proceedings of the International Kimberlite Conference* **6**, 649.

Vaganov, V.I., Golubev, Y.K., Rozova, Y.V. & D'yachkova, I.V., 1996. Mineralogical indicators for the presence of Mesozoic alkali-ultrabasic potentially diamond-bearing rocks in the north of the East European Platform. *Transactions (Doklady) of the Russian Academy of Sciences.Earth Science Sections* **344**, 42-46.

Vaganov, V.I., Golubev, Y.K. & Bogatykh, I.Y., 1999. Diamond complex in Russia and the world; state, problem and prospects. *Rudy i Metally* **1999**, 25-26.

Vaganov, V.I., Ilupin, I.P. & Kocherov, A.I., 1999. Chrome-spinelides as diamond indicators. *Rudy i Metally* **1999**, 35-47.

Valley, J.W. 2003, Oxygen isotopes in zircon; Zircon, In: Hanchar, JM and Hoskin, PWO (eds) Reviews in Mineralogy and Geochemistry **53**, 343-385.

Valley, J.W. 2006. Early Earth. *Elements* **2**, *201-204.*

Valley, J.W. 2006. M.S.A. presidential address: New tools for old minerals; Geological Society of America, 2006 annual meeting", Abstracts with Programs - *Geological Society of America* **38**, 377.

Valley, J.W. and Kita, N. 2005, Microanalysis of oxygen *isotopes; Abstracts of the 15th annual V. M. Goldschmidt conference", Geochimica et Cosmochimica Acta* **69**, 374.

Valley, J.W. & Lackey, *J.S.* 2005. Magmatic zircon, titanite, and garnet; oxygen isotope disequilibrium is good; Abstracts of the 15th annual V. M. Goldschmidt conference, *Geochimica et Cosmochimica Acta,* **69***,* 9.

Valley, J.W., Kitchen, N.E., Kohn, M.J., Niendorf, C.R. & Spicuzza, M.J., 1995. UWG-2, a garnet standard for oxygen isotope ratios: Strategies for high precision and accuracy with laser heating. *Geochim. Cosmochim. Acta* **59**, 5223-5221.

Valley, J.W., Bindeman, I.N., Peck, W.H., Humayun, M. & O'Neil, J.R., 2003. Empirical calibration of oxygen isotope fractionation in zircon; A special issue dedicated to Robert N. Clayton, *Geochimica et Cosmochimica Acta* **67**, 3257-3266.

Varshavsky, A.V., 1968. Anomaly birefringence and internal morphology of diamonds. Mosco, Nauka, 89 (in Russian).

Viljoen, K.S., 2002. An infrared investigation of inclusion-bearing diamonds from the Venetia kimberlite, Northern Province, South Africa: implications for diamonds from craton-margin settings. *Contributions to Mineralogy and Petrology* **144**, 98-108.

Viljoen, K.S., Swash, P.M., Otter, M.L., Schulze, D.J., & Lawless, P.J., 1992. Diamondiferous garnet harzburgite from the Finsch kimberlite, Northern Cape, South Africa. *Contrib. Mineral. Petrol.* **110**, 133-138.

Viljoen, K.S., Robinson, D.N., Swash, P.M., Griffin, W.L., Otter, M.L., Ryan, C.G. & Win, T.T., 1994. Diamond- and graphitebearing peridotite xenoliths from the Roberts Victor kimberlite, South Africa. In: Meyer, H. O. & Leonardos, O. H. (eds) Kimberlites, Related Rocks and Mantle Xenoliths. CPRM Special Publication 1A, 285–303.

Viljoen, K. S., Smith, C. B., and Sharp, Z. D., 1996. Stable and radiogenic isotope study of eclogite xenoliths from the Orapa kimberlite, Botswana. *Chem. Geol.* **131**, 235-255.

Viljoen, K.S., Phillips, D., Harris, J.W., & Robinson, D.N., 1999. Mineral inclusions in diamonds from the Venetia kimberlites, Northern Province, South Africa. *Proceedings of the 7th Kimberlite Conference*, 888-895.

Viljoen, K.S., Dobbe R., Smit, B., Thomassot, E., & Cartigny, P., 2004. Petrology and geochemistry of a diamondiferous lherzolite from the Premier diamond mine, South Africa. *Lithos* **77***,* 539-552.

Viljoen, K. S., Schulze, D. J., & Quadling, A. G., 2005. Contrasting group I and group II eclogite xenolith petrogenesis: Petrological, trace element and isotopic evidence from eclogite, garnet-websterite and alkremite xenoliths in the Kaalvallei kimberlite, South Africa. *J. Petrology* **46**,2059 - 2090.

Vischnevsky, A.A., 1991. Kelyphites on garnets in mantle xenoliths and kimberlites: compositions, genesis, petrological applications. *Extended Abstracts of 5th International Kimberlite Conference*, Araxa, Brasilia, 571-572.

Vladimirov, B.M., Volynuk, N.Y., & Ponomarenko, A.I., 1976. Deep-seated Inclusions from Kimberlites, Basalts and Kimberlitelike Rocks. Moscow, Nauka, 284 (in Russian).

Wagner, P.A., 1914. The Diamond Fields of Southern Africa. Transvaal Leader, Johannesburg, 347.

Walker, R. J., Carlson, R. W., Shirey, S. B., & Boy,d F. R., 1989. Os, Sr, Nd, and Pb isotope systematics of southern African peridotite xenoliths: Implications for the chemical evolution of sub-continental mantle. *Geochimica et Cosmochimica Acta* **53**, 1583-1595.

Walter, R. J., 1998. Melting of garnet peridotite and the origin of komatiite and depleted lithosphere. *Journal of Petrology* **39**, 29-60.

Walter, R. J., 1999. Melting residues of fertile peridotite and the origin of cratonic lithosphere. In: Fei, Y., Bertka, C., and Mysen, B. O. (eds.). Mantle Petrology: Field Observations and High Pressure Experimentation **6**, The Geochemical Society, 225-240.

Wang, W., 1998. Formation of diamond with mineral inclusions of "mixed" eclogite and peridotite paragenesis. *Earth and Planetary Science Letters* **160**, 831-843.

Wang, W. & Gasparik, T., 2001. Metasomatic clinopyroxene inclusions in diamonds from the Liaoning Province, China. *Geochimica et Cosmochimica Acta* ***65***, 611-620.

Wang, A., Dhamelincourt, P., Meyer, H.O.A., Guo, L. & Zhang, A., 1994, A carbon-rich multiphase inclusion in a Chinese diamond and its geochemical implications**.** *Contributions to Mineralogy and Petrology* **117,** 15-24.

Webb, K.J., Scott, Smith B.H., Paul, J.L. & Hetman, C.M., 2004. Geology of the Victor Kimberlite, Attawapiskat, Northern Ontario, Canada: Cross-cutting and nested craters. *Lithos* **76,** 29-50.

Weiss, Y., Griffin, W.L., Harris, J.W. & Navon, O., 2006. Pristine hydrous fluids in diamonds; direct analysis of major and trace elements in mantle fluids; AGU 2006 fall meeting, *EOS, Transactions, American Geophysical Union* **87**, @Abstract V41D-1759.

Westerlund, K.J. and Gurney, J.J., 2004. Silicate and oxide inclusion characteristics and infra-red absorption analysis of diamonds from the Klipspringer kimberlites, South Africa. *South African Journal of Geology* **107**, 131-146.

Westerlund, K. J., Shirey, S. B., Richardson, S. H., Carlson, R. W., Gurney, J. J. & Harris, J. W., 2006. A subduction wedge origin for Paleoarchean peridotitic diamonds and harzburgites from the Panda kimberlite, Slave craton: evidence from Re–Os isotope systematics, *Contributions to Mineralogy and Petrology* **152**, 275-294.

Wiesli, R.A., Taylor, L.A., Valley, J.W., Trommsdorff, V., & Kurosawa, M., 2001. Geochemistry of Eclogites and metapelites from Trescolmen, Central Alps, as observed from major and trace elements and oxygen isotopes. *International Geology Review* **43**, 95-119.

Williams, A.F., 1932. The Genesis of Diamond. London: Benn Ltd., 636.

Wirth, R., C. Vollmer, F. Brenker, S. Matsyuk, & F. Kaminsky. 2007. Inclusions of nanocrystalline hydrous aluminium silicate "phase egg" in superdeep diamonds from Juina (Mato Grosso state, Brazil). *Earth and Planetary Science Letters* **259**, 384-399.

Wood, B.J., 2000. Phase transformations and partitioning relations in peridotite under lower mantle conditions. *Earth Planet. Sci. Lett.* **174**, 341-351.

Wood, B.J., Bryndzia, L. T., & Johnson, K. E., 1990. Mantle oxidation state and its relationship to tectonic environment and fluid speciation. *Science* **248**, 337-345.

Woodland, A.B., 1998. The orthorhombic to high-P monoclinic phase transition in Mg-Fe pyroxenes: Can it produce seismic discontinuity? *Geophys. Res. Lett.* **25**, 1241-1244.

Wyman, D.A., Ayer, J.A., Conceição, R.V., Sage, R.P., 2006. Mantle processes in an Archean orogen: Evidence from 2.67 Ga diamond-bearing lamprophyres and xenoliths. *Lithos* **89**, 300-328.

Xu, S., Liu, Y., Chen, G., Ji, S., Ni, P., and Xlao, W., 2005. Microdiamonds, their classification and tectonic implications for the host eclogites from the Dabie and Su-Lu regions in central eastern China. *Mineralogical Magazine* **69**, 509-520.

Yamaoka, S., Kumar, M.D. S., & Kanda, H., 2002. Formation of diamond from CaCO in a reduced C–O–H fluid at HP–HT. *Diamond and Related Materials* **11**, 1496–1504.

Yaxley, G.M., Green, D.H., & Kamenetsky, V., 1998. Carbonatite metasomatism in the southern Australian lithosphere. *J. Petrol.* **39**, 1917-1930.

Yefimova, E.S., and Sobolev, N.V., 1977. Abundance of crystalline inclusions in diamonds of Yakutia. *Dokl. Akad. Nauk SSSR* **237**, 1475-1478 (in Russian).

Yefimova, E.S., Sobolev, N.V., & Pospelova, L.N., 1983. Sulfide inclusions in diamond and specific features of their paragenesis. *Zap. Vsesoy. Mineral. Obsh.* **112**, 300-310.

Yoder, H.S.J., 1976. Generation of Basaltic Magma. National Academy of Sciences. 265 pp.

Yoshioko, N., and Osagawara, Y., 2005. Cathodoluminescence of microdiamond in dolomite marble from Kokchetav massif–additional evidence for two-stage growth of diamond. *International Geology Review* **47,** 703-715.

Yurimoto, H., Yamashita, A., Nishida, N., & Sueno, S., 1989. Quantitative SIMS analysis of GSJ rock reference samples. *Geochem. Jour.* **23**, 215–236.

Zedgenizov, D.A., Logvinova, A.M., Shatskii, V S (Shatskiy,V.S.) & Sobolev, N.V., 1998. Inclusions in microdiamonds from some kimberlite diatremes of Yakutia, *Doklady Earth Sciences* **359**, 204-208.

Zedgenizov, D.A., Pokhilenko, N.P., Rylov, G.M., Milledge, J.H., Jones, A.P. & Parsons, I., 2002. Assorted diamond population from Snap Lake Mine (Canada); 18th general meeting of the International Mineralogical Association; mineralogy for the new millennium; programme with abstracts", vol. **18**, 116.

Zedgenizov, D.A., Reutsky, V.N., Shatsky, V.S., & Fedorova, E., 2002. A comparison of carbon isotope composition and impurity defects of microdiamonds of octahedral and cubic habit from Udachnaya kimberlite pipe (Yakutia); Abstracts of the 12th annual V. M. Goldschmidt conference. *Geochimica et Cosmochimica Acta* **66** (15A), 871.

Zedgenizov, D.A., Reutsky, V.N., Shatsky, V.S., & Fedorova, E., 2002. A comparison of carbon isotope composition and impurity defects of microdiamonds of octahedral and cubic habit from Udachnaya kimberlite pipe (Yakutia); Abstracts of the 12th annual V. M. Goldschmidt conference. *Geochimica et Cosmochimica Acta* **66** (15A), 871.

Zedgenizov, D.A., Kagi, H., Shatsky, V.S., & Sobolev, N.V., 2004. Carbonatitic melts in cuboid diamonds from Udachnaya kimberlite pipe (Yakutia): evidence from vibrational spectroscopy. *Mineralogical Magazine* **68**, 61-73.

Zedgenizov, D. A., Shiryaev, A.A., Shatsky, V.S., and Kagl, H., 2006. Water-related IR characteristics in natural fibrous diamonds. *Mineralogical Magazine* **70**, 219-229.

Zedgenizov, D.A., Harte, B., Shatsky, V.S., Politov, A.A., Rylov, G.M., & Sobolev, N.V., 2006. Directional chemical variations in diamonds showing octahedral following cuboid growth. *Contrib. Mineral. Petrol.* **151,** 45–57.

Zedgenizov, D. A., Rege, S., Griffin, W.L., Kagi, H., and Shatsky, V.S., 2007. Composition of trapped fluids in cuboid fibrous diamonds from the Udachnaya kimberlite: LAM-ICPMS analysis. *Chemical Geology* **240**, 151-162.

Zezin, R.B., Smirnova, E.P., Saparin, G.V., & Obyden, S.K., 1992. New growth features of natural diamond, revealed by color cathodoluminescence scanning electron microscope, CCL SEM technique. *Scanning* **14**, 3-10.

Zhuravlev, A. Z., Laz'ko, Y. Y., & Ponomarenko, A. I., 1991. Radiogenic isotopes and REE in garnet peridotite xenoliths from the Mir kimberlite pipe, Yakutia. *Geokhimiya* **7**, 982-994.

Zinchuk, N.N., Spetsius, Z.V., & Zuev, V.M., 1993. The Udachnaya kimberlite pipe. Novosibirsk: State University Press, 210 (in Russian).

Zindler, A., and Jagoutz, E., 1988. Mantle cryptology. *Geochim. Cosmochim. Acta.* **52**, 319-333.

Zonneveld, J.-P., Kjarsgaard, B.A., Harvey, S.E., Heaman, L.M., McNeil, D.H., & Marcia, K.Y., 2004. Sedimentologic and stratigraphic constraints on emplacement of the Star Kimberlite, east–central Saskatchewan. *Lithos* **76**, 115-138.